Stem Cell Biology and Regenerative Medicine

Series Editor
Kursad Turksen, Ph.D.

More information about this series at http://www.springer.com/series/7896

Barbara Zavan • Eriberto Bressan
Editors

Dental Stem Cells: Regenerative Potential

Humana Press

Editors
Barbara Zavan
Department of Biomedical Sciences
University of Padova
Padova, Italy

Eriberto Bressan
Department of Neurosciences
University of Padova
Padova, Italy

ISSN 2196-8985 ISSN 2196-8993 (electronic)
Stem Cell Biology and Regenerative Medicine
ISBN 978-3-319-33297-0 ISBN 978-3-319-33299-4 (eBook)
DOI 10.1007/978-3-319-33299-4

Library of Congress Control Number: 2016945174

Printed on acid-free paper

This Humana Press imprint is published by Springer Nature
The registered company is Springer International Publishing AG Switzerland

Contents

Contributors

Paul R. Cooper Oral Biology, School of Dentistry, University of Birmingham, Birmingham, UK

O.G. Davies Wolfson School of Mechanical and Manufacturing Engineering, Loughborough University, Loughborough, UK

Waruna Lakmal Dissanayaka Comprehensive Dental Care, Faculty of Dentistry, The University of Hong Kong, Hong Kong

Nileshkumar Dubey Discipline of Oral Sciences, Faculty of Dentistry, National University of Singapore, Singapore, Singapore

Henry F. Duncan Division of Restorative Dentistry and Periodontology, Dublin Dental University Hospital, Trinity College Dublin, Dublin, Ireland

Kitraki Efthymia Faculty of Dentistry, National and Kapodistrian University of Athens, Athens, Greece

Diamanti Evangelia Faculty of Dentistry, National and Kapodistrian University of Athens, Athens, Greece

Letizia Ferroni Department of Biomedical Sciences, University of Padova, Padova, Italy

Letizia Ferroni Department of Biomedical Sciences, University of Padova, Padova, Italy

Akhilesh K. Gaharwar Department of Biomedical Engineering, Texas A&M University, College Station, TX, USA

Department of Materials Science and Engineering, Texas A&M University, College Station, TX, USA

Chiara Gardin Department of Biomedical Sciences, University of Padova, Padova, Italy

Chiara Gardin Department of Biomedical Sciences, University of Padova, Padova, Italy

Paolo Ghensi Centre for Integrative Biology (CIBIO), University of Trento, Trento, Italy; Department of Neurosciences, University of Padova, Padova, Italy

Giovanna Iezzi Department of Medical, Oral, and Biotechnological Sciences, University of Chieti-Pescara, Chieti, Italy

Lucia Jimenez-Rojo Orofacial Development and Regeneration, Institute of Oral Biology, Centre for Dental Medicine, Medical Faculty, University of Zurich, Zurich, Switzerland

Amalia Kaparou Faculty of Dentistry, National and Kapodistrian University of Athens, Athens, Greece

Massimo Marrelli Department of Maxillofacial Surgery, Calabrodental, Crotone, Italy

Kyung-san Min Department of Conservative Dentistry, School of Dentistry, Chonbuk National University, Jeonju, South Korea

Thimios A. Mitsiadis Orofacial Development and Regeneration, Institute of Oral Biology, Centre for Dental Medicine, Medical Faculty, University of Zurich, Zurich, Switzerland,

Giovanna Orsini Department of Clinical Sciences and Stomatology, Polytechnic University of Marche, Ancona, Italy

Francesco Paduano Research and Development in Biomedicine – Stem Cells Unit, TECNOLOGICA Research Institute, Crotone (KR), Italy

Giorgio Pagni Private Practice, Florence, Italy

University of Milan, Milan, Italy

Adriano Piattelli Department of Medical, Oral, and Biotechnological Sciences, University of Chieti-Pescara, Chieti, Italy

Sara Ricci Department of Neurosciences, University of Padova, Padova, Italy

Vinicius Rosa Discipline of Oral Sciences, Faculty of Dentistry, National University of Singapore, Singapore, Singapore

B.A. Scheven School of Dentistry, College of Medical and Dental Sciences, University of Birmingham, Birmingham, UK

Jakub Suchánek Department of Dentistry, Faculty of Medicine in Hradec Králové, Charles University in Prague, Hradec Králové, Czech Republic

Fatima N. Syed-Picard Department of Bioengineering, University of Pittsburgh, Pittsburgh, PA, USA

Marco Tatullo Research and Development in Biomedicine – Stem Cells Unit, TECNOLOGICA Research Institute, Crotone (KR), Italy

G.K. Thakral Department of Dentistry, SGRR Institute of Medical & Health Sciences, Dehradun, India

Petridis Xenos Faculty of Dentistry, National and Kapodistrian University of Athens, Athens, Greece

Chengfei Zhang Comprehensive Dental Care, Faculty of Dentistry, The University of Hong Kong, Hong Kong

Dental Stem Cells (DSCs): Classification and Properties

Chiara Gardin, Sara Ricci, and Letizia Ferroni

Embryonic Origin and Components of the Tooth

A tooth is a complex structure made up of two major parts: the crown and the root. The crown of the tooth is what is visible in the mouth, whereas the root is the portion normally not visible in the mouth because it is anchored within the bone. Embryologically, mammalian teeth develop from sequential and reciprocal interactions between oral epithelium and neural crest-derived mesenchyme [1]. Most of the dental tissues have a cranial neural crest origin, namely dentin, dental pulp, cementum, periodontal ligament, and alveolar bone. The outer mineralized layer of enamel covering the tooth crown is the only component of ectodermal origin (Fig. 1).

Dentin comprises the main portion of the tooth. It is a mineralized connective tissue secreted by the odontoblasts, which are specialized cells located at the periphery of the dental pulp. The thickness of dentinal layer increases with age due to the deposition of secondary and tertiary (reparative) dentin, reducing the volume of the pulp chamber and the root canals [2]. It has been shown that the presence of dentin is essential for the differentiation of inner dental epithelial cells into ameloblasts [3]. Ameloblasts start to produce and secrete specific enamel matrix proteins, and soon after tooth eruption in the oral cavity, they completely disappear. During that time, tooth root develops, accompanied by cementum deposition and periodontium formation [4]. The cementum is a mineralized tissue covering the root of the tooth and produced by the cementoblasts. The main role of cementum is to serve as a medium by which the periodontal ligament can attach to the tooth for stability. At

C. Gardin • L. Ferroni (✉)
Department of Biomedical Sciences, University of Padova, via Ugo Bassi 58/B 35131, Padova, Italy
e-mail: letizia.ferroni@gmail.com

S. Ricci
Department of Neurosciences, University of Padova, via Giustiniani 5 35128, Padova, Italy

B. Zavan, E. Bressan (eds.), *Dental Stem Cells: Regenerative Potential*, Stem Cell Biology and Regenerative Medicine, DOI 10.1007/978-3-319-33299-4_1

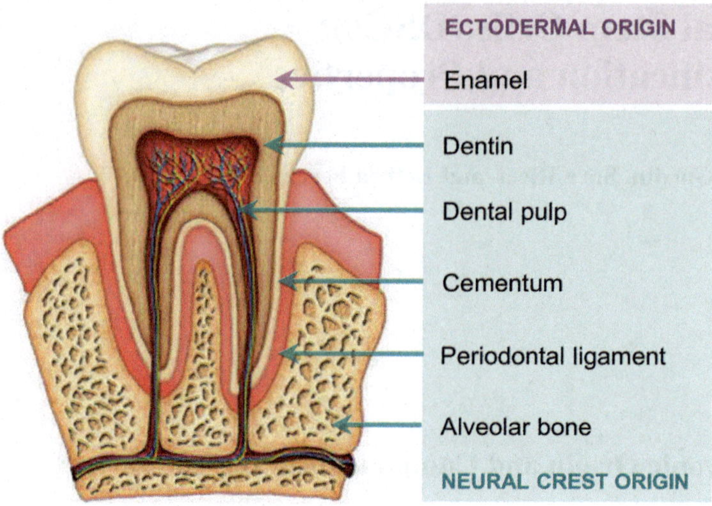

Fig. 1 Embryonic origin of the different tooth components. (Reproduced from www.bronxdental-spa.com/root-canal-bronx.htm)

the cementum-enamel junction, the cementum is acellular. This acellular type of cementum covers about two-third of the root; the apical portion of the root, that is more permeable, is covered by cellular cementum [5]. The periodontal ligament anchors the tooth to the alveolar bone, which is the bone in which teeth are encased. The outside wall of the alveolar bone is compact bone; the trabecular bone is inside and contains bone marrow. The number and the size of the trabeculae in this bone are determined by the function activity of the organ.

Dental pulp is the soft connective tissue of the tooth. It is located in its central cavity and is surrounded by the hard structures of enamel, dentin and cement. The dental pulp contains four layers. The external layer (*odontoblast layer*) is made up of odontoblasts producing dentin; the second layer (*cell-free zone*) is poor in cells and rich in collagen fibers; the third layer (*cell-rich zone*) contains fibroblasts and undifferentiated mesenchymal cells. From this layer, undifferentiated cells migrate to various districts where they can differentiate under different stimuli and make new differentiated cells and tissues. The innermost layer (*core of the pulp)* comprises blood vessels and nerves that enter the tooth mostly through the apical foramen. Other cells in the pulp include fibrocytes, macrophages, and lymphocytes [6].

Dental Stem Cells (DSCs)

Unlike other tissues such as bone, which have the ability to repair and remodel throughout life, human teeth have a very limited capacity to regenerate upon injury or disease [7]. Of all the dental components, the acellular enamel is incapable of regenerating its original structure, whereas the remaining dental tissues possess that capacity in varying degrees, dependent on multiple factors [8]. Stem cells have been opening a promising future in regenerative medicine because of their two remarkable features known as self-renewal and multilineage differentiation. Stem cells reside in a dynamic and specialized microenvironment denoted to as niche, which is composed of heterogeneous cell types, extracellular matrix (ECM), and soluble factors [9]. The niche regulates stem cells behavior, by maintaining a balance between quiescence, self-renewal, and differentiation [10]. Based on their origin, stem cells can be generally classified in Embryonic Stem Cells (ESCs), which differentiate into all cell types found in the human body, and Adult Stem Cells (ASCs), whose differentiation potential is restricted to certain cell lineages. ASCs have been identified in almost every adult tissues from both epithelial and mesenchymal origin, including skin [11], bone marrow [12], adipose tissue [13], peripheral blood [14], cartilage [15], intestine [16], and periosteum [17]. For this reason, ASCs are also referred to as postnatal stem cells. ASCs are more applicable than ESCs in stem cell-mediated therapies and regenerative medicine because these cells lack ethical concerns. In addition, they have low immunogenicity and less tumorigenic potency than their embryonic counterparts, being promising candidates for regenerative therapies [18].

Several populations of ASCs have been identified also in various dental tissues, and they are collectively referred to as Dental Stem Cells (DSCs) [19]. DSCs are considered a promising source of ASCs since they are easily accessible by tooth extraction with a local anesthetic or when a primary tooth is replaced. More interestingly, DSCs can also be harvested from inflamed or diseased dental tissues, and their properties are similar to those of DSCs obtained from healthy tissues [20–22]. Therefore, it is believed that DSCs could retain, at least to some extent, the stem cell properties and tissue regeneration potential, making them an important tool for future developments in regenerative medicine. Another advantage of teeth as a source of stem cells is that, due to their ectomesenchymal origins, DSCs may display characteristics of both mesoderm and ectoderm [23]. This fact is very important because the association of mesenchymal (that will form odontoblasts, cementoblasts, osteoblasts, and fibroblasts) and epithelial (that will form ameloblasts) stem cells it is necessary for regenerating or building a new tooth.

Classification and Properties of DSCs

To date, seven different human dental stem/progenitor cells have been isolated and characterized: Dental Pulp Stem Cells (DPSCs) [7]; Stem cells from Human Exfoliated Deciduous teeth (SHED) [24]; Periodontal Ligament Stem Cells (PDLSCs) [25]; Dental Follicle Progenitor Cells (DFPCs) [26]; and Stem Cells from Apical Papilla (SCAP) [27]. DPSCs, SHED, and SCAP are generally referred to as dental pulp-related stem cells, PDLSCs and DFPCs as periodontium-related stem cells [28]. Other dental-related stem cells have been identified later. These are

Fig. 2 Timeline in the history of identification of the different DSCs

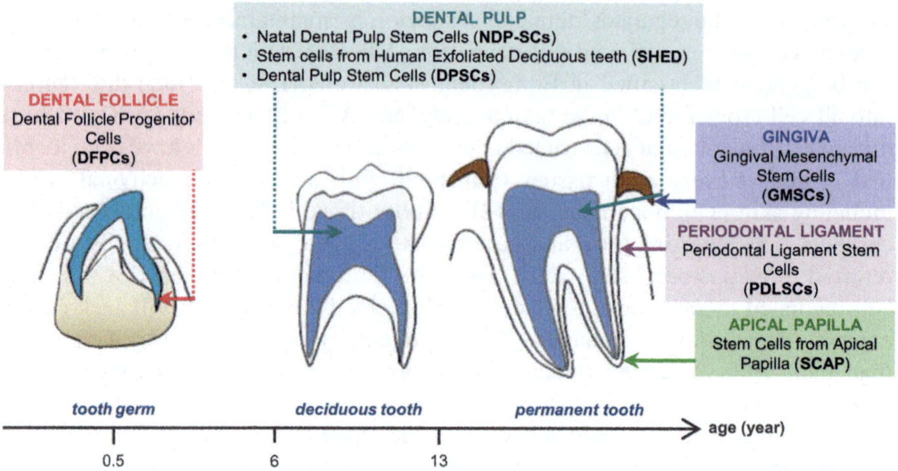

Fig. 3 Tooth developmental stages showing the anatomical localization of the different DSCs. (Reproduced from [31])

Gingival Mesenchymal Stem Cells (GMSCs) [29], and human Natal Dental Pulp Stem Cells (NDP-SCs) [30].

In this section, each type of DSCs will be described following the chronological order of their discovery (Fig. 2). Different biological aspects of DSCs will be discussed, starting from their first identification, the anatomical localization in the tooth (Fig. 3), the methods for their isolation, the peculiar expression of surface markers, and the differentiation potential in vitro and in vivo. The main biological properties of DSCs are summarized in Table 1.

Dental Pulp Stem Cells (DPSCs)

DPSCs Identification, Isolation, and Characterization

DPSCs from adult human dental pulp were first identified by Gronthos and colleagues in 2000 [7], even though the existence of stem cells in dental pulp has been reported by Yamamura in 1985 [58]. DPSCs were isolated on the basis of their high proliferation and frequency of colony formation that produced sporadic, but densely calcified nodules. The authors demonstrated that DPSCs could develop in vitro into odontoblasts, the cells that form the mineralized matrix of dentin. In addition, when

Table 1 Properties of DSCs

Type of DSCs	Location	Cell surface markers		Differentiation potential		References
		Positive	Negative	In vitro	In vivo	
DPSCs	Dental pulp of permanent tooth	CD9, CD10, CD13, CD29, CD44, CD59, CD73, CD90, CD105, CD106, CD146, CD166, STRO-1, Nanog, Oct-4, Sox-2	CD14, CD34, CD45, HLA-DR	Odontogenic Osteogenic Neurogenic Adipogenic Myogenic Chondrogenic	Dentin/pulp-like complex	[7, 23, 32–39]
SHED	Dental pulp of deciduous tooth	CD13, CD29, CD44, CD73, CD90, CD105, CD106, CD146, CD166, STRO-1, Nanog, Oct-4, SSEA-3, SSEA-4, TRA-1-60, TRA-1-81	CD14, CD34, CD45, HLA-DR	Odontogenic Osteogenic Neurogenic Adipogenic Myogenic Chondrogenic	Dentin formation New bone formation by recruiting host murine cells	[24, 40–43]

(continued)

Table 1 (continued)

Type of DSCs	Location	Cell surface markers		Differentiation potential		References
		Positive	Negative	In vitro	In vivo	
PDLSCs	Periodontal ligament	CD9, CD10, CD13, CD26, CD29, CD44, CD59, CD73, CD90, CD105, CD106, CD146, CD166, STRO-1, Nanog, Oct-4, Sox-2, Rex-1, SSEA-1, SSEA-3, SSEA-4, TRA-1-60, TRA-1-81	CD14, CD34, CD45, HLA-DR	Osteogenic Cementogenic Adipogenic Chondrogenic Insulin-producing cells	Cementum/periodontal ligament structure	[25, 33, 44–47]
DFPCs	Dental follicle of developing tooth	CD9, CD10, CD13, CD29, CD44, CD53, CD59, CD73, CD90, CD105, CD106, CD146, CD166, STRO-1, Nanog, Oct-4	CD34, CD45, CD117, HLA-DR	Osteogenic Adipogenic Neurogenic Chondrogenic	Mineralized tissue structure	[26, 28, 33, 48–51]

SCAP	Apical papilla of developing tooth	CD13, CD24, CD29, CD44, CD73, CD90, CD105, CD106, CD146, CD166, STRO-1, Nanog, Oct-4	CD18, CD34, CD45, CD150	Odontogenic Osteogenic Adipogenic Chondrogenic Neurogenic	Dentin/pulp-like complex	[27, 52, 53]
GMSCs	Gingiva	CD13, CD29, CD44, CD73, CD90, CD105, CD146, STRO-1, Oct-4, SSEA-4	CD14, CD34, CD45, HLA-DR	Osteogenic Adipogenic Chondrogenic Neurogenic Endothelial	Connective-like tissue	[29, 54–56]
NDP-SCs	Dental pulp of newborn	CD13, CD29, CD44, CD73, CD90, CD146, CD166, Nanog, Oct-4, Sox-2, Rex-1, FoxD3	CD14, CD34, CD45, HLA-DR	Osteogenic Adipogenic Chondrogenic Myogenic Neurogenic		[30, 57]

transplanted into immunocompromised mice, DPSCs mixed with hydroxyapatite/tricalcium phosphate (HA/TCP) were able to generate a dentin/pulp-like complex with a collagen matrix containing blood vessels and lined with odontoblasts [7].

The collection of stem cells from dental pulp is a common noninvasive surgical practice that can be performed in the adult during life. Dental pulp tissue from human third molars represents the most common source for DPSCs harvesting, since these teeth are often extracted and discarded. In order to isolate dental pulp, the tooth has to be sectioned at the cementum/enamel junction using dental instruments. Then, dental pulp tissue is gently separated from the crown and root, and minced into small fragments. From these, DPSCs can be isolated either by the enzymatic digestion (ED) method [7, 59] or by the outgrow (OG) method [60, 61]. Using the first approach, the dental pulp is digested in a solution of collagenase type I and dispase in order to obtain single cell suspensions. Conversely, the second technique is based on the outgrowth of stem cells from tissue fragments. Huang and colleagues compared the properties of DPSCs isolated with the two techniques [62], finding higher proliferation rate for the digested cells than for those isolated by the OG method. In addition, according to Karamzadeh and co-workers, DPSCs isolated by ED showed higher mineralization capacity than those obtained with the OG method [63]. In a later work of Hilkens and colleagues, it has been demonstrated that, when DPSCs were derived from the same donors and kept under the same culture conditions, there were no differences in cellular morphology, proliferation rate, stem cell marker expression and mesenchymal differentiation potential, regardless of the isolation method used [64].

Increasing evidence suggests that DPSCs can survive for long periods and can be passaged several times. Suchanek and colleagues showed that DPSCs achieved 60 population doublings in culture medium designed for bone marrow [65], whereas Laino and co-workers accomplished 80 passages by maintaining the DPSCs substrate interaction and cell-cell communication in the central region of the secreted ECM [66]. DPSCs can also be cryopreserved and stored for long periods without losing their multipotential differentiation ability [66].

In the dental pulp, stem cells are thought to reside in a perivascular niche within the tooth structure, as demonstrated by the expression of the endothelium-associated marker CD146 in blood vessel walls, but not in the surrounding fibrous tissue, odontoblast layer, and perineurium of the nerve [32]. DPSCs also express STRO-1, which is considered an early marker of different mesenchymal stem cells (MSCs). Positive staining for both STRO-1 and CD146 has been widely used to identify DSCs niches. In particular, STRO-1 and CD146 were found to co-localize on the outer walls of blood vessel in dental pulp, thus implying that the majority of DPSCs arise from the microvasculature [32]. Since DPSCs are considered a population of MSCs, markers that have been used for identifying MSCs are also used for DPSCs. Apart from CD146 and STRO-1, DPSCs result positive to other stromal-associated markers, such as CD9, CD10, CD13, CD29, CD44, CD73, CD90, CD105, CD106, and CD166, and negative to hematopoietic markers, such as CD14, CD34, and CD45, and HLA-DR [23, 33]. Nevertheless, different works have reported positive expression for CD34 in DPSCs [66, 67]. In humans, CD34 identifies a cell surface antigen expressed by the most primitive stromal other than hematopoietic stem cells. The controversy concerning CD34 expression among studies can be attributed to the different experimental conditions, mainly the high serum concentration in culture medium, which has been found responsible for the rapid loss of the CD34 antigen during expansion [68]. Undifferentiated DPSCs also express Oct-4, Nanog, and Sox-2, and present the cytoskeletal proteins nestin and vimentin, all of which are characteristic of undifferentiated ESCs. Apart from stemness markers, DPSCs exhibit expression for a variety of markers associated with bone, such as alkaline phosphatase (ALP), collagen type I (COL1), osteonectin (ON), osteopontin (OPN), osteocalcin (OCN), and bone sialoprotein (BSP); smooth muscle, for example α-smooth muscle actin; and fibroblasts, including collagen type III (COL3), and fibroblast growth factor 2 (FGF2) [7]. The lack of expression of the odontoblast-specific dentin sialoprotein (DSP), which is restricted to the outer pulpal layer containing differentiated odontoblasts, implies that DPSCs represent an immature pre-odontogenic population [32]. It is important to remind that, like all MSCs, DPSCs are heterogeneous and the various markers listed above may be expressed by subpopulations of these stem cells [23].

Differentiation Potential of DPSCs

DPSCs are considered similar to MSCs not only because they share a fibroblastic morphology with selective adherence to solid surfaces and good proliferative potential, but also for their ability to differentiate into multiple cell lineages in vitro. After their original characterization, DPSCs have been differentiated into odontogenic, osteogenic, neurogenic, adipogenic, myogenic, and chondrogenic tissues.

Human DPSCs were initially identified on the basis of their differentiation into osteoblasts and odontoblasts, as previously reported [7]. It has been shown that the potentiality of DPSCs differentiation into odontoblasts in vivo is reduced after passage 9, when these cells can only differentiate along the osteoblast lineage [69]. A selected subpopulation of human DPSCs, called Stromal Bone Producing DPSCs (SBP-DPSCs), has been further isolated by Laino and co-workers [66]. The authors obtained these cells from dental pulps of subjects between 30 and 45 years of age. The aged cells seemed not to be different from those found in younger patients, thus concluding that 30 years should not be considered as a critical age limit for DPSCs isolation. The authors found that SBP-DPSCs, which represent 10 % of dental pulp cells, displayed a great ability to differentiate into osteoblasts, producing a living autologous fibrous bone (LAB) tissue in vitro. When transplanted into immunocompromised rats, SBP-DPSCs formed a lamellar bone containing osteocytes. In this setting, SBP-DPSCs produced bone but not dentin, as shown by mRNA expression of bone markers including OCN, Runt-related transcription factor 2 (RUNX2), COL1, ALP, but not dentin sialophosphoprotein (DSPP), which is specific for dentin. In a later work of the same group, it has been observed that about 30 % of SBP-DPSCs become endothelial cells during their differentiation. These cells were found lining the vessel walls of the newly formed woven bone. Interestingly, a complete integration of vessels within bone chips took place after in vivo transplantation, leading to the formation of a vascularized bone tissue [34].

Although the majority of the studies have focused their attention on the ability of DPSCs to differentiate into odontoblasts or osteoblasts, further characterization revealed that DPSCs also possess adipogenic and neurogenic differentiation capacities [35, 36]. Remarkably, DPSCs are known to produce neurotrophic factors and even rescue motoneurons after spinal cord injury [37]. DPSCs were also found to undergo chondrogenic and myogenic differentiation in vitro. For example, Zhang and colleagues [38] demonstrated that human DPSCs were able to differentiate into neurogenic, osteogenic/odontogenic, adipogenic, myogenic, and chondrogenic lineages, when cultured under the appropriate inductive media. In a recent study of Paino and colleagues, it has been shown that DPSCs spontaneously differentiate in vitro toward the melanocytic lineage too [39]. The authors supposed that this spontaneous differentiation may be ascribed to the fact that both DPSCs and melanocytes arise from the neural crest cell population.

The ability of human DPSCs to differentiate into several cell lineages under defined conditions has been demonstrated also in a recent work [70]. DPSCs seeded onto a hyaluronan-based scaffold in the presence of a sequential cocktail of neuronal, glial, endothelial, and osteogenic factors proved to be a good strategy for the in vitro reconstruction of a dental pulp-like tissue. Once again, the plasticity and multipotential ability of DPSCs could be explained by the fact that dental pulp is made of both ectodermic and mesenchymal components, containing neural crest-derived cells.

Interestingly, Alongi and co-workers evaluated if DPSCs derived from Inflamed Pulps (DPSCs-IPs) could be used for dental tissue regeneration [20]. They found that DPSCs-IPs formed a dentin/pulp-like complex similar to healthy DPSCs when transplanted into immunocompromised mice. This is a very promising result since inflamed pulp, which is routinely discarded after pulpectomy, actually represents a possible source of stem cells.

Stem cells from Human Exfoliated Deciduous teeth (SHED)

SHED Identification, Isolation, and Characterization

Deciduous teeth is the formal name for what most people call *baby teeth* or, more colloquially, *milk teeth*. Children normally develop a set of 20 deciduous teeth, which appear after 6 months of life and generally are replaced, one tooth at a time, between age 6 and 13. The remaining crown of exfoliated deciduous teeth contains a living pulp remnant comprised of a normal dental pulp including connective tissue, blood vessels, and odontoblasts. Miura and colleagues were the first to identify in 2003 SHED from the remnant pulp in the crown of deciduous incisors of 7–8 years old children [24]. They isolated these clonogenic, proliferative cells using a technique very similar to the one Gronthos and co-workers used to isolate DPSCs

[7]. When cultured under a neuronal differentiation medium, SHED formed sphere-like clusters in which highly proliferative cells aggregated together in clusters that either adhered to the culture dish or floated freely in the culture medium [24]. The dissociation of these sphere-like clusters allowed cells to grow as individual fibroblastic cells. This phenomenon suggested a high proliferative capacity analogous to that of neural stem cells. In effect, SHED are characterized by a higher proliferation rate and increased cell population doublings, which are faster than those of Bone Marrow MSCs (BM-MSCs) [24].

SHED can be isolated using the same procedures described for DPSCs. These stem cells offer attractive advantages over other postnatal DSCs, as they are derived from a source which is non-invasive, readily accessible, naturally being disposed, and with very limited ethical or legal concerns [71]. Bakopoulou and colleagues compared the morphology, growth, immunophenotype, and in vitro osteo/odontogenic differentiation characteristics of SHED obtained using the two isolation methods of ED and OG [67]. They found that the enzymatically digested cells displayed a significant heterogeneity comprising different cell sizes and morphologies even within the same colonies; on the contrary, SHED isolated with the OG method showed a uniform fibroblast-like morphology. The two types of cultures showed differences also in the immunophenotypic profiles and in the mineralization rate, which was lower in SHED obtained with the OG method compared to the digested SHED. Despite these observations, no differences could be detected in the growth rates of SHED isolated using the two methods.

Similarly to DPSCs, SHED express the two early MSCs surface markers STRO-1 and CD146. Cells positive to STRO-1 and CD146 were found to be located around blood vessels of the remnant pulp, implying that SHED may originate from a perivascular microenvironment [24]. SHED have also been identified as Immature Dental Pulp Stem Cells (IDPSCs) by Kerkis and co-workers [40]. Besides confirming the findings reported above, the authors established that IDPSCs express the ESCs markers Oct-4, Nanog, stage specific embryonic antigens (SSEA-3, SSEA-4), and tumorigenic recognition antigens (TRA-1-60, TRA-1-81). Further characterization revealed that SHED were uniformly positive for CD13, CD29, CD44, CD73, CD90, CD105, and CD166; and negative for CD14, CD34, CD45, and HLA-DR [41]. SHED were also found positive to several neural and glial markers, such as nestin, βIII-tubulin, glutamic acid decarboxylase (GAD), neuronal nuclei (NeuN), glial fibrillary acidic protein (GFAP), neurofilament M (NFM), and 2′,3′-cyclic nucleotide-3′-phosphodiesterase (CNPase), possibly due to the neural-crest cell origin of dental pulp [72].

Differentiation Potential of SHED

Under appropriate culture conditions, SHED have demonstrated the ability to undergo differentiation not only into odontogenic, osteogenic, neurogenic, and adipogenic [24], but also myogenic and chondrogenic cell lineages [40, 42].

Miura and colleagues observed that, when SHED were transplanted with HA/TCP particles into immunocompromised mice, they were able to differentiate into odontoblasts and were immunoreactive to DSPP. Nevertheless, these cells failed to generate a complete dentin/pulp-like complex, as do DPSCs in vivo [24]. Additionally, the same authors observed that SHED could not be differentiated directly into osteoblasts; rather, they were able to induce new bone formation by recruiting host murine cells [24]. This is not a property attributed to DPSCs following transplantation in vivo. The osteoinductive potential of SHED was demonstrated in a further study, in which these cells were found able of repairing critical-size calvarial defects in immunocompromised mice [43]. Together, the potential for odontogenic differentiation and the high osteoinductive potential in vivo suggest that SHED may represent a population of multipotent stem cells that are more immature than DPSCs. This is not surprising since deciduous teeth are significantly different from permanent teeth with regards to their developmental processes, tissue structure, and function [42].

As reported above, SHED readily express a variety of neural cell markers under non-neuronal inductive conditions [24]. When stimulated with neurogenic medium, Miura and colleagues observed that expression of βIII-tubulin, GAD, and NeuN increased, whereas expression of the other neural markers remained unchanged. Under these conditions, SHED lost their typical fibroblastic morphology and developed multicytoplasmic processes. Neural developmental potential was further studied in vivo by injecting SHED into the dentate gyrus of the hippocampus of immunocompromised mice. These studies showed that SHED survived for more than 10 days inside the mouse brain microenvironment and continued to express both neuronal and glial cell markers [24]. Koyama and co-workers reported the differentiation of SHED into the adipogenic and chondrogenic lineages in vitro [42]. The adipogenic differentiation of SHED was detected by the in vitro accumulation of lipid droplets, and by the expression of adipogenic markers such as Peroxisome proliferator-activated receptor-gamma (PPARG), and Lipoprotein lipase (LPL). The comparison between SHED and DPSCs revealed a higher adipogenic differentiation capacity of SHED than DPSCs. SHED were also shown to express the chondrogenic markers Sox-9, collagen type II (COL2), and collagen type X (COL10) when cultured for 14 days with bone morphogenic protein 2 (BMP-2), a chondrogenic signaling protein in the transforming growth factor β (TGFβ) family [42].

The strong differentiation plasticity of SHED was demonstrated in engraftment studies in vivo. In particular, Kerkis and co-workers observed that, when IDPSCs were transplanted into immunocompromised mice via intraperitoneal injection, they engrafted into the lungs, liver, spleen, brain and kidney, producing a tissue that was indistinguishable from the host tissue for liver, spleen, brain and kidney [40].

All these observations suggest that deciduous teeth could represent an ideal source of stem cells for repairing damaged tooth structures, inducing bone regeneration, and possibly treating neurodegenerative diseases.

Periodontal Ligament Stem Cells (PDLSCs)

PDLSCs Identification, Isolation, and Characterization

The periodontal ligament, one of the highly specialized and complex connective tissues of the human body, is embedded between the cementum and the inner wall of the alveolar bone socket. Periodontal ligament not only has an important role in supporting teeth, but also contributes to tooth nutrition, homoeostasis, and repair of damaged tissue [73, 74]. Periodontal ligament contains heterogeneous cell populations that can differentiate into either cementoblasts or osteoblasts. The presence of multiple cell types within periodontal ligament has led to speculation that this tissue might contain progenitor cells responsible for maintenance of periodontal ligament homoeostasis and regeneration. In effect, PDLSCs were identified in the periodontal ligament of extracted teeth by Seo and co-workers in 2004 [25]. These cells have been reported to form adherent clonogenic clusters of fibroblast-like cells, and had a proliferation rate similar to DPSCs after 24 h in culture.

PDLSCs can be obtained from periodontal ligament tissue after its separation from the root surface of extracted third molars [25]. Tran and co-workers compared the biological properties of PDLSCs isolated using the OG or ED methods [75]. Cells showed a similar morphology, mainly polygonal with expansion of the cytoplasm, regardless of the isolation protocol used. Additionally, the two isolation methods did not influence the proliferation rates, as well as the immunophenotypic profiles of PDLSCs. Regarding the osteogenic differentiation potential, PDLSCs obtained with the OG method produced a minor total area of mineralized nodules compared to that generated by the enzymatically digested cells.

In the work of Wang and colleagues it has been found that periodontal ligament possesses asymmetrically distributed stem cells, and that PDLSCs properties vary depending on the harvest location [76]. Cells isolated from the alveolar bone surface exhibited strong proliferation capability, and they were shown able to better regenerate the alveolar bone when compared to cells obtained from the root surface.

PDLSCs are similar to other MSCs with respect to the expression of STRO-1 and CD146 markers, implying that these cells might also be derived from a population of perivascular cells. In effect, McCulloch and co-workers have previously identified progenitor cells in the perivascular region of the periodontal ligament of mice [77]. Later, Chen and colleagues localized small clusters of PDLSCs also in the extravascular areas and regions near the cementum [78]. Differently from DPSCs and SHED, PDLSCs were found to express high levels of scleraxis, a tendon-specific transcription factor associated with tendon cells [25]. This is not surprising as the dense collagen fiber structure of the periodontal ligament makes it similar to the tendon. Further characterization revealed that PDLSCs express the following cell surface markers: CD9, CD10, CD13, CD26, CD29, CD44, CD59, CD73, CD90, CD105, CD106, and CD166. Like DPSCs and SHEDs, PDLSCs do not express CD14, CD34, CD45, and HLA-DR [33]. Different works demonstrated that PDLSCs also express ESCs markers, including Oct-4, Nanog, Sox-2, Rex-1, SSEA-1, SSEA-3, SSEA-4, TRA-1-60, and TRA-1-81 [44, 45], and a subset of neural crest markers, such as nestin, slug, p75, and Sox-10 [46].

Differentiation Potential of PDLSCs

PDLSCs were shown to be multipotent, having the ability to differentiate into osteoblasts, cementoblast-like cells, adipocytes, chondrocytes, and collagen-forming cells [25].

The osteogenic potential of PDLSCs was demonstrated through their ability to form a mineralized matrix containing calcium deposits after 14 days of exposure to osteogenic differentiation medium [47]. However, unlike BM-MSCs and DPSCs, these deposits were sparsely distributed in the culture. Under osteogenic induction medium, PDLSCs showed increased protein expression of the osteoblastic/cementoblastic markers ALP, BSP, OCN, and TGFβ receptor I. Despite the expression of these osteogenic/cementoblastic markers in vitro, PDLSCs were unable to form dentin or bone in vivo [25]. Nevertheless, PDLSCs loaded onto a HA/TCP-scaffold could give rise to a cementum/periodontal ligament structure when transplanted into immunocompromised mice and rats [25]. In particular, a thin layer of cementum-like tissue was formed along with condensed collagen fibers with sparse cells resembling periodontal ligament structure. In 2012, Song and co-workers compared stem cells derived from the periodontal ligament of permanent teeth (pPDLSCs) to those obtained from deciduous teeth (dPDLSCs) [79]. There were no significant differences in the proliferation rate, expression of stem cell markers, or in vitro differentiation between the two cell populations. Nevertheless, pPDLSCs generated more cementum/periodontal ligament-like structures in vivo than dPDLSCs, becoming better candidates for use in periodontium regeneration. In a recent work of Park and colleagues, PDLSCs from human healthy and inflamed periodontal ligament tissues were isolated and compared for their regenerative potential [22]. The proliferative potential did not differ between the two cell populations, and both of them were successfully able to differentiate under osteogenic/cementogenic and adipogenic culture conditions. Both cell types exhibited new cementum-like tissue and related periodontal ligament fiber regeneration after in vivo transplantation.

Although several cell types reside in the periodontal ligament tissue, adipocytes and chondrocytes have not been reported as native components. Nevertheless, PDLSCs were capable of differentiation into adipocytes after 25 days of culture in adipogenic inducing medium as demonstrated by the formation of lipid droplets, and the upregulation of adipocyte specific markers [25, 47]. Gay and colleagues also showed that PDLSCs could undergo chondrogenic differentiation after 21 days culture in vitro [47]. They not only observed cells with a chondrocyte cell morphology, but also the expression of chondrogenic markers, such as COL2 and proteoglycans. Another work demonstrated that PDLSCs differentiate into insulin producing cells, implying that these cells could also be able to generate cells of the endodermal lineage [46]. All these findings seem to suggest the pluripotency of PDLSCs and their ability to differentiate into cells from all three germ layers.

Dental Follicle Progenitor Cells (DFPCs)

DFPCs Identification, Isolation, and Characterization

The dental follicle is the ectomesodermal connective tissue surrounding the enamel organ and the dental papilla of the developing tooth germ prior to eruption [80]. Its main biological function is the coordination of tooth eruption through the regulation of the osteoclastogenesis and osteogenesis processes [81]. The dental follicle contains progenitor cells that form cementoblasts, periodontal ligament, and osteoblasts. It is thought that the dental follicle cells near the forming root differentiate into cementoblasts, whereas the cells towards the alveolar bone differentiate to osteoblasts. The dental follicle cells located between the cementoblast and osteoblast precursor cells develop into fibroblasts producing the ECM of the periodontal ligament [80].

Morsczeck and co-workers identified DFPCs in the dental follicle area of impacted wisdom teeth in 2005 using the same protocol that Gronthos and colleagues used five years earlier to discover DPSCs [26]. For the isolation of DFPCs, attached dental follicles are separated from the mineralized tooth, then minced and digested with an enzymatic solution [26]. It has to be reminded that DFPCs are only available from patients during wisdom tooth eruption, usually between 15 and 28 years of age [28]. DFPCs have been isolated by their ability to form clonogenic, fibroblastic-like colonies in culture that adhered to plastic [26]. DFPCs were found positive for nestin and notch-1 even after longer periods of propagation in culture. In addition, DFPCs expressed COL1, BSP, OCN, and FGF receptor 1-IIIC [26]. These cells also showed positive expression for STRO-1 and CD146, the two early MSCs markers also present on DPSCs, SHED, and PDLSCs. Furthermore, DFPCs have been reported to express CD9, CD10, CD13, CD29, CD44, CD53, CD59, CD73, CD90, CD105, CD106, and CD166 surface markers; these cells were negative for the hematopoietic antigens CD34 and CD45, CD117, and HLA-DR [28, 33, 48–50].

In order to establish if dental follicle contains heterogeneous cell populations, Luan and co-workers cloned 3 cell lines starting from murine DFPCs [48]. They found significant differences between the 3 dental follicle-derived cell lines in terms of proliferation rates, cell cycle distribution patterns, and cell surface markers. One cell line had high proliferation activity but it did not display any mineralization behavior, suggesting that it might be related to a periodontal ligament-type lineage. The second line was similar to the first one regarding proliferation potential but featured high ALP activity indicative of a highly undifferentiated state. The third cell line exhibited mineralization characteristics, indicating that it might be of cementoblastic or alveolar bone osteoblastic lineage.

Differentiation Potential of DFPCs

DFPCs have shown the potential to undergo osteogenic, adipogenic, chondrogenic, and neurogenic differentiation in vitro in both mice and rats [48, 51].

After 5 weeks of induction with osteogenic differentiation medium containing dexamethasone, human DFPCs at passage 6 were able to form a membrane-like structure. The structure of this membrane could be compared to that of the periodontal ligament, consisting of a fibrous tissue with parallel or net-like arranged collagen fibers and fibroblastic cells covered by a cell layer [26]. In contrast, when DFPCs at passage 17 were cultured under the same conditions for 4 weeks, they were able to produce compact calcified nodules. These globular or nodular elements were interconnected by connective tissue strands and surrounded by a connective tissue capsule. These results together would suggest that dental follicle derived from human wisdom teeth represents a source of precursor cells for periodontal ligament-fibroblasts, cementoblasts, and osteoblasts.

Morsczeck and co-workers also investigated osteogenic potential in vivo by transplanting DFPCs in conjunction with hydroxyapatite powder (HAP) into immunocompromised mice [26]. After 8 weeks from transplantation, the cells generated a structure lining the surfaces of the HAP particles, comprised of fibrous or rigid connective tissues. Nevertheless, transplanted DFPCs were unable to build a matrix for cementum or bone formation. More recently, the hard tissue-forming potential of human DFPCs has been evaluated by Yagyuu and co-workers after transplanting these cells with ceramics discs into immunocompromised rats [49]. After 8 weeks, a mineralized tissue structure containing cementocyte/osteocyte cells was formed in vivo, but the exact identity of the tissue type could not be determined as dentin, cementum or bone.

The multilineage potential of murine DFPCs has been assessed in the work of Luan and colleagues [48]. They obtained dental follicle tissues from mouse molar developing tooth organs of 8 days postnatal mice. The isolated DFPCs were induced to differentiate into adipogenic, chondrogenic, or osteogenic lineages using different induction media. Lipid droplets were observed in culture after induction of DFPCs with adipogenic medium for 6 days. A glycosaminoglycan (GAG)-rich fibrous cartilage-type ECM was generated following chondrogenic induction for 21 days; whereas the presence of a mineralized tissue was detected after placing DFPCs in osteogenic medium for 21 days.

In the study of Yao and colleagues, dental follicles were surgically removed from the first mandibular molars of rat pups at days 5–7 postnatally [51]. The isolated DFPCs were then subjected to osteogenic, adipogenic, or neurogenic differentiation in vitro. When placed in an osteogenic differentiation medium for 2 weeks, osteoblasts/cementoblasts formed and developed mineralization nodules, as revealed by Alizarin Red S and von Kossa staining. Adipogenic differentiation potential of DFPCs was revealed by the presence of Oil Red O stained cells. The incubation of DFPCs with a neuronal induction medium for 24 h resulted in the development of cells that resembled multipolar neurons. The neural differentiation in vitro has been demonstrated by Morsczeck and colleagues in a later work [50]. After stimulation with different serum-replacement media for 7 days, the authors observed that, like SHED, human DFPCs expressed neural stem cells-associated markers Sox-2, nestin, and ATP binding cassette, subfamily G, member 2 (ABCg2). Nevertheless, SHED and DFPCs did not have the same neural differentiation potential in vitro, as demonstrated by the absence of neurosphere-like cell clusters in DFPCs differentiated cultures. In addition, the authors excluded a glial cell differentiation potential of DFPCs as revealed by the weak expression of the glial cell marker GFAP.

Stem Cells from Apical Papilla (SCAP)

SCAP Identification, Isolation, and Characterization

Sonoyama and co-workers were the first to isolate SCAP from the apical papilla of impacted wisdom teeth of 18-20 years old patients in 2006 [27]. The apical papilla is the soft tissue found at the apex of the developing permanent tooth that eventually becomes the pulp tissue in the mature tooth [82]. A further study conducted by Sonoyama and colleagues demonstrated that apical papilla appears to be histologically distinct from pulp and contains unique potent MSCs [52]. In particular, histological characterization of the apical papilla showed that there is an apical cell-rich zone lying between the apical papilla and the pulp. In comparison to the dental pulp, apical papilla seems to contain less blood vessels and cellular components.

The isolation of SCAP is achieved by gentle separation of root apical papilla from the surface of the root with immature apex during the extraction of third molars. Apical papilla is minced, then digested with an enzymatic solution to obtain single cell suspensions [27]. SCAP have been identified as adherent clonogenic cell clusters of fibroblast-like cells. These cells were shown to have a greater proliferation rate and population doubling than DPSCs isolated from the same tooth [27]. In particular, over 70 population doublings were measured for SCAP. Furthermore, it has been reported that SCAP display higher levels of survivin and telomerase, both of which are markers associated with cell proliferation, when compared to DPSCs from the same tooth [27].

SCAP show expression of the same surface markers as the other DSCs, that is STRO-1, CD13, CD24, CD29, CD73, CD90, CD105, CD106, CD146, and CD166. At the same time, these cells are negative for CD18, CD34, CD45, and CD150 antigens. Interestingly, SCAP express a unique marker, not detectable in other DSCs: CD24 [27, 53].

Differentiation Potential of SCAP

SCAP have shown the ability to undergo odontoblastic/osteoblastic, adipogenic, chondrogenic, and neural differentiation in vitro [27, 52, 53].

The potential of human SCAP to undergo odontoblastic/osteoblastic differentiation was evaluated after 4 weeks of induction with an osteogenic differentiation medium [27]. SCAP were able to form small round Alizarin Red-positive nodules, indicating calcium accumulation in vitro. Interestingly, the expression of the SCAP-specific marker CD24 was downregulated in response to osteogenic stimulation.

However, the biological significance of this finding is currently unclear. Similarly to DPSCs and SHED, SCAP have shown odontogenic potential in vitro. Nevertheless, SCAP express lower levels of DSP, matrix extracellular phosphoglycoprotein (MEPE), TGFβ receptor II, Flt-1 (VEGF receptor 1), Flg (FGFR1), FGFR3, and melanoma-associated glycoprotein (MUC18) compared to DPSCs [52]. When transplanted into immunocompromised mice, human SCAP mixed with HA/TCP particles underwent in vivo differentiation into odontoblasts, regenerating a dentin/pulp-like structure and connective tissue similarly to DPSCs [27]. Although SCAP are comparable to DPSCs, they represent a distinct source of dental stem/progenitor cells. It has been speculated that SCAP appear to be the source of primary odontoblasts responsible for the formation of root dentin, whereas DPSCs are likely the source of replacement odontoblasts that form reparative dentin [23]. In order to play a functional role in vivo, the root has to connect with the periodontal ligament. In a preliminary study using minipigs, co-transplantation of swine SCAP and PDLSCs onto HA/TCP particles produced the formation of dentine and periodontal ligament, respectively [27]. These findings suggest that SCAP, together with PDLSCs, could be used to create a biological root that could be used in a similar way as a metal implant, by capping with an artificial dental crown.

SCAP also demonstrated the ability to undergo adipogenic and chondrogenic differentiation following induction in vitro [27, 52, 53]. The adipogenic differentiation potential of SCAP was similar to that of DPSCs but much weaker than that of BM-MSCs. The chondrogenic differentiation was observed as the cells formed a pellet with a spherical appearance and expressed the cartilage extracellular protein COL2 [53].

An interesting feature of SCAP is that they express several neural markers even when not exposed to neurogenic stimulation, implying their possible origin from the neural crest [53]. However, when stimulated, more neural markers are expressed, such as βIII-tubulin, GFAP, GAD, nestin, NeuN, NFM, neuron-specific enolase (NSE), and CNPase [52].

Considering all these observations, SCAP seem to be more embryonic-like than other sources of DSCs [83]. Consequently, SCAP could have potential for applications not only in dentistry, but also in the treatment of neurodegenerative and ischemic diseases, diabetes research, and other applications in the field of regenerative medicine [84].

Gingival Mesenchymal Stem Cells (GMSCs)

GMSCs Identification, Isolation, and Characterization

Gingiva is a unique oral tissue attached to the alveolar bone of tooth sockets, recognized as a biological mucosal barrier and a distinct component of the oral mucosal immunity [29]. One of the most striking characteristics of gingiva is its remarkable regenerative and wound healing capacity, characterized by a rapid reconstitution of tissue architecture and little evidence of scarring, in contrast to the common scar formation present in skin [85]. Histologically, gingiva is composed of three layers: an epithelial layer, a basal layer, and a lower spinous layer that is similar to the dermis of the skin. Zhang and co-workers were the first to report the isolation of GMSCs from the spinous layer of human gingiva in 2009 [29]. They used healthy gingival tissues obtained from remnant or discarded tissues following routine dental procedures. In this context, human gingiva represents an interesting alternative to the other dental tissues as source of DSCs. The major advantage of using gingiva is the possibility to achieve a large quantity of stem cells without the need to sacrifice a tooth irreversibly to obtain its pulp, periodontal ligament, or dental follicle. Furthermore, gingival tissues can be obtained from minimally invasive procedures at any time in life.

For the isolation of GMSCs, gingival tissues are generally incubated with dispase to separate the epithelial and lower spinous layer. The tissues are then minced and enzymatically digested with collagenase [29, 54]. Other methods described for the isolation of GMSCs involved the use of collagenase and dispase at the same time, followed by the removal of the first digested cell suspension for excluding epithelial cells, and by additional incubations with the same enzymatic solution [55, 86].

GMSCs were identified as plastic-adherent cells with a uniform fibroblast-like morphology [29]. These cells have been found to display stable phenotype, and to maintain normal karyotype and telomerase activity in long-term cultures [55].

Regarding the expression of surface markers, GMSCs result positive to CD13, CD29, CD44, CD73, CD90, CD105, CD146, STRO-1, Oct-4, and SSEA-4 antigens; at the same time, they lack expression of the hematopoietic markers CD14, CD34, and CD45, and HLA-DR [29, 54, 55].

Interestingly, GMSCs have been found to display immunomodulatory properties; in particular, they inhibit lymphocytes proliferation and express a wide range of immunosuppressive factors, such as Interleukin-10 (IL-10), IDO, inducible NO synthase (iNOS), and cyclooxygenase 2 (COX-2) in response to the inflammatory cytokine Interferon γ (IFNγ) [29].

Differentiation Potential of GMSCs

GMSCs demonstrated a multilineage differentiation potential into adipogenic, osteogenic, and chondrogenic cell lineages in vitro [29, 54, 56].

Under adipogenic stimulation, single colony-derived GMSCs could differentiate into adipocytes as demonstrated by Oil Red O staining, and increased expression of specific adipogenic markers, including PPARG and LPL [29]. The osteogenic differentiation was evaluated after incubation of GMSCs with an osteogenic medium. GMSCs were able to form mineralized nodules as determined by Alizarin Red S staining, and they showed increased expression of several osteogenic markers [29, 55]. When GMSCs were cultured as pelleted micromass under a chondrogenic differentiation medium, they were able to synthesize a multilayered sulphated proteglycan-rich matrix, and express chondrocyte specific genes, such as aggrecan, Sox-9, and COL2 [55]. It has also been reported that GMSCs undergo endothelial and neural differentiation in vitro when stimulated with proper culture medium [29]. Endothelial differentiation was observed by the expression of the endothelial marker CD31; whereas neural differentiation was ascertained by the positivity for the neural markers GFAP, NFM, and βIII-tubulin.

The in vivo differentiation potential of human GMSCs has been evaluated by transplanting these cells with HA/TCP particles into immunocompromised mice [29]. GMSCs were able to regenerate a connective-like tissue containing collagen fibers; however, osteogenic differentiation of the transplanted GMSCs was not observed. The in vivo bone regeneration potential of GMSCs was demonstrated in a another study, in which GMSCs were seeded onto HA/TCP particles for 24 h, incubated in osteogenic medium for 10 days, then implanted subcutaneously in immunocompromised mice [55]. After 10 weeks, the retrieved implants showed the presence of a highly mineralized tissue.

Overall, these findings seem to indicate that human gingiva is an interesting source of DSCs, since GMSCs have shown ability of multilineage differentiation in vitro, and tissue regeneration in vivo.

Natal Dental Pulp Stem Cells (NDP-SCs)

NDP-SCs Identification, Isolation, and Characterization

Natal teeth are deciduous teeth that arise in newborn. These teeth are smaller than primary teeth and have little or no root development. Natal teeth are very rare since very few newborns are born with teeth, that is approximately one in every 2000–3000 births [87].

Human NDP-SCs are a unique type of dental pulp-related stem cells isolated only from the pulp of newborn teeth [30]. Karaöz and co-workers isolated NDP-SCs from 2 natal teeth obtained from a healthy newborn female in 2010 [30]. After fracturing of the dental crown, the dental pulp was uncovered and digested using collagenase to generate single cell suspensions. A small number of NDP-SCs adhered to plastic in culture, and displayed a fibroblast-like morphology that became flattened at later passages. When compared to SHED and DPSCs, NDP-SCs had higher proliferation rate [57]. NDP-SCs express CD13, CD29, CD44, CD73, CD90, CD146, and CD166 markers, but do not express CD14, CD34, CD45, and HLA-DR antigens, in a similar way to the other DSCs [30, 57]. Interestingly, NDP-SCs express detectable levels of the ESCs markers Nanog, Rex-1, and Oct-4, as well as the transcription factors Sox-2 and FoxD3, suggesting that these cells display some of the characteristics for pluripotency. Surprisingly, when growth in standard medium, NDP-SCs show a positive immune reaction for the osteogenic markers COL1, OCN, ON, OPN, BMP-2, and BMP-4. Furthermore, these cells express some myogenic markers, such as desmin and myogenin, and the chondrogenic marker COL2 under the same culture conditions. NDP-SCs were also found positive to neural markers, including nestin, vimentin, GFAP, and βIII-tubulin [30]. In a recent study of Akpinar and co-workers the protein expression profile of human NDP-SCs was analyzed and compared to that of human DPSCs and SHED [57]. They found that NDP-SCs cell proteome was more similar to the SHED proteome than the DPSCs proteome, and they identified 61 proteins that were predominantly expressed by all the 3 DSCs. Further analysis on the molecular function of the identified proteins revealed that most of them played roles in cellular architecture. Proteins involved in the folding machinery were also described, as well as transcription, protein biosynthesis, and degradation related proteins, thus implying the presence of cellular self-renewal and proliferation.

Differentiation Potential of NDP-SCs

The multilineage differentiation potential of NDP-SCs has been demonstrated by their differentiation into osteoblasts, adipocytes, chondroblasts, myoblasts, and neuro-glial-like cells in vitro [30, 57].

Adipogenic differentiation of human NDP-SCs was confirmed by the formation of lipid droplets that enlarged and invaded the entire cytoplasm during cell differentiation. Following culture in osteogenic medium, it was possible to identify calcium deposits by Alizarin Red S staining. Chondrogenic differentiation was evidenced by the presence of round cells which resembled hyaline chondrocytes, and expressed GAG. When treated with 5-azacytidine, NDP-SCs acquired a myocyte phenotype in culture and, within 8 days following seeding, the myoblasts fused to form small myotubes. These immature myotubes had clustered nuclei that differentiated during the next 2–3 days into large multinucleate syncytia. The neuronal differentiation of NDP-SCs was demonstrated by the formation of neuron-like cells that displayed different morphologies, ranging from simply bipolar to branched multipolar cells, and expressed several neural-specific markers.

Conclusions

The observations reported in this chapter seem to indicate that DSCs are a viable alternative to other adult MSCs for regenerative medicine. The main advantage in using DSCs is that they are easily obtainable from several dental tissues which are often discarded. In addition, DSCs have been described as highly clonogenic and proliferative cells, possessing multilineage differentiation potential in vitro, and regenerative properties in vivo. A better comprehension of the biology of DSCs would result in significant benefits for the management of diseases affecting human dental tissues as well as other organs.

References

1. Miletich I, Sharpe PT (2004) Neural crest contribution to mammalian tooth formation. Birth Defects Res C Embryo Today 72(2):200–212
2. Zavan B, Bressan E, Sivolella S et al (2011) Dental pulp stem cells and tissue engineering strategies for clinical application on odontoiatric field. In: Pignatello R (ed) Biomaterials science and engineering, Intech. doi:10.5772/24871
3. Jimenez-Rojo L, Ibarretxe G, Aurrekoetxea M et al (2010) Epiprofin/Sp6: a new player in the regulation of tooth development. Histol Histopathol 25(12):1621–1630
4. Wright T (2007) The molecular control of and clinical variations in root formation. Cells Tissues Organs 186(1):86–93
5. Friedman A, Friedman G (2013) Plasma-assisted healing and treatment of diseases. In: Plasma medicine. Wiley, Oxford. doi:10.1002/9781118437704.ch9
6. D'Aquino R, De Rosa A, Laino G et al (2009) Human dental pulp stem cells: from biology to clinical applications. J Exp Zool B Mol Dev Evol 312B(5):408–415
7. Gronthos S, Mankani M, Brahim J et al (2000) Postnatal human dental pulp stem cells (DPSCs) in vitro and in vivo. Proc Natl Acad Sci U S A 97(25):13625–13630
8. Inanç B, Elçin YM (2011) Stem cells in tooth tissue regeneration-challenges and limitations. Stem Cell Rev 7(3):683–692
9. Schofield R (1978) The relationship between the spleen colony-forming cell and the haemopoietic stem cell. Blood Cells 4(1–2):7–25
10. Morrison SJ, Spradling AC (2008) Stem cells and niches: mechanisms that promote stem cell maintenance throughout life. Cell 132(4):598–611
11. Toma JG, Akhavan M, Fernandes KJ et al (2001) Isolation of multipotent adult stem cells from the dermis of mammalian skin. Nat Cell Biol 3(9):778–784
12. Friedenstein AJ, Chailakhjan RK, Lalykina KS (1970) The development of fibroblast colonies in monolayer cultures of guinea pig bone marrow and spleen cells. Cell Tissue Kinet 3(4):393–403
13. Zuk PA, Zhu M, Ashjian P et al (2002) Human adipose tissue is a source of multipotent stem cells. Mol Biol Cell 13(12):4279–4295
14. Zvaifler NJ, Marinova-Mutafchieva L, Adams G et al (2000) Mesenchymal precursor cells in the blood of normal individuals. Arthritis Res 2(6):477–488
15. Alsalameh S, Amin R, Gemba T et al (2004) Identification of mesenchymal progenitor cells in normal and osteoarthritic human articular cartilage. Arthritis Rheum 50(5):1522–1532
16. Wong MH (2004) Regulation of intestinal stem cells. J Investig Dermatol Symp Proc 9:224–228
17. De Bari C, Dell'Accio F, Luyten FP (2001) Human periosteum derived cells maintain phenotypic stability and chondrogenic potential throughout expansion regardless of donor age. Arthritis Rheum 44(1):85–95
18. Potdar PD, Jethmalani YD (2015) Human dental pulp stem cells: applications in future regenerative medicine. World J Stem Cells 7(5):839–851
19. Karamzadeh R, Eslaminejad MB (2013) Dental-related stem cells and their potential in regenerative medicine. In: Andrades JA (ed) Regenerative medicine and tissue engineering, InTech, doi:10.5772/55927
20. Alongi DJ, Yamaza T, Song Y et al (2010) Stem/progenitor cells from inflamed human dental pulp retain tissue regeneration potential. Regen Med 5(4):617–631
21. Sun HH, Chen B, Zhu QL et al (2014) Investigation of dental pulp stem cells isolated from discarded human teeth extracted due to aggressive periodontitis. Biomaterials 35(35): 9459–9472
22. Park JC, Kim JM, Jung IH et al (2011) Isolation and characterization of human periodontal ligament (PDL) stem cells (PDLSCs) from the inflamed PDL tissue: in vitro and in vivo evaluations. J Clin Periodontol 38(8):721–731

23. Huang GT, Gronthos S, Shi S (2009) Mesenchymal stem cells derived from dental tissues vs. those from other sources: their biology and role in regenerative medicine. J Dent Res 88(9):792–806
24. Miura M, Gronthos S, Zhao M et al (2003) SHED: stem cells from human exfoliated deciduous teeth. Proc Natl Acad Sci U S A 100:5807–5812
25. Seo BM, Miura M, Gronthos S et al (2004) Investigation of multipotent postnatal stem cells from human periodontal ligament. Lancet 364(9429):149–155
26. Morsczeck C, Götz W, Schierholz J et al (2005) Isolation of precursor cells (PCs) from human dental follicle of wisdom teeth. Matrix Biol 24(2):155–165
27. Sonoyama W, Liu Y, Fang D et al (2006) Mesenchymal stem cell-mediated functional tooth regeneration in swine. PLoS One 1, e79
28. Morsczeck C, Schmalz G, Reichert TE et al (2008) Somatic stem cells for regenerative dentistry. Clin Oral Investig 12(2):113–118
29. Zhang Q, Shi S, Liu Y et al (2009) Mesenchymal stem cells derived from human gingiva are capable of immunomodulatory functions and ameliorate inflammation-related tissue destruction in experimental colitis. J Immunol 183(12):7787–7798
30. Karaöz E, Doğan BN, Aksoy A et al (2010) Isolation and in vitro characterisation of dental pulp stem cells from natal teeth. Histochem Cell Biol 133(1):95–112
31. Abdullah MF, Ponnuraj KT, Mokhtar KI (2013) DPSCs and SHED in tissue engineering and regenerative medicine. Open Stem Cell J 4:1–6
32. Shi S, Gronthos S (2003) Perivascular niche of postnatal mesenchymal stem cells in human bone marrow and dental pulp. J Bone Miner Res 18(4):696–704
33. Lindroos B, Mäenpää K, Ylikomi T et al (2008) Characterisation of human dental stem cells and buccal mucosa fibroblasts. Biochem Biophys Res Commun 368(2):329–335
34. D'Aquino R, Graziano A, Sampaolesi M et al (2007) Human postnatal dental pulp cells co-differentiate into osteoblasts and endotheliocytes: a pivotal synergy leading to adult bone tissue formation. Cell Death Differ 14(6):1162–1171
35. Gronthos S, Brahim J, Li W et al (2002) Stem cell properties of human dental pulp stem cells. J Dent Res 81(8):531–535
36. Arthur A, Rychkov G, Shi S et al (2008) Adult human dental pulp stem cells differentiate toward functionally active neurons under appropriate environmental cues. Stem Cells 26(7):1787–1795
37. Nosrat IV, Widenfalk J, Olson L et al (2001) Dental pulp cells produce neurotrophic factors, interact with trigeminal neurons in vitro, and rescue motoneurons after spinal cord injury. Dev Biol 238(1):120–132
38. Zhang W, Walboomers XF, Shi S et al (2006) Multilineage differentiation potential of stem cells derived from human dental pulp after cryopreservation. Tissue Eng 12(10):2813–2823
39. Paino F, Ricci G, De Rosa A et al (2010) Ecto-mesenchymal stem cells from dental pulp are committed to differentiate into active melanocytes. Eur Cell Mater 20:295–305
40. Kerkis I, Kerkis A, Dozortsev D et al (2006) Isolation and characterization of a population of immature dental pulp stem cells expressing OCT-4 and other embryonic stem cell markers. Cells Tissues Organs 184(3-4):105–116
41. Suchánek J, Visek B, Soukup T et al (2010) Stem cells from human exfoliated deciduous teeth–isolation, long term cultivation and phenotypical analysis. Acta Medica (Hradec Kralove) 53(2):93–99
42. Koyama N, Okubo Y, Nakao K et al (2009) Evaluation of pluripotency in human dental pulp cells. J Oral Maxillofac Surg 67(3):501–506
43. Seo BM, Sonoyama W, Yamaza T et al (2008) SHED repair critical-size calvarial defects in mice. Oral Dis 14(5):428–434
44. Kawanabe N, Murata S, Murakami K et al (2010) Isolation of multipotent stem cells in human periodontal ligament using stage-specific embryonic antigen-4. Differentiation 79(2):74–83

45. Trubiani O, Zalzal SF, Paganelli R et al (2010) Expression profile of the embryonic markers nanog, OCT-4, SSEA-1, SSEA-4, and frizzled-9 receptor in human periodontal ligament mesenchymal stem cells. J Cell Physiol 225(1):123–131
46. Huang CY, Pelaez D, Dominguez-Bendala J et al (2009) Plasticity of stem cells derived from adult periodontal ligament. Regen Med 4(6):809–821
47. Gay IC, Chen S, MacDougall M (2007) Isolation and characterization of multipotent human periodontal ligament stem cells. Orthod Craniofac Res 10(3):149–160
48. Luan X, Ito Y, Dangaria S et al (2006) Dental follicle progenitor cell heterogeneity in the developing mouse periodontium. Stem Cells Dev 15(4):595–608
49. Yagyuu T, Ikeda E, Ohgushi H et al (2010) Hard tissue-forming potential of stem/progenitor cells in human dental follicle and dental papilla. Arch Oral Biol 55(1):68–76
50. Morsczeck C, Völlner F, Saugspier M et al (2010) Comparison of human dental follicle cells (DFCs) and stem cells from human exfoliated deciduous teeth (SHED) after neural differentiation in vitro. Clin Oral Investig 14(4):433–440
51. Yao S, Pan F, Prpic V et al (2008) Differentiation of stem cells in the dental follicle. J Dent Res 87(8):767–771
52. Sonoyama W, Liu Y, Yamaza T et al (2008) Characterization of the apical papilla and its residing stem cells from human immature permanent teeth: a pilot study. J Endod 34(2):166–171
53. Abe S, Yamaguchi S, Amagasa T (2007) Multilineage cells from apical pulp of human tooth with immature apex. Oral Sci Int 4:45–58
54. Gao Y, Zhao G, Li D et al (2014) Isolation and multiple differentiation potential assessment of human gingival mesenchymal stem cells. Int J Mol Sci 15(11):20982–20996
55. Tomar GB, Srivastava RK, Gupta N et al (2010) Human gingiva-derived mesenchymal stem cells are superior to bone marrow-derived mesenchymal stem cells for cell therapy in regenerative medicine. Biochem Biophys Res Commun 393(3):377–383
56. Mitrano TI, Grob MS, Carrión F et al (2010) Culture and characterization of mesenchymal stem cells from human gingival tissue. J Periodontol 81(6):917–925
57. Akpinar G, Kasap M, Aksoy A et al (2014) Phenotypic and proteomic characteristics of human dental pulp derived mesenchymal stem cells from a natal, an exfoliated deciduous, and an impacted third molar tooth. Stem Cells Int 2014:457059
58. Yamamura T (1985) Differentiation of pulpal cells and inductive influences of various matrices with reference to pulpal wound healing. J Dent Res 64:530–540
59. Bressan E, Ferroni L, Gardin C et al (2012) Donor age-related biological properties of human dental pulp stem cells change in nanostructured scaffolds. PLoS One 7(11), e49146
60. Park SH, Hsiao GY, Huang GT (2004) Role of substance P and calcitonin gene-related peptide in the regulation of interleukin-8 and monocyte chemotactic protein-1 expression in human dental pulp. Int Endod J 37(3):185–192
61. Spath L, Rotilio V, Alessandrini M et al (2010) Explant-derived human dental pulp stem cells enhance differentiation and proliferation potentials. J Cell Mol Med 14(6B):1635–1644
62. Huang GT, Sonoyama W, Chen J et al (2006) In vitro characterization of human dental pulp cells: various isolation methods and culturing environments. Cell Tissue Res 324(2):225–236
63. Karamzadeh R, Eslaminejad MB, Aflatoonian R (2012) Isolation, characterization and comparative differentiation of human dental pulp stem cells derived from permanent teeth by using two different methods. Vis Exp 69:4372
64. Hilkens P, Gervois P, Fanton Y et al (2013) Effect of isolation methodology on stem cell properties and multilineage differentiation potential of human dental pulp stem cells. Cell Tissue Res 353(1):65–78
65. Suchánek J, Soukup T, Ivancaková R et al (2007) Human dental pulp stem cells–isolation and long term cultivation. Acta Medica (Hradec Kralove) 50(3):195–201
66. Laino G, D'Aquino R, Graziano A et al (2005) A new population of human adult dental pulp stem cells: a useful source of living autologous fibrous bone tissue (LAB). J Bone Miner Res 20(8):1394–1402

67. Bakopoulou A, Leyhausen G, Volk J et al (2011) Assessment of the impact of two different isolation methods on the osteo/odontogenic differentiation potential of human dental stem cells derived from deciduous teeth. Calcif Tissue Int 88(2):130–141
68. Simmons PJ, Torok-Storb B (1991) CD34 expression by stromal precursors in normal human adult bone marrow. Blood 78(11):2848–2853
69. Yu J, He H, Tang C et al (2010) Differentiation potential of STRO-1+ dental pulp stem cells changes during cell passaging. BMC Cell Biol 11:32
70. Ferroni L, Gardin C, Sivolella S et al (2015) A hyaluronan-based scaffold for the in vitro construction of dental pulp-like tissue. Int J Mol Sci 16(3):4666–4681
71. Brar GS, Toor RS (2012) Dental stem cells: dentinogenic, osteogenic, and neurogenic differentiation and its clinical cell based therapies. Indian J Dent Res 23(3):393–397
72. Chai Y, Jiang X, Ito Y et al (2000) Fate of the mammalian cranial neural crest during tooth and mandibular morphogenesis. Development 127(8):1671–1679
73. Beertsen W, McCulloch CA, Sodek J (1997) The periodontal ligament: a unique, multifunctional connective tissue. Periodontol 13:20–40
74. Shimono M, Ishikawa T, Ishikawa H et al (2003) Regulatory mechanisms of periodontal regeneration. Microsc Res Tech 60(5):491–502
75. Tran Hle B, Doan VN, Le HT et al (2014) Various methods for isolation of multipotent human periodontal ligament cells for regenerative medicine. Vitro Cell Dev Biol Anim 50(7):597–602
76. Wang L, Shen H, Zheng W et al (2011) Characterization of stem cells from alveolar periodontal ligament. Tissue Eng Part A 17(7-8):1015–1026
77. McCulloch CA (1985) Progenitor cell populations in the periodontal ligament of mice. Anat Rec 211(3):258–262
78. Chen SC, Marino V, Gronthos S et al (2006) Location of putative stem cells in human periodontal ligament. J Periodontal Res 41(6):547–553
79. Song JS, Kim SO, Kim SH et al (2012) In vitro and in vivo characteristics of stem cells derived from the periodontal ligament of human deciduous and permanent teeth. Tissue Eng Part A 18(19–20):2040–2051
80. Ten Cate AR (1997) The development of the periodontium-a largely ectomesenchymally derived unit. Periodontol 13:9–19
81. Wise GE, Frazier-Bowers S, D'Souza RN (2002) Cellular, molecular, and genetic determinants of tooth eruption. Crit Rev Oral Biol Med 13(4):323–334
82. Linde A, Goldberg M (1993) Dentinogenesis. Crit Rev Oral Biol Med 4(5):679–728
83. Volponi AA, Pang Y, Sharpe PT (2010) Stem cell-based biological tooth repair and regeneration. Trends Cell Biol 20(12):715–722
84. Mayo V, Sawatari Y, Huang CY et al (2014) Neural crest-derived dental stem cells–where we are and where we are going. J Dent 42(9):1043–1051
85. Irwin CR, Picardo M, Ellis I et al (1994) Inter- and intra-site heterogeneity in the expression of fetal-like phenotypic characteristics by gingival fibroblasts: potential significance for wound healing. J Cell Sci 107:1333–1346
86. Jin SH, Lee JE, Yun JH et al (2015) Isolation and characterization of human mesenchymal stem cells from gingival connective tissue. J Periodontal Res 50(4):461–467
87. Leung AK, Robson WL (2006) Natal teeth: a review. J Natl Med Assoc 98(2):226–228

Protocols for Dental-Related Stem Cells Isolation, Amplification and Differentiation

Jakub Suchánek, Klára Zoe Browne, Tereza Suchánková Kleplová, and Yvona Mazurová

If I have seen further, it is by standing on the shoulders of giants. Isaac Newton

Election of Research Material

To elect research material which would appropriately accommodate individual dental-related stem cell experiments it is essential to understand both the genesis of specific tissue-stem cell lineages and their consequent inherent properties such as population curve in time and behavior under various environmental conditions in vivo (in an intact organism) as well as in vivo modified (transplanted as modified autologous or heterologous material) and in vitro (outside of an organism) [1]. The below chapters provide comprehensive overview of the up-to-date scientific knowledge in this field of study concerning all the tissues of tooth and the tissues immediately adjoined.

Firstly, we outline dental-related tissues genesis and anatomical location, and thereafter the quality and quantity of their stem cell populations. We consider the complete range of postnatal dental-related tissues-stem cell populations, starting from the incident predeciduous dentition [2] to the primary and secondary dentition [3].

Specifically, the feasible dental-related postnatal sources of stem cells are all the mature tissues of all dentitions (natal/neonatal dental pulp [4], immature dental pulp [5], deciduous teeth dental pulp [6], adult dental pulp [7]; apical papilla [8], periodontal ligament [9] and gingiva [10] plus the immature tissues of the secondary dentition dental follicle [11] and dental papilla).

J. Suchánek, Ph.D. (✉) • T.S. Kleplová, M.D.
Department of Dentistry, Faculty of Medicine in Hradec Králové, Charles University in Prague, Sokolská 581, 500 05 Hradec Králové, Czech Republic
e-mail: SuchanekJ@lfhk.cuni.cz

K.Z. Browne, M.Sc. • Y. Mazurová, C.Sc., M.D.
Department of Histology and Embryology Faculty of Medicine in Hradec Kralove, Charles University in Prague, Sokolská 581, 500 05 Hradec Králové, Czech Republic

B. Zavan, E. Bressan (eds.), *Dental Stem Cells: Regenerative Potential*, Stem Cell Biology and Regenerative Medicine, DOI 10.1007/978-3-319-33299-4_2

Dental-Related Tissues' Genesis

Dental-related tissues—lifecycle spans from early stages on intrauterine life and depending on tissue type can last till the host's organism demise. These tissues histogenesis and odontogenesis start when the cells of the ectoderm and mesoderm germ layers interact, differentiate and form increasingly specialised tissues in the area of mouth cavity. The process does not stop until the tooth organ maturation and eruption by when all the dental-related tissues are mature. Seeing that different dentitions (predeciduous, primary, secondary) start forming at different times from tissues of different level of maturation, it is sensible that their dental-related tissues quality differs as well. The following overview of human teeth histogenesis and odontogenesis describes the developments in the dental-related tissues-quality over time.

Histogenesis

All the cells which form dental-related tissues originate in adjoined leaf-like layers of ectoderm and mesoderm. These germ layer-cells move around and create tissues which take on different morphological characteristics and purpose throughout the process of odontogenesis and regeneration.

While the tissues originating from mesoderm germ layer gives rise to oral mesenchyme (the source of all connective and supporting soft tissues and vessels composing the tooth and tooth socket), the ectoderm germ layer gives rise to two original types of tissues: the stomodeal ectoderm (oral epithelium) and the cranial neural-crest-derived ectomesenchyme (neural-crest cells which have undergone an epithelial to mesenchymal transition and manifest mesenchymal phenotype [12]); which together construct all the hard tissues of the tooth.

Organogenesis/Odontogenesis

The complex process of tooth organogenesis/odontogenesis starts at the 6th week of the human embryogenesis and, depending on the type of a tooth, lasts up to 18 years. Odontogenesis is divided into four stages: (1) *the initiation stage*, (2) *the bud stage*, (3) *the cap stage*, and (4) *the bell stage*.

The initiation stage starts with the migration of oral epithelium cells into the underlying mesenchyme and ectomesenchyme of the stomodeum where they form the primary (general) dental lamina, the first signaling centre of tooth development which gives rise to all the teeth buds of primary dentition (20 deciduous teeth) and 12 molars of secondary dentition. Subsequently, around 12th week, the secondary (successional) dental lamina evolves, as a lingual extension of the primary lamina. Finally, at 14th week of embryonic development, the secondary dental lamina gives rise to the first germs of permanent teeth with predecessors (20 teeth)/at the primary dentition bell stage.

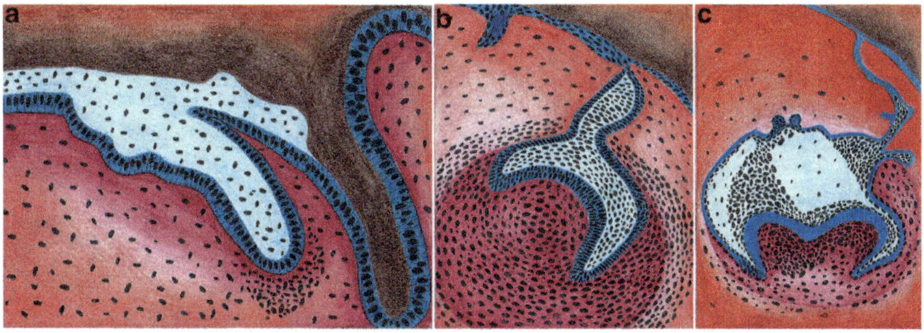

Fig.1 (**a**) developing tooth in a bud stage, (**b**) developing tooth in a cap stage, (**c**) developing tooth in a bell stage

The bud stage (Fig. 1a) is the first stage of the actual tooth germ development (at about 8th week for deciduous teeth). The ectodermal (epithelial) solid bud induces the proliferation and condensation of the underlying mesenchymal and ectomesenchymal cells to form the dental papilla which inductive capacity is essential for further development of ectodermal epithelial bud.

During *the cap stage* (Fig. 1b) (at about 9th–10th week), 3 basic structures develop, i.e. the enamel organ, dental papilla and dental follicle.

The bell stage (Fig. 1c) (approx. 11th–12th week) is primarily characterized by the activation of the enamel organ and the formation of the dentino-enamel junction. The enamel organ is composed of 4 epithelial layers—the outer enamel epithelium (OEE), reticular epithelium, intermediate stratum and inner enamel epithelium (IEE). The IEE starts to differentiate into preameloblasts which induce the proliferation of ectomesenchymal cells in the outer layer of dental papilla and their differentiation to (pre)odontoblasts. Odontoblasts start to secrete predentin which induces the process of preameloblasts maturation into ameloblasts, and so the secretion of the enamel commences. When the amelogenesis (i.e. the production and full mineralization of the enamel) is finished, the ameloblasts degenerate and disappear.

Later in the bell stage, the cervical loop (an area of the enamel organ located on the future crown-root border, composed of the IEE and the OEE) elongates apically through proliferation towards the epithelial diaphragm and eventually becomes Hertwig's epithelial root sheath (HERS).)The HERS then initiates the tooth root formation and activates the production of root dentin (in the same manner as in the crown region). After the activation of the local odontoblasts the root sheath disintegrates, this way the newly generated dentin becomes exposed to the ectomesenchymal cells in the inner layer of dental follicle, which eventually induces the differentiation of these follicular cells into cementoblasts and so the production of precementum. The exposed dentin further induces the differentiation of fibroblasts from the middle layer of dental follicle and so eventually the creation of periodontal ligament, and the differentiation of osteoblasts of alveolar socket leading to the creation of the outer layer of the dental follicle.

Eventually, the mature soft dental-related tissues are dental pulp, periodontal ligament, gingiva and apical papilla, and the mature hard dental-related tissues are enamel (oral epithelium), dentine and cementum (neural-crest-derived ectomesenchyme).

Dental-Related Tissues' Stem. Cell Lineages Differentiation Potential and Anatomic Location

Tissues' stem cell populations quality and quantity vary with their origin and stage of maturation. At maturation, the hard dental tissues posses no stem cell populations, while the soft dental-related tissues do and facilitate both self-regeneration as well as regeneration of them adjoined hard tissues [13, 14]. The only hard dental tissue which lacks the possibility to repair is enamel for the absence of communication with any soft tissue.

For the summary of up-to-date differentiation potential of dental-related stem cell lineages see Table 1.

Tooth Organ-Stem Cell Lineages

Immature Tooth Organ-Stem Cell Lineages

Dental Papilla Stem Cells: iTDPSC

Immature Tooth Dental Papilla is an engineering tissue of mesenchymal origin enclosed by the enamel organ on the top and by the dental follicle on the bottom, which eventually converts to pulp tissue. It plays central role in epithelial-mesenchymal interactions responsible for tooth morphogenesis. It is firstly responsible for the crown morphogenesis and later, when it becomes apical to dental pulp (a converted top part of dental papilla, after the onset of dentinogenesis/bell stage) it becomes secondly also responsible for the morphogenesis of the tooth roots (in the form of Apical Papilla) [54–56]. The most suitable teeth for obtaining dental papilla are third molars where the roots are not developed at all, this condition is fulfilled before the 14 years of age (Figs. 2 and 3a).

The up-to-date knowledge of the differentiation potential of the Dental Papilla Stem Cells is minimal and calls for further research. Most recently, in 2014, Vishwakarma et al. have stated that the dental papilla stem cells "can induce both enamel-dentine and dentine-cementum production", but only under certain specific conditions and combined with the appropriate cell populations.

Apical Papilla Stem Cells/Immature Root Papilla Stem Cells: SCAP/iRPSCs

The root formation of developing tooth starts when the cervical loop-epithelial cells proliferate apically and induce the differentiation of dental follicle mesenchymal cells into odontoblasts and cementoblasts [57] This apically extending epithelial

Table 1 Differentiation potential of dental-related tissues stem cells lineages

Differentiation capacity	DPSC	iRPSC (SCAP)	iTDPSC(germ)	NDP-SCs	SHED	DFSCs	PDLSCs	GMSCs
Adipogenic	Struys [15]	Sonoyama [16]	Yalvac [17]	Akpinar [18]	Miura et al. [6]	Kémoun [19]	Xu [20]	Gao et al. [21]
Chondrogenic	Koyama [22]	Wang [23]	Demerci [24]	Akpinar [18]	Kerkis [5]	Kémoun [19]	Xu [20]	El-Sayed [25]
Osteogenic	Mori [26]	Bakopoulou et al. [27]	Yalvac [17]	Akpinar [18]	Miura et al. [6]	Morsczeck [11]	Xu [20]	Gao et al. [21]
Neurogenic	Osathanon [28]	Sonoyama [29]	Yalvac [17]	Karaöz [30]	Miura et al. [6]	Morsczeck [11]	Shi [31]	Zhang et al. [32]
Myogenic	Zhang [33]	Abe et al. [34]	Tasli [35]	Karaöz [30]	Kerkis [5]	d'Aquino [36]	Shi [31]	Ansari [37]
Endothelial cells	Hilkens [38]	Bakopoulou [39]	Yalvac [17]		Cordeiro [40]	Dögan [41]		Zhang et al. [32]
Odontogenic	Huang [42]	Sonoyama [29]	Demerci [24]		Miura et al. [6]	Chen [43]	Shi [31]	Gao et al. [21]
Hepatogenic	Ishkitiev [44]	Patil [45]	Ikeda [46]		Ishkitiev [47]	Patil [45]	Kawanabe [48]	
Cementogenic	Kim [49]	Kim [49]				Kémoun [19]	Seo [9]	
Melanocytes	Paino [50]							
Cardiomyocytes	Armiñán [51]							
Epithelial cells			Dögan [41]					
Endodermal cells								Zhang et al. [10]
Periodontal ligaments						Morsczeck [11]		
Pancreatic islets	Ishkitiev [52]							
Cornea cells	Gomes [53]							

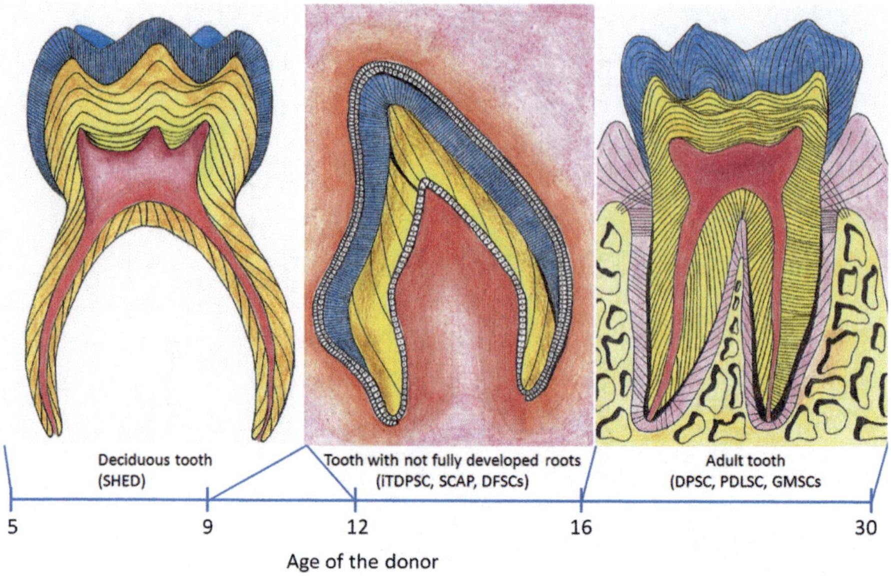

Fig. 2 Dental related stem cells lineages and the most common age of the donors of the different stem cells populations

wall forms Hertwig's epithelial root sheath (HERS) which guides the shape of the tooth roots. It is at this time dental papilla becomes apical to dental pulp tissue (hence apical papilla) and forms apical cell-rich zone as a divide between the two [55]. Overall, apical papilla is histologically different from dental pulp of immature tooth in multiple ways, according to Sonoyama et al., its stem cells proliferate 2 to 3 times quicker than those of the adjoined immature dental pulp, it contains less cellular and vascular components and it further contains unique potent MSCs. As apical papilla is a tissue derived from the neural crest, its stem cells express numerous neurogenic markers [58].

As SCAP are isolated from the apical papilla of a tooth with not yet fully developed roots, the most suitable teeth to extract are the first premolars of up to 12 years old patients and the third molars of 12–16 years old patients (Figs. 2 and 3b–d).

Mature Tooth Organ-Stem Cell Lineages

Dental Pulp

Dental pulp originates in dental papilla and is made up of connective tissue and four distinct cell-zones. The dental pulp connective tissue consist of an extracellular matrix (ECM) composed of ground substance with relatively high content of glycosaminoglycans, proteoglycans and other adhesion molecules (fibronectin, laminin, etc.) and the sparsely distributed type III collagen fibres which form a rich network only around vessels and nerves. The outmost cell-zone of the dental pulp is the odontoblast layer-zone (gives rise to the primary "predentine", secondary and tertiary

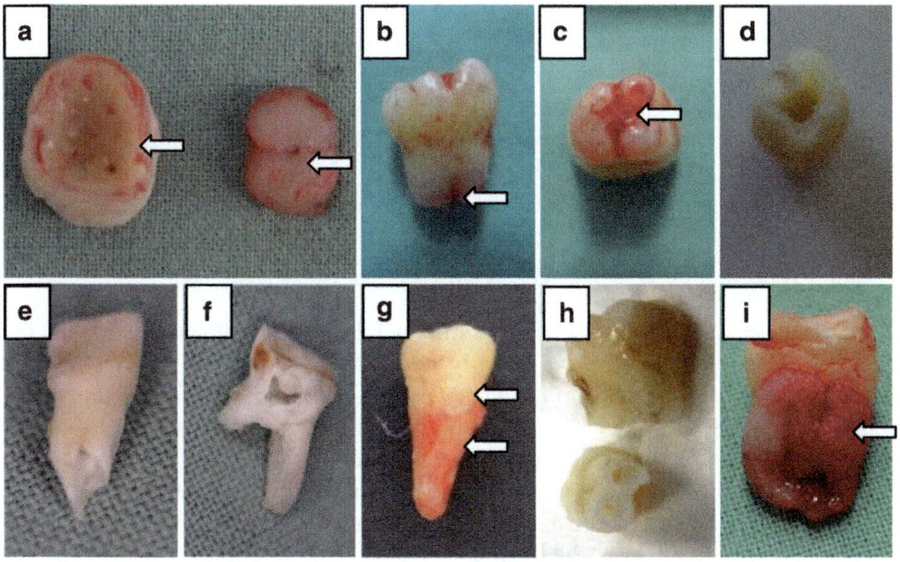

Fig. 3 (**a**) Extracted 3rd lower molar in a bell stage (germectomy) on the left side and its pulp on the right side. (**b**) Extracted 3rd lower molar with not fully developed roots. (**c**) Extracted 3rd lower molar with not fully developed roots with the apical papilla covering the root orificium. (**d**) Extracted 3rd lower molar with not fully developed roots from the (**c**) after extraction of the pulp tissue. (**e**) Extracted upper deciduous first incisor where the root apex is reabsorbed and the root canal widely opened. (**f**) Extracted second lower deciduous molar after splitting the crown using the bur to open the pulp chamber. (**g**) Extracted upper 3rd molar, the lower arrow indicates the area which is covered by the periodontal ligaments and presents the suitable source of the periodontal ligament stem cells, the *upper arrow* indicates the cement-enamel junction. (**h**) Two extracted upper first premolars after the isolation of the dental pulp. (**i**) Extracted lower 3rd molar with dental follicle tissue marked with the *arrow*

"reparative" dentine), followed inwards by a cell-free-zone called the Weil basal layer (contains dense capillary and nerve network), then cell-rich-zone, and finally arriving at the pupal core. The last two zones share similar structure as they are composed of larger vessels, nerves and great amount of cells, specifically: fibroblasts, cells belonging to the defense system, and undifferentiated mesenchymal stem cells.

Dental pulp stem cells lineages' quality and characteristics differ with dentition (predeciduous, primary, secondary) and level of their maturation.

Predeciduous Dentition Dental Pulp Stem Cells/Natal, Neo-Natal Dental Pulp Stem Cells NDP-SCs

The natal (erupted before birth) and neonatal (erupted within the days after birth) teeth are rarity found in some newborn babies within the first few days after birth. Natal teeth are more frequent than neonatal teeth, with the ratio being approximately 3:1 [59]. These teeth are usually indicated for extraction due to the danger connected to their inhalation, the risk that they could cause deformity and mutilation of tongue, dehydration, malnutrition, growth retardation of the following dentitions and due to the health risks they can cause to the mother during breastfeeding. Histologically, the natal teeth have only thin or completely absent layer of both enamel and cementum. The predentin layer is often of various thickness compiled of irregular dentinal tubules. Furthermore, unlike dental pulp of teeth of the primary and secondary dentitions, the natal dental pulp has extra inflammatory areas and does not have the Weil basal layer and cell-rich-zone. These teeth tend to have a wide pulp chamber and underdeveloped roots which makes isolation of their dental pulp easier.

Primary Dentition Dental Pulp Stem Cells: SHED

The amount of stem cells from human exfoliated deciduous teeth decreases with the receding dental pulp (which is gradually replaced by gum). This implies that the deciduous teeth that fall out spontaneously are likely to have little to no pulp, therefore little to no SHED population. To ensure a sufficient amount of stem cells will be isolated, the rule of thumb is to focus on teeth which have at least 1/3 of the original root length (after the onset of primary dentition root resorption), and for the multirooted teeth it is best to isolate teeth with the furcation area still present (Fig. 3e, f).

The ideal time to extract the deciduous teeth for their dental pulp stem cells is between the ages of 5 and 9. Beware; dental pulp of teeth affected by dental carries is not fitting for use in research.

Secondary Dentition Dental Pulp Stem Cells: DPSCs

The teeth most commonly used for isolation of the dental pulp stem cells are the first premolars and the third molars (Fig. 3g, h), as they frequently attract preliminary extraction at early age of the dental tissue. The first premolars are usually extracted shortly after eruption, due to the orthodontic reasons, around the age of 12. The "youth" of their soft dental-tissues, not fully developed root, and very low probability of carries makes these teeth the ideal stem cell source. The third molars are often extracted to prevent health complication such as the surrounding tissue inflammation, as well as due to the orthodontic reasons. While the extraction for orthodontic reasons comes around the age of 16, the extraction due to tissue inflammation occurs later, usually between the ages of 16–30, hence involves at least partially-erupted and often fully developed tooth.

Tooth-Supportive Tissues-Stem Cell Lineages

Immature Tooth-Supportive Tissues-Stem Cell Lineages

Dental Follicle Stem/Progenitor Cells: DFSCs

The dental follicle is a loose connective tissue that surrounds the developing tooth and is separated from dentin by an epithelial layer (Hertwing's sheet). Once this epithelial layer disintegrates and dental follicle touches the dentin, it is prompt to differentiate into periodontium including the alveolar bone, cementum, and periodontal ligament [60]. According to Vishwakarma et al. [61], the tissue contains at least three distinct stem cell populations (hDF1, hDF2, hDF3) with distinct morphologies, gene expressions and differentiation potentials.

Most commonly, the unerupted third molars are used for isolation of these stem cells, usually before 14 years of age (Fig. 3i).

Mature Tooth-Supportive Tissues-Stem Cell Lineages

Periodontal Ligament Stem Cells: PDLSCs

The periodontal ligament connects alveolar bone and cementum to support teeth in situ and preserve the hard tissues homoeostasis [9]. Periodontal ligament stem cells are commonly isolated from the soft tissue adjoined to the root below its enamel-cementum border (Fig. 3g); they form adherent clonogenic population of fibroblastic-like cells, forming flat and loose aggregates [62]. In 2010, Kawanabe et al. [62] have shown that PDLSCs express embryonic stem cell-associated antigen SSEA-4, implying these stem cells are great source of building material for regenerative medicine, being capable of differentiating into cells of all three germ layers.

The periodontal ligament tissue is usually isolated from around the first premolars and third molars from donors who have no history of periodontal disease and healthy periodontium.

Overall, PDLSCs were demonstrated to be able to differentiate into adipocytes, collagen-forming cells, cementoblast-like cells [9, 63], chondrocytes, neurons, hepatocytes [62], and osteoblasts [32] in vitro; and into cementum- and periodontal ligament-like tissues in vivo modified [9].

Gingiva-Derived Mesenchymal Stem Cells: GMSCs

The gingiva is a mucosal soft tissue which forms a shield-like physical protection of the adjoined dental-related tissues by sealing the gap between tooth's exposed enamel and periodontal ligament protected cementum; it further exhibits both immunomodulatory and anti-inflammatory capacities, which allow it to help the adjoined tissues to regenerate [10].

The GMSCs are usually derived from the spinous layer of human gingiva [64]. This highly homogenous population of stem cells carries not only the usual mesenchymal stem cell markers but also a positive expression of extracellular matrix proteins [10]. GMSCs have been successfully differentiated into osteoblasts, chondrocytes, adipocytes.

Donor

Theoretically any vital tooth and its enveloping tissues provide possible source of dental-related stem cells. However, it is important to narrow the pool of possible donors to those who are most likely to donate viable and research feasible stem cell populations, will be the least impacted by the extraction procedure, populations, while undergo the least impactful extraction procedure. Eventually, electing the right research material is directly related to choosing a fitting donor.

The three basic recommendations in choosing the right donor:

1. *Dental-related tissues to be extracted*
   - Different stem cells lineages can only be isolated from donors of certain age/development status
2. *Health status of the donor*
   - General medical history—as certain illnesses and diseases (e.g. genetic predispositions) can cause researcher bias as they may influence the qualities of the research material
   - Possible impact of tissue extraction on the donor's health
3. *Clinical status of the tissue*
   - The general condition of the tissue and its immediate environment represent yet another research bias (e.g. caries lesions represent risk of bacterial transmission, inflammation of the soft tissue can adversely impact the research material, erupted tooth is contaminated by the oral micro microflora, fully developed roots complicate the process of the dental pulp isolation)

Election of Research Method

Bioethics, Legislation and Regulation of Human Stem Cell Research

The principal goal of human biomedical stem cell research is to relieve and avert suffering caused by both genetic and acquired impediments to health condition [65].

Due to the interdisciplinary character of the knowledge base and the worldwide network of research which aims to build upon each other, while oftentimes towards different goals, it is eminently important to unify the standards for information integrity, full disclosure, and respect for research participants—their health and freedoms [65]. While the research focused on the unipotent, multipotent and pluripotent stem cells avoids the major ethical and legal issues connected to the totipotent embryonic stem cells and oocytes, there are other concerns which remain standing.

The international consensus over the stem cell research guidelines is conveyed by the *ISSCR—International Society for Stem Cell Research*, established independent non-profit organisation founded in 2002, which releases yearly updated guidelines for stem cell research and clinical trials and provides templates of legal documents connected to biomaterial donation and transfer [65]. The ICSCN—*International Consortium of Stem Cell Networks* then provides the contacts and means to bridge varying national stem cell legislation and regulation to foster best practice international research collaborations. The consortium is managed by a secretariat comprising of nationally funded representatives from Australia, Canada, Germany and Scotland and meets annually in conjunction with ISSCR.

Beyond the international organisations, the research ethics and regulations are further narrowed by supra-national consortia such as the European *EuroStemCell* and the American USFDA—*Food and Drug Administration of the United States of America*; national legislation and regulations issued by National Academies; and eventually by the institutional SCRO—*Stem Cell Oversight Committee* [66]. The SCRO must respectively consist of proxies of both scientific and ethical communities (also referred to as: IRE—Institutional Review Board/IEC—Independent Ethics Committee/ERB—Ethical Review Board or REB—Research Ethics Board), from which a part is non-affiliated. The ethical community further demands membership of representatives of ethical expertise and those representing the lay public [67]. Finally, the text summarising the institutional professional standards and self-regulations, issued by the institutional SCRO, deals with all the above issues in two separate parts: Part 1: *Respect for research participants* and Part 2: *Integrity of the research enterprise* [67]. The SCRO standards for non-totipotent stem cells research should, according to the ISSCR 2015 Draft Guidelines [65] and Hyun [68], address the following:

1. *Respect for Research Participants*

   The respect for research participants remains accompanied with the issue of donor/patient consent to share genetic information and the issue of donor/patient mental and physical state now and in the future; which prohibits scientists to treat the biomaterial freely and subject it to downstream research which has not been approved by the SRCO and the donor in advance [68]. The above issues require written and legally valid informed consent to donate (if the donor is not sui juris, their legal representative has to subscribe the informed consent instead), which according to the ISSCR Draft 2015, should address both the donor and research organisations' rights and obligations towards each other, specifically:

   (a) The provision of accurate information on risks to donor health connected to tissue retrieval.
   (b) The provision of accurate information on risks, limitations, possible benefits, and available alternatives to patient treatment.
   (c) The donated tissue treatment and storage—genetic modification, recipient profile, discard protocol, ad cetera.
   (d) The genetic and disease screening of the donated material.
   (e) The full disclosure of the derived information.
   (f) The donor/patient private information (identification, genomic information, ad cetera) treatment and discard protocol.
   (g) The future accessibility of the donor/patient to obtain additional consent, additional material or information.
   (h) The commercial potential of the donated material and intellectual property rights of the researchers/donor.
   (i) The treatment of incidental findings.
   (j) The treatment and legal liability of the researchers and medical staff in case of unforeseeable impacts on patient health [65].

2. Integrity of the Research Enterprise
   The treatment of the biological material founded upon valid, timely, reliable and reproducible protocol; executed following the GLP—Good Laboratory Practices; and reported in transparent, reliable manner and accessible format ensures integrity of research enterprise. All of the above is pertinent for independent peer review and downstream research.

The US FDA provides a step-by-step toolkit to institutional SCRO organisation, operation, administration and reporting which can be used as the building blocks of the institutional standards [65].

Dental-Related Tissue Extraction and Transport

Dental-Related Tissues Extraction

The extraction of tissues for stem cell research is performed under the same condition as an ordinary prescribed extraction, while we are interested in preserving the biological material as intact as possible, the health of the donor/patient comes first. Fundamentally, there are two types of extraction, the simple extraction and the surgical extraction; where both are performed under local anesthesia.

The simple extraction is adopted in the case of fully erupted tooth and is executed using various dental tools including elevator and forceps to break the tooth' bonds to periodontal ligament and to widen its alveolar bone socket, and so allow for smooth removal of intact tooth.

The surgical extraction is adopted in case of an impacted or not yet fully erupted tooth (the most common source of dental-related stem cell material) (Fig. 4).

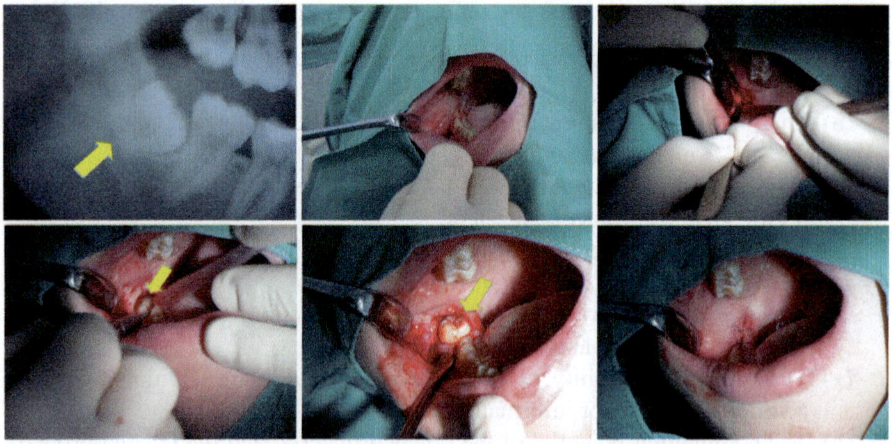

Fig. 4 The procedure of complicated extraction of lower 3rd molar. First picture represent the cut of the ortopantomogram. Second and third pictures show the moving of the mucoperiostal flap. On the fourth picture the bone covering the tooth is removed and on the fifth picture we can see the extraction using the levers

This procedure involves rising of mucoperiosteal flap/tissue flap on the intra-oral side of gums covering the tooth (for esthetic reasons) and trimming the exposed bone present around the tooth roots to allow for smooth root removal (here the prime consideration is to limit the amount of bone being removed). At this point, depending on situation, the surgeon either retains the tooth as a whole, which is preferable, or sections it to further ease its extraction. Once sectioned, however; the dental-related tissues are compromised together with their value in research.

Immediately after the extraction it is recommended to disinfect the tooth using any disinfectant solution commonly used in the dental cavity, to decrease the possible amount of bacterial presence. Then, the tooth needs to be fully immersed in the Hanks balanced salt solution enhanced by antibiotics and antifungals (composed of 1 ml of Hanks balanced salt solution, 9 ml water for inj., 200 μl/10 ml gentamycin, 200 μl/10 ml streptomycin, 200 μl/10 ml and 200 μl/10 ml penicillin) to fully eliminate any remaining bacterial presence.

Dental-Related Tissues Transport

The temperature of the solution during transport is highly recommended to be kept at 4 °C. Even though stem cells are theoretically resistant to hypoxia and survive longer than somatic cells, it is absolutely essential they are isolated from the extracted tissues as soon as possible (no later than 24 h after extraction). For this reason it is necessary to invest in proper preliminary planning to avoid any wastage.

Isolation of the Dental-Related Tissues

Tooth-Organ Tissues Extraction

Immature Tooth-Organ Tissues Extraction

Dental Papilla and Apical Papilla Extraction

Dental papilla tissue is extracted from tooth without developed roots, hence sourced from very young donors/patients. Such extraction is called germectomy (Fig. 3a) (an identical procedure is used to obtain dental follicle tissue (Fig. 3i). The most common source of the dental papilla tissue for stem cell research are third molars of donors/patients aged 10–14 years, as those teeth are commonly indicated for extraction. After the onset of root dentine formation and root development, the dental papilla splits into the soft tissues of dental pulp and apical papilla.

Apical papilla is localized at the area of root growth, below dental pulp (Fig. 3b–d). The dental pulp and apical papilla are clearly separated by apical papilla adjoined apical cell-rich zone which however does not stick to dental pulp and so allows smooth separation of the two tissues.

Whilst dental and apical papilla tissues can be obtained from immature teeth of both dentitions, it is rare to find immature deciduous teeth indicated for extraction which have no caries lesion; therefore the SCAP and iTDPSC are widely considered as stem cell lineages of secondary dentition.

Mature Tooth-Organ Tissues Extraction

Dental Pulp Extraction

Dental pulp consists of connective tissue and odontoblasts locked together inside the dental pulp chamber. During the process of tooth maturation, dental pulp tissue becomes increasingly enclosed in hard dental-related tissue, until only a tiny opening at the roots' apex (physiological apical foramen) is left (0.25–0.35 mm wide at maturation) to allow the nervous connections and blood vessels. To ensure the extracted dental pulp tissue is as intact as possible, we aim to use the least destructive method which varies with the tooth level of maturation (including all postnatal immature and mature dental pulps of all dentitions).

The three most commonly used methods of dental pulp extraction are: (1) Extraction of the dental pulp through the physiological apical foramen, (2) Splitting the tooth using forceps and (3) Splitting the tooth using burr (rotary file).

The first method is regarded as the easiest, fastest and the most qualitatively efficient of all; however, it can only be performed on teeth with wide apical foramen (approx. $x > 2$ mm), therefore reserved exclusively for teeth with roots which are not yet developed (isolating immature dental pulp tissue of primary and secondary dentition) or for teeth which roots are resorbing (isolating primary dentition dental pulp tissue). The first step of this extraction process is to gently detach the dental pulp from the adjoined dentine (commonly performed using a needle). The following step is to pull the pulp out downwards using tweezers (commonly performed using anatomical stainless steel hard tweezers, model ADSON) (Fig. 5a).

When the physiological foramen is too narrow to use as the extraction canal, we are left with other methods which involve mechanical disintegration of the hard dental tissue surrounding the pulp, opening the dental chamber from the top. To access the soft dental pulp, we tend to aim to split the tooth at the enamel-cementum junction (cervical area) using either forceps (crashing the junction) (Fig. 5b) or a burr (employing a diamond burr on a fast micromotor handpiece—red stripe, about 200,000 rpm; cooling the tooth with water spray; performing horizontal cut at the junction (Fig. 5c) (it is also possible to employing a disc shaped burr and performing vertical cut, though this approach is regarded as very risky as far as the pulp integrity and quality is concerned (Fig. 5d)). It is also important to mention that the burr method has potentially detrimental impact on the dental pulp due to the heat emitted during filing. If successful the dental pulp can be pulled out upwards using the same method as described above. While these mechanical methods are relatively complicated, slow and less efficient in the amount of tissue harvested (in comparison with the first method), they allow researchers access to much greater number of donors/donated tissues.

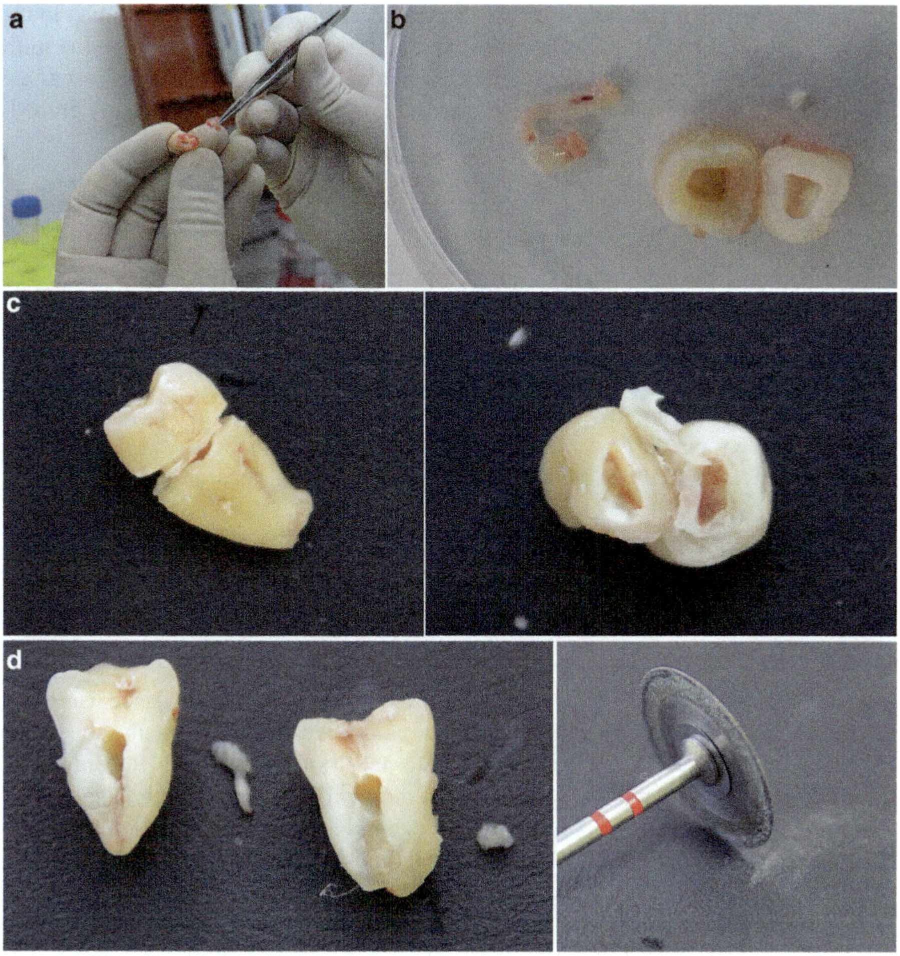

Fig. 5 (**a**) Extraction of the pulp tissue from the tooth with not fully developed roots, where the pulp chamber is widely opened. (**b**) Extracted lower 3rd molar where the crown was separated from the root in the using the forceps and the isolated dental pulp tissue. (**c**) Extracted upper 3rd molar where the crown was separated from the root in the using the diamond burr on the *left side* and the same tooth still containing the pulp tissue on the *right side*. (**d**) Extracted upper 3rd molar where the crown was separated from the root in the using the forceps soflex discs (the bur is on the *right picture*)

Tooth-Supportive Soft Tissues Extraction

Immature Tooth-Supportive Tissues Extraction

Dental Follicle Extraction

Dental follicle is loose connective tissue surrounding the developing tooth germ (Fig. 3i) which degenerates after the tooth erupts and finishes its development. It's harvest after tooth organ extraction is relatively easy as it is not firmly adjunct to the tooth. Due to this timing issue, the most frequent source of this tissue are the third molars extracted from the young donors (around 10–14 years old). These teeth are frequently extracted for orthodontic reasons and are commonly completely embedded in the jaw bone.

Mature Tooth-Supportive Tissues Extraction

Periodontal Ligament and Gingiva Extraction

Periodontal ligament is specialized connective tissue localized between the alveolar bone and root surface, and is hidden under gingiva. While not all extracted teeth carry a research-sufficient amount of gingiva tissue, as the gingiva is commonly raised from the tooth surface preceding the tooth organ extraction, it commonly carries enough periodontal ligament tissue. Both the tissues are harvested off the tooth root surface by scraping.

Dental-Related Stem Cells Isolation

In general, dental-related stem cells are isolated from their maternal tissue using one of the two following approaches: (1) Enzymatic Digestion (ED) of tissues or (2) Spontaneous Outgrowth (OG) from the maternal tissue mince [64]. Recent scientific research suggests that isolation method most probably impacts the qualities of the isolated stem cell lineages [27, 69]. For example, while some researchers, comparing the impact of the above isolation methods on the resulting dental-related stem cell populations, concluded that OG isolated stem cell population had weaker stem cell-markers expression (SHED-ED × SHED-OG [70]) and lower proliferation rate (DPSC-ED × DPSC-OG [69], other studies concerning non-dental-related stem cells report contradicting results, mostly arriving to the decision to opt out from using the collagenase enzyme as it was believed to alter the stem cell phenotype [38, 71]. Ergo, it is important to keep an eye on the upcoming research in this field to help us narrate the most probably bias-free methodology.

Enzymatic Digestion/Dissociation: ED

During enzymatic digestion the extracted tissue is submerged in digestive enzymes solution (most commonly Hank's balanced Salt Solution (HBSS) carrying diluted collagenase I (3 mg/ml), dispase (4 mg/ml) and phosphate-buffered saline (PBS) (buffer solution) in a ratio of 1:1:1:1) for about 30–60 min, depending on the size of pieces of the tissue, at 37 °C. The result of ED is a single-cell suspension which contains many different cell types and remnants of vessels and extracellular matrix (Fig. 6). While it is not imperative to remove the remnants from the suspension, it

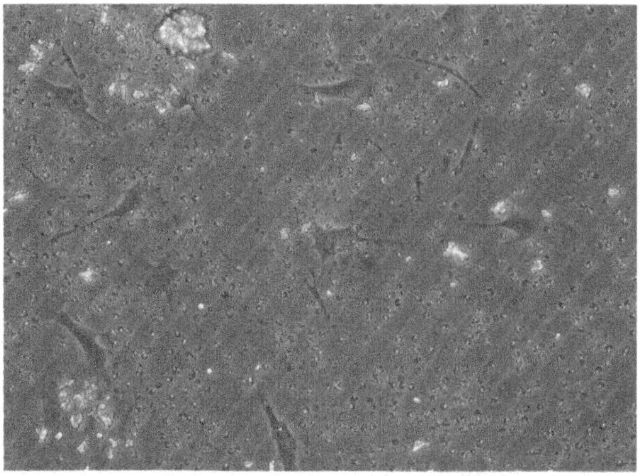

Fig. 6 Dental pulp stem cells in the prime culture 6 days after seeding with the remnants of the soft tissue and the erythrocytes

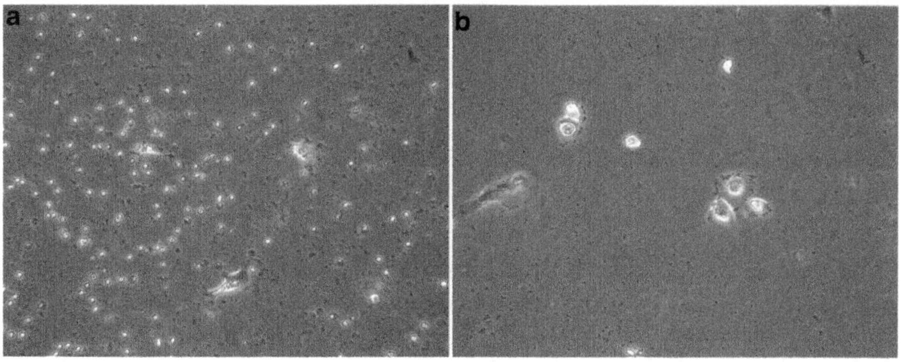

Fig. 7 (**a**, **b**) The same lineage of dental pulp stem cells before and after changing the media

can be done by either sieving through a strainer (3 μm followed by 20 μm), by magnetic-activated cell sorting (MACS) or by stem cell colony cultivation [64]. Whether researchers decide to work with suspension containing remnants or not, the single-cell suspension is centrifuged (600 g (g-force), 2000 rpm for 5 min) and the cell pellet is finally seeded into cultivation dishes where it is resuspended in amplification media. After 7 days of amplification the stem cells adhere to the surface of culture dish. At this point all the other material is washed away and new amplification media is added (Fig. 7a, b).

Spontaneous Outgrowth/Explant Culture: OG

The outgrowth method is technically easier and faster than ED. The soft tissue is minced to pieces smaller than 2 mm (explants) and planted on cultivation dish, fully immersed in amplification media. After 3 to 5 days of amplification we can observe stem cells adherent to the culture dish everywhere under and around the minced tissue (Fig. 8) [64].

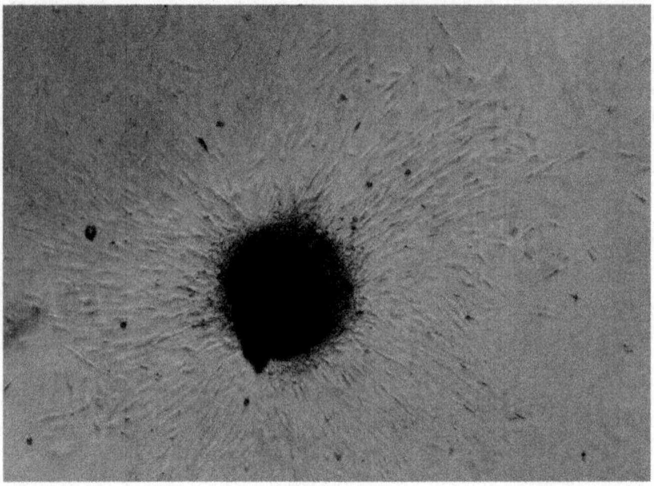

Fig. 8 Part of the dental pulp tissue with outgrowing stem cells. The photo was taken 12 days after seeding

Dental-Related Stem Cells Amplification and Differentiation Culture Systems

After we have successfully isolated our stem-cell lineages we aim to expand their population to research-relevant amounts (their amount at isolation varies by maternal tissue and method of extraction [72]. Depending on the future use of the amplified population it is important to choose the right amplification conditions and environment (physical characteristics of the culture system (adherent, non-adherent/sphere-forming), amplification media sera (xenogeneic, autologous serum, serum-free) and co-culture (epithelial/ectodermal-mesenchymal interaction) [64]) as they do affect the quality of the resulting stem cell population [73]. Especially, when human regenerative medicine is concerned, the researchers should avoid using xenogeneic sera as their use carries multiple legal, ethical and medical issues ranging from research results bias, to transfer of zoonoses (interspecies infectious agents); viral, bacterial and fungal contamination; and permanent alteration to the stem cell population genetic code [74].

Amplification and Differentiation Culture Systems

Generally, dental-related stem cells are amplified and differentiated at 37 °C at 5 % CO_2 atmosphere. To maintain high proliferation rate, cultivated stem cells should be passaged when the population reaches 70 % confluence (confluence >70 % leads to decrease in proliferation due to the physiological contact inhibition). The scientific community recognises two basic physical cultivation environments for stem cell amplification and differentiation: the (1) *Adherent Culture Systems* and the (2) *Non-adherent (Suspension/Sphere-Forming) Culture Systems*; plus one extra type of differentiation environment, the (3) *Co-Culture Systems*. Overall, while the non-adherent culture systems are easier for passaging and the growth of cells is not limited by the surface area of the cultivation dishes, the adherent culture systems allow for easier visual inspection and provide less opportunities for cell heterogeneity [75]. The co-culture systems then provide "site-specific" lineages interaction which leads to differentiation of specific progenitor/adult cells [64].

Adherent Culture Systems

The adherent system consists of a dish which allows the cultured stem cells to adhere to it. At the end of cultivation, the dish is washed with phosphate-buffered saline, which safely removes all cultivation media. Once the stem cells are the only content of the dish, it's whole surface is covered by trypsin which detaches the stem cell off the surface of the dish. The trypsin is neutralised after 5 min by solution of α-MEM and serum (in concentration of 20 %). The resulting suspension with cultivated cells is centrifuge (600 g, at 2000 rpm for 5 min). After centrifugation, the supernatant is removed and the resulting cell pellet is resuspended in new cultivation media and placed on new culture dish, in concentration of 2500 cells per cm^2. This culture system is the default system in stem cell research, except for cultivation and differentiation of non-adhesive cell lines (e.g. neural and hematopoietic cells).

Non-adherent/Suspension/Sphere-Forming Culture Systems

Spheres are free-floating cultures of stem cells. The non-adherent sphere-forming culture is generally serum-free and contains epidermal growth factor (EGF), basic fibroblast growth factor (bFGF) and leukemia inhibitory factor (LIF) and rests upon either super-hydrophilic (non-adherent) plates or in a culture flask [76, 77]. While this system has been widely applied for assaying neural stem cells up to neurogenesis [78] it has been also studied an effective system for generation of mineralized tissues (from SCAP) [34].

Jensen and Parmar's report on the applicability of this system raises a few important issues, they argue that the cultured cell types properties appeared to be sensitive to the culturing method used, variation in cell density in the system, as well as the concentrations of factors in the media, method and frequency of passaging and the number of passages after isolation [78]. Furthermore, multiple researchers report that the sphere-forming cells become increasingly heterogeneous with further passage as the growing spheres contain cells at various stages of differentiation, from stem cells to progenitor cells and specific cell types [75, 79].

Co-culture Systems

Co-culturing of various dental-related stem cells is not used for their amplification as much as it is used for their differentiation. These co-cultivation systems applied in the dental-related stem cell research are based on our knowledge of the dental-tissue genesis (epithelial-mesenchymal co-culture/ectodermal-mesenchymal co-culture). For example, Bai et al. [80] have demonstrated that co-cultivation of DFSCs and SCAP (rat) can lead to formation of bone-like particles and Xiao and Tsutsui [81] demonstrated that co-cultivation of oral epithelial cells (OECs) and DPSCs can build an epithelium invagination like model. These systems are very promising tool for future research as they replicate the original in vivo environment more accurately than the above single-cell systems.

Amplification Media

The cultivation media, as well as any component of extracellular environment affects the properties of cultivated cells. Due to the vast heterogeneity of media composition used through dental-related stem cell studies, it becomes challenging to make any generally applicable judgements about the media on the amplified or differentiated cell populations. Therefore it is imperative that when outlining a research method, one should critically and attentively evaluate the validity of preliminary research' methods and their results to eventually arrive at scientifically credible conclusion and results in their work, results which can be build upon. The three general groups of media are: (1) *Serum-Free Medium,* (2) *Allogeneic-Serum Medium,* and (3) *Xenogeneic-Serum Medium.* The serum-free media are currently the most preferable solution due to absence of risks associated with the xenogeneic and allogeneic media (e.g. transfer of viruses and zoonoses).

Serum-Free Media

The serum-free medium, most frequently used in dental-related stem cell research, is a combination of the following additives: Dulbecco's modified Eagle's medium, insulin, transferrin, sodium selenite supplement, Embryotrophic Factor [82]. While the serum-free media are most possibly the future of stem cell research, they are not yet universally effective when used for various types of stem cell lineages, and their efficiency in boosting stem cell proliferation and accessibility remains low [83]. Today, the serum-free media are used as a complementary media to amplification and differentiation processes powered primarily by either allogeneic or xenogeneic media.

According to a study, conducted by Tarle et al. [84], on use of serum-free media for amplification of PDLSCs and SHED; the stem cells cultivated partially under serum-free condition express higher positivity of stem cells markers then when amplified in xenogeneic medium alone. They also manifested similar differentiation capacity, while their proliferation rate was significantly slower [84].

Allogeneic Serum-Containing Amplification Media

The most commonly used allogeneic sera are human plasma (HP) or platelet rich plasma (PRP) [85–89]. While HP does deliver on its promise, it is proven that PRP, due to its boosted contents of platelets/platelet lysate, is more potent in maintaining the cultured dental-related stem cell properties and its use results in greater cells' proliferation rate, immunosuppressive activity and migration activity [87, 90, 91].

The disadvantages of these sera are the hazards involved in transferring heterologous material which can carry diseases (e.g. HIV, Hepatitis), drug metabolites, and depending on the donor also irregular concentration of growth factors and other components which vary within physiological borders. Nonetheless, the disease and drug metabolite hazards can be managed and the concentrations of the components can be successfully averaged by pooling material sources from multiple donors (>4) [92].

Xenogeneic Serum-Containing Amplification Media

The xenogeneic media contain around 2–20 % of fetal calf serum/bovine serum, where concentrations of >10 % are the most common. The advantages of this serum include its high content of growth factors and low concentration of inhibitory factors. The disadvantage is; however, that the concentration of those, together with the serum nutrients, enzymes and proteins, is not stable and varies sample to sample, within the species physiological borders. In general, the higher concentration of the bovine serum in media the better cell adherence, the more colony forming units, and the greater proliferation potential. Nevertheless, these research beneficial outcomes are offset by research detrimental impacts on the cultivated cells such as chromosomal instability, phenotype amalgamation, malignant transformations, and at concentrations >20 % even toxicity.

When attempting to minimise the above mentioned detrimental effects of high concentrations of xenogeneic serum on cultured cells while retaining the high proliferation rate and good adherence, it is recommended to use media with <10 % bovine serum and to make up for the higher concentrations by adding growth factors like platelet derived growth factor and epidermal growth factor to boost its performance. According to some studies the addition of insulin-transferrin-sodium selenite solution results in higher proliferation rate as well [92].

Nevertheless, there exists no solution for one of its major shortcomings, its xenogeneic origin, which brings a whole lot of possible hazards to the table. Eventually, its application can lead to internalization of bovine serum proteins by the cultured stem cells and finally stimulation of the recipient immune system and unwanted immunological reaction [93–95], or to transmission of the zoonoses [96, 97], and induction of metabolic and morphological changes in the cultured cells [98, 99]. Overall, while this serum is easily obtainable and relatively cheap, it is recommended it is phased out off the stem cell scientific research as it provides biased results which are not applicable for human medicine [100, 101].

Dental-Related Stem Cells Differentiation and Evaluation

According to their origin dental-related stem cells are pre-programmed to differentiate into certain mature cell lineages and it is harder, although not impossible, to induce them to differentiate into foreign lineages. To prove that the amplified dental-related stem cells are indeed dental-related stem cells we would test their capacity to differentiate into osteoblasts, chondroblasts and adipocytes, which capacity is characteristic for all mesenchymal stem cells. Yet again, the results of the differentiation process are highly dependent on the artificial extracellular environment and hence this step attracts good planning and thorough preliminary research. Besides choosing the basic culture system, as per chapter *i. Amplification and Differentiation Culture Systems*, we have to choose the right differentiation method (there are multiple approaches to the same end) and follow its evaluation process to produce valid results. (*Interesting new fact, in* 2015, Zhang et al. *have proven that LMHF mechanical vibration promotes PDLSCs osteogenic differentiation. This discovery implies that vibration may be one of the mechanisms of external environment which impacts the biological processes, hence should be considered as a variable when planning and reporting research results.*)

The below overview lists the most commonly used differentiation protocols in dental-related stem cell research:

Osteo-Differentiation Protocol and Evaluation

The dental-stem cells undergoing osteogenic differentiation are cultivated in adherent cultivation dish in a monolayer and are supposed to become almost 100 % confluent. The differentiation is induced by osteogenic differentiation media composed of α-MEM, 0.2 mM ascorbic acid 2-phosphate, 2 mM L-glutamine, 100 U/ml penicillin, 100 μg/ml streptomycin, 10 mM β-glycerophosphate, 0.1 μM dexamethasone and 10 % fetal calf serum and the process takes >4 weeks.

To prove a successful differentiation we use standard histological methods like Masson's blue trichrome and Von Kossa stain, to identify calcium deposits and collagen I fibres; or immunodetection of osteonectin, osteopontin, or procollagen I. Karbanová et al. [102] states that, by the week 3–4, we should be able to visually identify small extracellular aggregates of bone tissue.

Chondro-Differentiation Protocol and Evaluation

The dental-related stem cell chondro-differentiation can be successfully performed in either an adherent system (monolayer cultivation dish) and a non-adherent system (sphere-forming system). The chondrogenic differentiation medium is composed of high glucose DMEM, 0.2 mM ascorbic acid 2-phosphate, 0.1 μM dexamethasone, 1 % ITS + 1 supplement, 100 U/ml penicillin; 100 μg/ml streptomycin, 40 μg/ml proline, 100 μg/ml sodium pyruvate, 10 ng/ml TGF-β3. Four weeks into the chondro-differentiation the acid mucins within the extracellular matrix can by visualized by alcian blue staining [102], proving the success of the process.

Adipo-Differentiation Protocol and Evaluation

Nowadays, adipogenic differentiation is most often induced using copyrighted adipogenic media. While the precise formula is not public, it is usually some mixture of dexamethasone, 3-isobutyl-1-methyl-xanthine, insulin, and indomethacin and standard cultivation media [103]. The results can be evaluated 4 weeks after the onset of the differentiation and the proof is provided by 0.18 % Oil red O staining of the intracellular lipid droplets in mature adipocytes with.

Neuro-Differentiation Protocol and Evaluation

The dental-related stem cells differentiation into neuro-like cells can be done, alike the chondro-differentiation, in both adherent and non-adherent systems using neurogenic differentiation medium composed of 50 ml DMEM, 50 ml F12 HAM, 2 ml B27 Supplement, 1 ml N2 Supplement, 10 ng/ml bFGF, 20 ng/ml EGF, 1 ml L-Glutamine, 1 ml Penicilin. The positive differentiation is proven using immunodetection methods to detect the markers typical for neural tissue—β-III tubulin, FORSE-1, GFAP, Nestin and PanNF.

Myogenic-Differentiation Protocol and Evaluation

Myogenic differentiation can be induced by treating the dental-related stem cell population with 3 μM 5-azacytidine in serum-free medium DMEM for the duration of 24 h. After this treatment, cells are washed with PBS and maintained in DMEM containing 10 % bovine serum [104]. To prove a successful myogenic-differentiation we should be able to visually identify myocyte markers using immunofluorescence-labeled antibodies against human cardiac actin, cardiac actinin, cardiac troponin C, cardiac troponin T, connexin 43, sarcomeric actin).

Hepatogenic-Differentiation Protocol and Evaluation

The dental-related stem cells undergo hepato-differentiation in media composed of 20 ng/ml recombinant human hepatocyte growth factor and 2 % bovine serum, for the first 5 days, and then a mixture of 10 ng/ml Oncostatin M, 10 nM Dexamethasone and 1 % Insulin-Transferrin-Selenium-X (ITS-X), for another 15 days [47].

The success of the differentiation is evaluated by either using imunohistochemistry detection of albumin, α-fetoprotein, hepatic nuclear factor 4α and insulin like growth factor, or periodic-acid-shift staining for visualising the stored glycogen.

Dental-Related Stem Cells Cryopreservation

SHED and third molar DPSCs represent the majority of dental-related stem cells material which is cryopreserved for either research or future personal-medical use. There is no clear winner amongst the cryopreservation methods available today as there are important questions pending to be answered: (1) What is the optimal cryoprotective agent and in which concentration should it be used to reduce or eliminate the side effect of cryopreservation?; (2) How many cells should be cryopreserved to obtain viable lineage after thawing?; and (3) What is the optimal process of cryopreservation, storage time and storage temperature to maintain maximum amount of viable cells?

Preceding cryopreservation the stem cell population should be harvested from the cultivation system. After centrifugation the cell pellet is resuspended using the chosen cultivation media and the amount of viable cells is measured. It is recommended to cryopreserve stem cell populations composed of at least one million cells. To avoid damage to the cells inflicted by the cryopreservation itself we expose the cells to cryoprotectants (e.g. glycerol or dimethyl sulfide—DMSO) in the concentration of maximum 10 % (as DMSO is cytotoxic)at room temperature for about 30 min, to allow the protectant to enter the cells but not damage them. The DMSO concentrate should be premixed before it touches the cells as it releases heat during the process of dilution. Finally, the treated cells should be fastly frozen using liquid nitrogen.

Good luck researchers!

References

1. The Stem Cell (2009) Retrieved September 28, 2015, from http://stemcells.nih.gov/info/scireport/pages/chapter1.aspx
2. Malik M, Saini RS, Jain R, Laller S, Bahadurgarh H (2014) A comprehensive update: predecidous teeth (natal and neonatal). J Adv Med Dental Sci Res 2:2
3. Nelson SJ (2014) Wheeler's dental anatomy, physiology and occlusion. Elsevier Health Sciences and occlusion. Elsevier Health Sciences, St. Louis, Missoury
4. Karaöz E, Doğan BN, Aksoy A, Gacar G, Akyüz S, Ayhan S, Sarıboyacı AE (2010) Isolation and in vitro characterisation of dental pulp stem cells from natal teeth. Histochem Cell Biol 133(1):95–112
5. Kerkis I, Kerkis A, Dozortsev D, Stukart-Parsons GC, Gomes Massironi SM, Pereira LV, Caplan AI, Cerruti HF (2006) Isolation and characterization of a population of immature dental pulp stem cells expressing OCT-4 and other embryonic stem cell markers. Cells Tissues Organs 184(3–4):105–116

6. Miura M, Gronthos S, Zhao M, Lu B, Fisher LW, Robey PG, Shi S (2003) SHED: stem cells from human exfoliated deciduous teeth. Proc Natl Acad Sci 100(10):5807–5812
7. Gronthos S, Mankani M, Brahim J, Robey PG, Shi S (2000) Postnatal human dental pulp stem cells (DPSCs) in vitro and in vivo. Proc Natl Acad Sci 97(25):13625–13630
8. Sonoyama W, Liu Y, Fang D, Yamaza T, Seo BM, Zhang C, Wang S (2006) Mesenchymal stem cell-mediated functional tooth regeneration in swine. PloS One 1(1):e79
9. Seo BM, Miura M, Gronthos S, Bartold PM, Batouli S, Brahim J, Young M, Gehron Robey P, Wang CY, Shi S (2004) Investigation of multipotent postnatal stem cells from human periodontal ligament. Lancet 364(9429):149–155
10. Zhang Q, Shi S, Liu Y, Uyanne J, Shi Y, Shi S, Le AD (2009) Mesenchymal stem cells derived from human gingiva are capable of immunomodulatory functions and ameliorate inflammation-related tissue destruction in experimental colitis. J Immunol 183(12):7787–7798
11. Morsczeck C, Götz W, Schierholz J, Zeilhofer F, Kühn U, Möhl C, Sippel C, Hoffmann KH (2005) Isolation of precursor cells (PCs) from human dental follicle of wisdom teeth. Matrix Biol 24(2):155–165
12. Huang X, Saint-Jeannet JP (2004) Induction of the neural crest and the opportunities of life on the edge. Dev Biol 275(1):1–11
13. Handschel J, Meyer T, Wiesmann HP (2009) Fundamentals of tissue engineering and regenerative medicine. Springer, Heidelberg, p 360
14. Menicanin D, Mrozik KM, Wada N, Marino V, Shi S, Bartold PM, Gronthos S (2013) Periodontal-ligament-derived stem cells exhibit the capacity for long-term survival, self-renewal, and regeneration of multiple tissue types in vivo. Stem Cells Dev 23(9):1001–1011
15. Struys T, Moreels M, Martens W, Donders R, Wolfs E, Lambrichts I (2010) Ultrastructural and immunocytochemical analysis of multilineage differentiated human dental pulp-and umbilical cord-derived mesenchymal stem cells. Cells Tissues Organs 193(6):366–378
16. Sonoyama W, Liu Y, Fang D, Yamaza T, Seo BM, Zhang C et al (2006) Mesenchymal stem cell-mediated functional tooth regeneration in swine. PLoS One 1(1):e79
17. Yalvac ME, Ramazanoglu M, Rizvanov AA, Sahin F, Bayrak OF, Salli U et al (2010) Isolation and characterization of stem cells derived from human third molar tooth germs of young adults: implications in neo-vascularization, osteo-, adipo-and neurogenesis. Pharmacogenomics J 10(2):105–113
18. Akpinar G, Kasap M, Aksoy A, Duruksu G, Gacar G, Karaoz E (2014) Phenotypic and proteomic characteristics of human dental pulp derived mesenchymal stem cells from a natal, an exfoliated deciduous, and an impacted third molar tooth. Stem Cells Int 2014:457059
19. Kémoun P, Laurencin-Dalicieux S, Rue J, Farges JC, Gennero I, Conte-Auriol F et al (2007) Human dental follicle cells acquire cementoblast features under stimulation by BMP-2/-7 and enamel matrix derivatives (EMD) in vitro. Cell Tissue Res 329(2):283–294
20. Xu J, Wang W, Kapila Y, Lotz J, Kapila S (2008) Multiple differentiation capacity of STRO-1+/CD146+ PDL mesenchymal progenitor cells. Stem Cells Dev 18(3):487–496
21. Gao Y, Zhao G, Li D, Chen X, Pang J, Ke J (2014) Isolation and multiple differentiation potential assessment of human gingival mesenchymal stem cells. Int J Mol Sci 15(11):20982–20996
22. Koyama N, Okubo Y, Nakao K, Bessho K (2009) Evaluation of pluripotency in human dental pulp cells. J Oral Maxillofac Surg 67(3):501–506
23. Wang JJ, Dong R, Wang LP, Wang JS, Du J, Wang SL et al (2015) Histone demethylase KDM2B inhibits the chondrogenic differentiation potentials of stem cells from apical papilla. Int J Clin Exp Med 8(2):2165
24. Demirci S, Doğan A, Şişli B, Sahin F (2014) Boron increases the cell viability of mesenchymal stem cells after long-term cryopreservation. Cryobiology 68(1):139–146
25. El-Sayed KMF, Paris S, Graetz C, Kassem N, Mekhemar M, Ungefroren H et al (2015) Isolation and characterisation of human gingival margin-derived STRO-1/MACS+ and MACS– cell populations. Int J Oral Sci 7(2):80–88

26. Mori G, Brunetti G, Oranger A, Carbone C, Ballini A, Muzio LL et al (2011) Dental pulp stem cells: osteogenic differentiation and gene expression. Ann N Y Acad Sci 1237(1):47–52
27. Bakopoulou A, Leyhausen G, Volk J, Tsiftsoglou A, Garefis P, Koidis P, Geurtsen W (2011) Comparative analysis of in vitro osteo/odontogenic differentiation potential of human dental pulp stem cells (DPSCs) and stem cells from the apical papilla (SCAP). Arch Oral Biol 56(7):709–721
28. Osathanon T, Sawangmake C, Nowwarote N, Pavasant P (2014) Neurogenic differentiation of human dental pulp stem cells using different induction protocols. Oral Dis 20(4):352–358
29. Sonoyama W, Liu Y, Yamaza T, Tuan RS, Wang S, Shi S, Huang GTJ (2008) Characterization of the apical papilla and its residing stem cells from human immature permanent teeth: a pilot study. J Endod 34(2):166–171
30. Karaöz E, Doğan BN, Aksoy A, Gacar G, Akyüz S, Ayhan S et al (2010) Isolation and in vitro characterisation of dental pulp stem cells from natal teeth. Histochem Cell Biol 133(1):95–112
31. Shi S, Bartold PM, Miura M, Seo BM, Robey PG, Gronthos S (2005) The efficacy of mesenchymal stem cells to regenerate and repair dental structures. Orthod Craniofac Res 8(3):191–199
32. Zhang C, Lu Y, Zhang L, Liu Y, Zhou Y, Chen Y, Yu H (2015) Influence of different intensities of vibration on proliferation and differentiation of human periodontal ligament stem cells. Arch Med Sci 11(3):638
33. Zhang W, Walboomers XF, Van Kuppevelt TH, Daamen WF, Van Damme PA, Bian Z, Jansen JA (2008) In vivo evaluation of human dental pulp stem cells differentiated towards multiple lineages. J Tissue Eng Regen Med 2(2–3):117
34. Abe S, Hamada K, Yamaguchi S, Amagasa T, Miura M (2011) Characterization of the radioresponse of human apical papilla-derived cells. Stem Cell Res Ther 2(1):2
35. Taşlı PN, Doğan A, Demirci S, Şahin F (2016) Myogenic and neurogenic differentiation of human tooth germ stem cells (hTGSCs) are regulated by pluronic block copolymers. Cytotechnology 68(2):319–329
36. d'Aquino R, Tirino V, Desiderio V, Studer M, De Angelis GC, Laino L et al (2011) Human neural crest-derived postnatal cells exhibit remarkable embryonic attributes either in vitro or in vivo. Eur Cell Mater 21:304–316
37. Ansari S, Chen C, Xu X, Annabi N, Zadeh HH, Wu BM et al (2016) Muscle tissue engineering using gingival mesenchymal stem cells encapsulated in alginate hydrogels containing multiple growth factors. Ann Biomed Eng 44:1908–20
38. Hilkens P, Gervois P, Fanton Y, Vanormelingen J, Martens W, Struys T, Politis P, Lambrichts I, Bronckaers A (2013) Effect of isolation methodology on stem cell properties and multilineage differentiation potential of human dental pulp stem cells. Cell Tissue Res 353(1):65–78
39. Bakopoulou A, Kritis A, Andreadis D, Papachristou E, Leyhausen G, Koidis P et al (2015) Angiogenic potential and secretome of human apical papilla mesenchymal stem cells in various stress microenvironments. Stem Cells Dev 24(21):2496–2512
40. Cordeiro MM, Dong Z, Kaneko T, Zhang Z, Miyazawa M, Shi S et al (2008) Dental pulp tissue engineering with stem cells from exfoliated deciduous teeth. J Endod 34(8):962–969
41. Doğan A, Demirci S, Şahin F (2015) In vitro differentiation of human tooth germ stem cells into endothelial-and epithelial-like cells. Cell Biol Int 39(1):94–103
42. Huang GJ, Gronthos S, Shi S (2009) Mesenchymal stem cells derived from dental tissues vs. those from other sources: their biology and role in regenerative medicine. J Dent Res 88(9):792–806
43. Chen G, Sun Q, Xie L, Jiang Z, Feng L, Yu M et al (2015) Comparison of the odontogenic differentiation potential of dental follicle, dental papilla, and cranial neural crest cells. J Endod 41(7):1091–1099

44. Ishkitiev N, Yaegaki K, Imai T, Tanaka T, Nakahara T, Ishikawa H et al (2012) High-purity hepatic lineage differentiated from dental pulp stem cells in serum-free medium. J Endod 38(4):475–480
45. Patil R, Kumar BM, Lee WJ, Jeon RH, Jang SJ, Lee YM et al (2014) Multilineage potential and proteomic profiling of human dental stem cells derived from a single donor. Exp Cell Res 320(1):92–107
46. Ikeda E, Yagi K, Kojima M, Yagyuu T, Ohshima A, Sobajima S et al (2008) Multipotent cells from the human third molar: feasibility of cell-based therapy for liver disease. Differentiation 76(5):495–505
47. Ishkitiev N, Yaegaki K, Calenic B, Nakahara T, Ishikawa H, Mitiev V, Haapasalo M (2010) Deciduous and permanent dental pulp mesenchymal cells acquire hepatic morphologic and functional features in vitro. J Endod 36(3):469–474
48. Kawanabe N, Murata S, Murakami K, Ishihara Y, Hayano S, Kurosaka H et al (2010) Isolation of multipotent stem cells in human periodontal ligament using stage-specific embryonic antigen-4. Differentiation 79(2):74–83
49. Kim BC, Bae H, Kwon IK, Lee EJ, Park JH, Khademhosseini A, Hwang YS (2012) Osteoblastic/cementoblastic and neural differentiation of dental stem cells and their applications to tissue engineering and regenerative medicine. Tissue Eng Part B Rev 18(3):235–244
50. Paino F, Ricci G, De Rosa A, D'Aquino R, Laino L, Pirozzi G et al (2010) Ecto-mesenchymal stem cells from dental pulp are committed to differentiate into active melanocytes. Eur Cell Mater 20:295–305
51. Armiñán A, Gandía C, Bartual M, García-Verdugo JM, Lledó E, Mirabet V et al (2009) Cardiac differentiation is driven by NKX2. 5 and GATA4 nuclear translocation in tissue-specific mesenchymal stem cells. Stem Cells Dev 18(6):907–918
52. Ishkitiev N, Yaegaki K, Kozhuharova A, Tanaka T, Okada M, Mitev V et al (2013) Pancreatic differentiation of human dental pulp CD117+ stem cells. Regen Med 8(5):597–612
53. Gomes JAP, Melo GB, Monteiro BG, Maranduba CCM, Fonseca SAS, Cerruti H et al (2007) Differentiation of immature adult stem cells isolated from human dental pulp into corneal ephitelium in an animal model of total limbal stem cell deficiency. Invest Ophthalmol Vis Sci 48(13):4436
54. Harris P, Nagy S (2009) Mosby's dictionary of medicine, nursing and health professions. Elsevier, Australia
55. Sonoyama W, Liu Y, Yamaza T, Tuan RS, Wang S, Shi S, Huang GTJ (2008) Characterization of the apical papilla and its residing stem cells from human immature permanent teeth: a pilot study. J Endod 34(2):166–171
56. Tziafas D, Kodonas K (2010) Differentiation potential of dental papilla, dental pulp, and apical papilla progenitor cells. J Endod 36(5):781–789
57. D'souza R (2002) Development of the pulpodentin complex. Seltzer and Bender's dental pulp. Quintessence publishing Co, Inc., Hanover Park, pp 13–40
58. De Berdt P, Vanacker J, Ucakar B, Elens L, Diogenes A, Leprince JG, des Rieux A (2015) Dental apical papilla as therapy for spinal cord injury. J Dental Res 94(11):1575–1581
59. Mhaske S, Yuwanati MB, Mhaske A, Ragavendra R, Kamath K, Saawarn S (2013) Natal and neonatal teeth: an overview of the literature. ISRN Pediatrics
60. Rad MR (2014) Characteristics of dental follicle stem cells and their potential application for treatment of craniofacial defects (Doctoral dissertation, Faculty of the Louisiana State University and Agricultural and Mechanical College in partial fulfillment of the requirements for the degree of Doctor of Philosophy in The Interdepartmental Program in Veterinary Medical Science through The Department of Comparative Biomedical Sciences by Maryam Rezai Rad DDS, Tehran University of Medical Sciences)
61. Vishwakarma A, Sharpe P, Shi S, Ramalingam M (eds) (2014) Stem cell biology and tissue engineering in dental sciences. Academic Press, London, UK
62. Kawanabe N, Murata S, Murakami K, Ishihara Y, Hayano S, Kurosaka H, Kamioka H, Takano-Yamamoto T, Yamashiro T (2010) Isolation of multipotent stem cells in human periodontal ligament using stage-specific embryonic antigen-4. Differentiation 79(2):74–83

63. Mrozik K, Gronthos S, Shi S, Bartold PM (2010) A method to isolate, purify, and characterize human periodontal ligament stem cells. In: Oral biology. London, UK: Humana Press, pp 269–284
64. Karamzadeh R, Eslaminejad MB (2013) Dental-related stem cells and their potential in regenerative medicine pp 95–116 http://dx.doi.org/10.5772/55927
65. Kimmelman J, Daley G (2015) Guidelines for stem cell science 2 and clinical translation-draft 2015. In: International Society for Stem Cell Research. Retrieved September 28, 2015. http://www.isscr.org/docs/default-source/2015-am-stockholm/draft-guidelines-for-stem-cell-science-and-clinical-translation-2015.pdf?sfvrsn=2
66. Lo B, Parham L (2013) Ethical issues in stem cell research. Endocr Rev 30(3):204–213
67. US Food and Drug Administration (2013) Code of federal regulations—title 21—food and drugs. Part 56: Institutional
68. Hyun I (2010) The bioethics of stem cell research and therapy. J Clin Invest 120(1):71
69. Huang GTJ, Sonoyama W, Chen J, Park SH (2006) In vitro characterization of human dental pulp cells: various isolation methods and culturing environments. Cell Tissue Res 324(2):225–236
70. Karamzadeh R, Eslaminejad MB, Aflatoonian R (2012) Isolation, characterization and comparative differentiation of human dental pulp stem cells derived from permanent teeth by using two different methods. J Vis Exp 69, e4372
71. Busser H, De Bruyn C, Urbain F, Najar M, Pieters K, Raicevic G, Meuleman N, Bron D, Lagneaux L (2014) Isolation of adipose-derived stromal cells without enzymatic treatment: expansion, phenotypical, and functional characterization. Stem Cells Dev 23(19): 2390–2400
72. Mokry J, Soukup T, Micuda S, Karbanova J, Visek B, Brcakova E, Suchanek J, Bouchal J, Vokurkova D, Ivancakova R (2010) Telomere attrition occurs during ex vivo expansion of human dental pulp stem cells. J Biomed Biotechnol 2010:673513, Epub 2010 Oct 4
73. Morsczeck C, Völlner F, Saugspier M, Brandl C, Reichert TE, Driemel O, Schmalz G (2010) Comparison of human dental follicle cells (DFCs) and stem cells from human exfoliated deciduous teeth (SHED) after neural differentiation in vitro. Clin Oral Investig 14(4):433–440
74. Ducret M, Fabre H, Degoul O, Atzeni G, McGuckin C, Forraz N, Alliot-Licht B, Mallein-Gerin F, Perrier-Groult E, Farges JC (2015) Manufacturing of dental pulp cell-based products from human third molars: current strategies and future investigations. Front Physiol 6:213
75. Suslov ON, Kukekov VG, Ignatova TN, Steindler DA (2002) Neural stem cell heterogeneity demonstrated by molecular phenotyping of clonal neurospheres. Proc Natl Acad Sci 99(22):14506–14511
76. Sasaki R, Aoki S, Yamato M, Uchiyama H, Wada K, Okano T, Ogiuchi H (2008) Neurosphere generation from dental pulp of adult rat incisor. Eur J Neurosci 27(3):538–548
77. Techawattanawisal W, Nakahama K, Komaki M, Abe M, Takagi Y, Morita I (2007) Isolation of multipotent stem cells from adult rat periodontal ligament by neurosphere-forming culture system. Biochem Biophys Res Commun 357(4):917–923
78. Jensen JB, Parmar M (2006) Strengths and limitations of the neurosphere culture system. Mol Neurobiol 34(3):153–161
79. Parmar M, Sjöberg A, Björklund A, Kokaia Z (2003) Phenotypic and molecular identity of cells in the adult subventricular zone: in vivo and after expansion in vitro. Mol Cell Neurosci 24(3):741–752
80. Bai Y, Bai Y, Matsuzaka K, Hashimoto S, Kokubu E, Wang X, Inoue T (2010) Formation of bone-like tissue by dental follicle cells co-cultured with dental papilla cells. Cell Tissue Res 342(2):221–231
81. Xiao L, Tsutsui T (2012) Three-dimensional epithelial and mesenchymal cell co-cultures form early tooth epithelium invagination-like structures: expression patterns of relevant molecules. J Cell Biochem 113(6):1875–1885

82. Hirata TM, Ishkitiev N, Yaegaki K, Calenic B, Ishikawa H, Nakahara T, Haapasalo M (2010). Expression of multiple stem cell markers in dental pulp cells cultured in serum-free media. J Endod 36(7):1139–1144
83. Brunner D, Frank J, Appl H, Schöffl H, Pfaller W, Gstraunthaler G (2010) Serum-free cell culture: the serum-free media interactive online database. Altex 27(1):53
84. Tarle SA, Shi S, Kaigler D (2011) Development of a serum-free system to expand dental-derived stem cells: PDLSCs and SHEDs. J Cell Physiol 226(1):66–73
85. Kobayashi T, Watanabe H, Yanagawa T, Tsutsumi S, Kayakabe M, Shinozaki T, Higuchi H, Takagishi K (2005) Motility and growth of human bone-marrow mesenchymal stem cells during ex vivo expansion in autologous serum. J Bone Joint Surg 87(10):1426–1433
86. Lin HT, Tarng YW, Chen YC, Kao CL, Hsu CJ, Shyr YM, Ku H-H, Chiou SH (2005) Using human plasma supplemented medium to cultivate human bone marrow–derived mesenchymal stem cell and evaluation of its multiple-lineage potential. Transplant Proc 37:4504–4505
87. Mizuno N, Shiba H, Ozeki Y, Mouri Y, Niitani M, Inui T, Inui T, Kurihara H (2006) Human autologous serum obtained using a completely closed bag system as a substitute for fetal calf serum in human mesenchymal stem cell cultures. Cell Biol Int 30(6):521–524
88. Shahdadfar A, Frønsdal K, Haug T, Reinholt FP, Brinchmann JE (2005) In vitro expansion of human mesenchymal stem cells: choice of serum is a determinant of cell proliferation, differentiation, gene expression, and transcriptome stability. Stem Cells 23(9):1357–1366
89. Stute N, Holtz K, Bubenheim M, Lange C, Blake F, Zander AR (2004) Autologous serum for isolation and expansion of human mesenchymal stem cells for clinical use. Exp Hematol 32(12):1212–1225
90. Bernardo ME, Avanzini MA, Perotti C, Cometa AM, Moretta A, Lenta E, Locatelli F (2007) Optimization of in vitro expansion of human multipotent mesenchymal stromal cells for cell-therapy approaches: further insights in the search for a fetal calf serum substitute. J Cell Physiol 211(1):121–130
91. Kilian O, Flesch I, Wenisch S, Taborski B, Jork A, Schnettler R, Jonuleit T (2004) Effects of platelet growth factors on human mesenchymal stem cells and human endothelial cells in vitro. Eur J Med Res 9(7):337–344
92. Suchanek J, Soukup T, Visek B, Ivancakova R, Kucerova L, Mokry J (2009) Dental pulp stem cells and their characterization. Biomed Papers 153(1):31–35
93. Heiskanen A, Satomaa T, Tiitinen S, Laitinen A, Mannelin S, Impola U, Laine J (2007) N-glycolylneuraminic acid xenoantigen contamination of human embryonic and mesenchymal stem cells is substantially reversible. Stem Cells 25(1):197–202
94. Martin MJ, Muotri A, Gage F, Varki A (2005) Human embryonic stem cells express an immunogenic nonhuman sialic acid. Nat Med 11(2):228–232
95. Spees JL, Gregory CA, Singh H, Tucker HA, Peister A, Lynch PJ, Hsu S, Smith J, Prockop D (2004) Internalized antigens must be removed to prepare hypoimmunogenic mesenchymal stem cells for cell and gene therapy. Mol Therapy 9(5):747–756
96. Aguzzi A, Barbara J, Brown P, Budka H, Diringer H, Dormont D, Heymann DL (1997) Medicinal and other products and human and animal transmissible spongiform encephalopathies: memorandum from a WHO meeting
97. Halme DG, Kessler DA (2006) FDA regulation of stem-cell–based therapies. N Engl J Med 355(16):1730–1735
98. Bruinink A, Tobler U, Hälg M, Grünert J (2004) Effects of serum and serum heat-inactivation on human bone derived osteoblast progenitor cells. J Mater Sci Mater Med 15(4):497–501
99. Dictus C, Tronnier V, Unterberg A, Herold-Mende C (2007) Comparative analysis of in vitro conditions for rat adult neural progenitor cells. J Neurosci Methods 161(2):250–258
100. Dimarakis I, Levicar N (2006) Cell culture medium composition and translational adult bone marrow-derived stem cell research. Stem Cells 24(5):1407–1408

101. Sekiya I, Larson BL, Smith JR, Pochampally R, Cui JG, Prockop DJ (2002) Expansion of human adult stem cells from bone marrow stroma: conditions that maximize the yields of early progenitors and evaluate their quality. Stem Cells 20(6):530–541
102. Karbanová J, Soukup T, Suchánek J, Pytlík R, Corbeil D, Mokrý J (2011) Characterization of dental pulp stem cells from impacted third molars cultured in low serum-containing medium. Cells Tissues Organs 193(6):344–365
103. Vater C, Kasten P, Stiehler M (2011) Culture media for the differentiation of mesenchymal stromal cells. Acta Biomat 7:463–477
104. Wang HS, Hung SC, Peng ST, Huang CC, Wei HM, Guo YJ, Fu Y, Lai M, Chen CC (2004) Mesenchymal stem cells in the Wharton's jelly of the human umbilical cord. Stem Cells 22(7):1330–1337

Isolation and Cryopreservation of Stem Cells from Dental Tissues

O.G. Davies and B.A. Scheven

Introduction

One of the principal aims of cell-based therapies is to deliver personalised medicine for the repair and regeneration of tissues lost to accidents or disease. To achieve this aim sub-zero temperatures (−196°) are applied that halt biological activity, thus preserving the cells for future clinical applications. The idea of banking stem cells as a means of '*biological insurance*' has seen a recent rise in popularity that is at least in part due to increased media attention and a greater public awareness of regenerative medicine. Consequently, several companies now exist offering individuals the opportunity to store their own multipotent cells, with the aim of future therapeutic application to restore or regenerate a multitude of tissues throughout the body.

In order for cellular therapies to be effective and widely available a number of biological and pragmatic considerations need to be taken into account. Firstly, the cryopreservation procedure employed must assure the quality and safety of the final biological product. This is of particular importance when the banked cells are intended for clinical use. In order to ensure that cryopreserved stem cells are suitable for clinical application standardised protocols and procedures are required that adhere to current good manufacturing practice (cGMP) [1, 2]. This means that cryo-profiles including type, concentration, equilibration, and dilution of cryoprotective agent (CPA) as well as freezing rate and thawing conditions must be optimised and

O.G. Davies (✉)
Wolfson School of Mechanical and Manufacturing Engineering,
Loughborough University, Loughborough, Leicestershire LE11 3TU, UK
e-mail: O.G.Davies@lboro.ac.uk

B.A. Scheven
School of Dentistry, College of Medical and Dental Sciences,
University of Birmingham, St Chad's Queensway, Birmingham B4 6NN, UK

B. Zavan, E. Bressan (eds.), *Dental Stem Cells: Regenerative Potential*, Stem Cell Biology and Regenerative Medicine, DOI 10.1007/978-3-319-33299-4_3

validated for particular cells/tissues [3]. Additional factors to be considered when banking stem cells for clinical applications are: potential cytotoxicity of the CPA and the immunogenicity of animal or human serum components often incorporated within the cryopreservation medium (CPM). Implementation of these standards will help ensure that cell viability, proliferation, genetic and epigenetic stability, and differentiation are not compromised, allowing banked stem cells to be as clinically effective as freshly isolated cells. One final point concerns the more pragmatic aspects of space, cost-effectiveness, and the establishment of GMP standards governing freezing protocols and the use of CPAs [4]. The scaling of stem cell therapies to meet clinical demand is by no means a simple task and the future success of cellular therapies is equally likely to depend on advances in the field of healthcare engineering as well as in basic science and medicine [5]. This concise review will focus on issues surrounding the cryopreservation of dental tissues, in particular stem cells isolated from dental pulp, which are an emerging and promising source for a wide range of regenerative therapies.

Principles of Cryopreservation

Cryostorage causes a physical insult to cells which could result in structural and molecular changes. The principal aim of any cryopreservation procedure is to provide adequate protection from damage caused by the formation of intracellular and extracellular ice crystals during both freezing and thawing stages [6]. In order to achieve this, CPAs are incorporated in the freezing medium. An ideal CPM is one in which cells maintain their stem cell characteristics over long periods of time (durability), with no loss in biological viability or vitality. CPAs work by penetrating the cell membrane where they bind to water molecules in solution. This acts to prevent an efflux of water from the cytoplasm during freezing, avoiding cellular dehydration and maintaining stable intracellular salt concentrations and pH [7]. To date, a number of different CPMs have been tested for the storage of mesenchymal stem cells (MSCs) derived from a variety of tissues, with the most common solution incorporating dimethyl sulfoxide (DMSO) as a CPA [8]. In addition to the choice and concentration of CPA used, the rate of freezing is also important for maintaining stem cell viability. Cells may incur damage caused by the process of rapid cooling (cold shock) or by the low temperatures used to enable long term preservation (chilling injury). The optimal cooling rate is likely to be specific for each tissue, and is largely determined by the ratio of cell volume to surface area, and by the permeability of the membrane to water and the CPA. The optimal cooling rate is also likely to depend on the concentration of CPA within the cryopreservation medium. Ideally, a CPA solution should be non-toxic for cells and patients, non-antigenic, chemically inert, ensure high cell viability after thawing and allow quick and simple transplantation without the need for additional washing steps [9]. However, producing a freezing environment that is optimal for both cells and patient will require further research, and it is likely that a compromise will have to be reached.

Dental Stem Cell Banking

This chapter focuses on the long-term storage of dental tissues, in particular stem cells derived from the vital central gelatinous tissue called the dental pulp. Dental pulp stem cells (DPSCs) were first identified at the turn of the twenty-first century and have since been shown to have the capacity to differentiate down multiple lineages [10]. DPSCs originate from the neural crest and are responsible for the formation of dentine producing cells called odontoblasts [11].

Barlett and Reade were the first scientists to cryopreserve dental tissue [12]. However, at that time dental cryopreservation was intended for the storage of whole teeth required for auto-transplantation rather than for stem cell isolation [13]. Recently the focus has shifted away from auto-transplantation towards the use of adult dental stem cells as a means of restoring or replacing lost or decayed dental tissue [10]. Indeed, adult stem cells are becoming increasingly popular as a cell source for regenerative medicine since they are not subject to the same ethical concerns as embryonic stem cells. Teeth in particular contain a significant proportion of adult mesenchymal stem cells (MSCs) in relation to their tissue mass [14]. Dental stem cells from extracted teeth have further appeal due to the fact that the process of isolating these multipotent cells is relatively simple and avoids invasive procedures associated with other popular sources of MSCs, such as bone marrow or fat. One of the most promising sources of dental stem cells resides within the primary dentition, which becomes lost during maturation. When human deciduous teeth are lost and replaced by permanent teeth, stem cells located within the pulp chamber can be harvested and stored as a convenient and pain-free source for regenerative medicine. The stem cells from human exfoliated deciduous teeth, or SHED cells, are a rapidly proliferating and multipotent source that are often favoured over other bankable dental stem cells located within the pulp of wisdom teeth or orthodontically extracted teeth [15]. The process of isolating and storing cells from teeth is relatively cost-effective when compared with MSCs derived from cord blood—a popular source for banking [16]. Additionally, the efficacy of these cryopreserved dental MSCs has been validated by a number of studies showing that post-thaw viability and differentiation following short and long term storage remains unaffected [17, 18]. These data highlight the promise of dental MSCs as a source of regenerative cells for long-term banking and clinical application. However, for the creation of cell banks the question of whether to bank intact whole teeth or the isolated stem cell population needs to be further considered.

Banking Intact Whole Teeth

Frequently stem cells are isolated, purified and expanded (to approximately 80–90 % confluence) prior to cryopreservation and the majority of this chapter will primarily focus on this method. However, one should be aware that this method requires resources for stem cell culture and scaling to be available upfront, thereby increasing workload and subsequent cost, with the further caveat of introducing possible phenotypic changes in the cells during expansion. It has been argued that a more rational and cost effective approach would instead be to bank whole teeth, allowing for the isolation and expansion of stem cells only when required. At present research into whole tooth banking for stem cell isolation is limited, and several groups attempting to cryopreserve intact teeth have shown a 25 % incidence of longitudinal fractures after freezing, which may compromise cell viability [19]. Methods for the banking of whole teeth have also encountered difficulties in enabling the CPA to reach the central pulpal region where the MSCs are localised. This consequently leads to problems with pulp necrosis, and can significantly reduce the viability, and hence the availability and clinical application of the cells [20, 21]. Importantly, current evidence demonstrates that MSC immunophenotype and differentiation is unaffected following whole tooth banking [22]. However, there are mixed reports concerning the effects of whole tooth banking on cell proliferation rates and morphologies [19, 22–24]. In order to improve the efficacy of whole tooth banking modern methods such as laser-piercing have been used to introduce microchannels, allowing CPA to reach the central pulp [21]. However, the procedure of whole tooth banking for MSC extraction is still in its infancy, and further advancements will be required if such a method is to be applied on a global scale.

International Licensed Dental Stem Cell Banks

To deliver effective autologous regenerative therapies dental stem cells need to be stored in a state of stasis. To achieve this purpose tooth banks are being developed that provide members of the public the opportunity to bank these clinically useful cells for an annual fee. Currently, the number of international licensed dental stem cell and tooth banks is low. However, due to increased media and public awareness of regenerative medicine and the potentially revolutionary impact of these technologies on the healthcare sector the number is likely to increase. Currently, a number of international laboratories offer collection and storage of dental stem cells, and the ease of the process typically makes it simpler and cheaper than banking cord blood stem cells. Historically, the world's first tooth bank was developed in Japan at Hiroshima University in 2005 [25]. The country's national tooth bank was opened soon after in 2008—a collaboration between Hiroshima University and Taipei Medical University. A similar development has since taken place in Norway as collaboration between the Norwegian Institute of Public Health and the University of Bergen [26]. However, tooth banking is not restricted to academic institutions, with a number of commercially licensed tooth banks existing in developed countries such as the UK and USA, as well as in developing countries such as India. At present the USA dominates the commercial tooth banking market with several companies such as BioEden, Pharmacells, Store-A-Tooth and StemSave expanding throughout North America, Europe and the rest of the world.

Cryopreservation Procedure

If banked stem cells are intended for therapeutic use it is important that current standards of good manufacturing practice are implemented throughout the entire process, from cell collection to freezing, storage and post-thaw assessment. The development of a standardised cryopreservation protocol for dental stem cells will require specification of the following:

Tooth Collection

One of the most significant problems related to the cryopreservation of dental tissue is the time elapsed between tissue harvesting and banking [27]. As soon as the tooth is lost or removed the vital pulp will begin to degrade, causing irreparable damage to the cellular component. A previous study has shown that in order to maintain DPSC viability, cell isolation and cryopreservation must occur within 120 h of tooth extraction [23]. Currently, to minimise tissue degradation, teeth are placed in sterile hypotonic phosphate buffered saline solution, which prevents dehydration. This can be done by the individual, the individual's parents/guardians (in the case of SHED cells) or a dental health professional. The tooth is then transferred into a vial, which is carefully sealed and placed into a thermette (a temperature phase change carrier) and transported to the banking facility in an insulated metal transport vessel—these materials are often provided by the tooth bank [16]. The procedure is designed to maintain the sample in a hypothermic state, and the process is referred to as sustentation. It is of critical importance that at the time of extraction, the central pulp region appears red in colour, indicating sufficient blood flow at the time of isolation. Teeth should be excluded if the central pulp appears grey, as this indicates pulp necrosis. The most appropriate source of deciduous teeth for isolation and banking are the canines and incisors, since these teeth often contain a healthy source of pulp that is starting to loosen. Primary incisors and canines with no pathology and at least one third of the root in-tact can also be banked [28]. Primary molars are not recommended for SHED banking as they take a longer time to resorb, which can have a negative impact on the pulp chamber and consequently on the viability of the stem cells [29].

Cell Isolation and Expansion

The majority of tooth banks isolate and expand the cells prior to freezing. However, before the dental pulp can be isolated the tooth must undergo a rigorous disinfection procedure. Sterilization can be achieved by washing the cells with PBS and povidone iodine [16]. Washes with commercially available bactericidal solutions (e.g.

chlorhexidine) may also be used to reduce the number of microorganisms and help further decrease the risk of cross-contamination [23]. After the tooth has been washed and disinfected DPSCs are isolated using protocols originally developed by Gronthos et al. in the following way [30]:

1. Pulp tissue is removed from the tooth manually using sterile forceps or a dental excavator. The pulp may also be removed by flushing the pulp chamber with sterile salt water or saline solution.
2. The excavated pulp is placed in a sterile petri dish and washed several times with PBS to further reduce the risk of bacterial contamination.
3. Pulp tissue is digested with a Collagenase type-I (3 mg/mL)/Dispase II (4 mg/mL) solution or 0.5 % Trypsin/EDTA solution for a period of 1 h at 37 °C.
4. Digested pulp tissue is passed through a 70 μm cell strainer to obtain single cell suspensions.
5. Cell suspensions are cultured at 37 °C/5 % CO_2 in growth medium; Alpha-modified Eagle's medium (α-MEM) supplemented with 20 % foetal bovine serum (FBS), 2 mM L-glutamine, 100 μM L-ascorbate-2-phosphate, 50 U/mL penicillin, and 50 μg/mL streptomycin.
6. At this stage there is the option to use the expanded MSCs to obtain differentiated cell lines such as odontogenic [31], adipogenic [32] or neural [33] by making adjustments to the culture medium.

In order to characterise or improve the homogeneity of MSC cultures fluorescence activated- (FACS) or magnetic activated- cell sorting (MACS) can be applied to select for Stro-1^+ or $CD146^+$ cells. Previous studies have demonstrated that DPSCs can be accurately identified through the presence of these cell surface antigens [34]. However, it should be noted that purification of MSC cell populations may not always be beneficial as it can adversely affect their differentiation and therapeutic potential [32].

Introduction of a Cryopreservation Solution

Incorporation of a CPA within the freezing medium prevents the formation of intracellular ice crystals that damage the cytoplasmic membrane during freezing. Protocols utilising DMSO as a CPA are widely used both in research and the clinic, since this method consistently allows high post-thaw recovery and differentiation [35]. However, there are a number of documented issues concerning the use of DMSO. Firstly, DMSO is recognised as a cytotoxic chemical that can cause adverse side effects such as sickness if introduced to a patient during cell therapy [36]. Secondly, DMSO has been reported to cause epigenetic changes and cytotoxicity at levels routinely used for MSC cryopreservation [37, 38]. Surprisingly, given the known adverse effects of DMSO there are currently no regulations or guidelines governing the use of this chemical for cell banking. In fact, a review of the current literature has identified that 10 % DMSO is most commonly incorporated within the freezing medium—currently used by approximately 80 % of cell therapy centres [39]. However, due to the documented negative effects of this CPA on both cells and patient there is a need to find effective alternative methods. Reducing the concentration of DMSO has been postulated as one way of reducing associated cellular toxicity [8]. However, reports have shown that the survival and number of colonies formed by MSCs was significantly decreased as a consequence, and that the magnitude of this decrease was inversely proportional to DMSO concentration [23]. Furthermore, concentrations below 7 % have been shown to cause a significant decline in DPSC viability [2, 23]. Therefore, it was proposed that an alternative approach that employs a simple post-thaw washing step may represent an effective way of reducing the patient-associated effects of this CPA without affecting cell recovery [40]. However, this approach fails to take into account any epigenetic changes associated with the incorporation of 10 % DMSO. Furthermore, if applying such an approach, one has to be careful not to trigger a damaging osmotic shock [23]. To avoid triggering an osmotic shock a similar result might achieved by avoiding post-thaw washes and instead simply fractionating the delivery of stem cells [41].

Alternatively, one could reduce the concentration of DMSO in the CPM by incorporating biocompatible CPAs that reduce the risk of cellular toxicity and accompanying side effects upon cell transplantation. It is important that these non-toxic alternatives are able to provide a level of efficacy similar to DMSO without the documented side-effects. Lowering the concentration of DMSO by incorporating sugar derivatives such as hydroxyethyl starch or trehalose has been shown to provide comparable levels of protection from freezing-induced cell damage to traditional 10 % DMSO-based solutions [9, 42]. The incorporation of CPAs such as methylcellulose or polyethylene glycol has also shown promise, reducing the concentration of DMSO required in the CPM. Importantly, these freezing mediums provide a level of protection that virtually eliminates the need to incorporate serum, thereby reducing potential immunological complications resulting from cell transplantation. Indeed, the formulation of an effective serum-free CPM is likely to represent an important step required for the clinical translation of cellular therapies [43–46].

Complete replacement of DMSO with non-toxic alternatives has proven a more difficult task. For instance CPAs such as glycerol and proline are effective when used in combination with DMSO but provide inadequate protection when used independently [47]. In fact only a small number of CPAs have been identified that provide adequate cryoprotection without the need to incorporate DMSO. Currently, only CPMs containing polyvinylpyrollidone (PVP) or ectoin have been shown to achieve levels of MSC post-thaw viability and differentiation comparable to DMSO-based solutions [45, 47]. However, at present these studies are not comprehensive enough to draw any significant conclusions, and the efficacy of these CPAs will need to be evaluated in long-term studies before their clinical applicability can be comprehensively assessed. In conclusion, significant progress is being made in the identification of serum-free freezing mediums that contain non-toxic CPAs for the cryoprotection of MSCs. However, at present no studies have analysed the effects of novel biocompatible CPAs on multipotent dental stem cells. Therefore, considerable further investment is required into research that comprehensively examines the efficacy of non-toxic CPAs for dental stem cell banking.

Freezing Protocol

Controlled-Rate Freezing

Damage caused by cell dehydration and ice crystal formation can be minimised by controlling the freezing rate. A slow and controlled freezing rate of 1–2 °C per minute is generally considered optimal for maintaining MSC viability during cryopreservation [48, 49]. Controlled rate freezers (CRFs) are frequently used to accurately control the freezing rate, and can be either manual or programmable [50, 51]. Although controlled-rate freezing has the advantage of reducing the likelihood of cellular damage occurring through ice crystal formation, one of the fundamental issues with CRFs is that they cannot be scaled to provide uniformity of temperature to all vials during large scale banking [52]. This represents a significant problem, and if cell therapies are to be applied on a national or international scale advances in CRF technology will be required. Furthermore, CRF development, along with vapour phase storage in liquid nitrogen is costly, and could impact the clinical translation of regenerative medicine. Ultra-slow programmable freezing systems (USPF) represent an alternative method that apply a highly controlled freezing rate of −0.3 to −0.6 °C per minute, allowing comparatively high recovery rates and increased levels of efficiency that are in compliance with cGMP requirements [4, 53]. However, a significant caveat is that USPFs are more complex and costly than traditional slow rate freezing (SRF) [54]. Much like CRFs, scaling is also a significant pragmatic consideration when applying USPFs for the high-volume banking of dental tissue. Ultimately, the choice between programmable or manual freezing will be dependent on the resources available and the experience of the scientists or technicians on site.

Uncontrolled-Rate Freezing

An alternative approach where samples are first cooled at −4 °C and then directly deposited into a freezer at −80 °C or liquid phase nitrogen can also be applied [55]. Uncontrolled-rate freezing represents a comparatively cheap and simple way of cryopreserving cells that requires no specialist training. This method has been most frequently applied for the preservation of peripheral blood stem cells with consistent results [56, 57]. Initial studies have also shown that uncontrolled freezing can be successfully used for the long-term cryopreservation of dental pulp stem cells [58]. Based on the parameters of the study the authors found that this method was as effective at maintaining cell viability, proliferation, and differentiation as CRFs with subsequent liquid nitrogen storage. However, to date uncontrolled-rate freezing has only been tested in the presence of 10 % DMSO, and studies will be required to test the efficacy of this method for dental stem cell preservation in combination with low DMSO concentrations or non-toxic DMSO replacements.

Rapid Freezing (Vitrification)

Vitrification represents a viable alternative to controlled-rate freezing for the preservation of tissue and stem cells [59, 60]. The process of vitrification and its application for the cryopreservation of biological material was first described by Luyet during the early part of the twentieth century [61]. This approach applies a high concentration of cryoprotectant and rapid cooling rate to produce an arrested liquid state and a glass-like solidification [53]. Faster cooling is considered advantageous because it reduces the toxic effects of the CPA and minimises the formation of ice crystals during freezing and thawing [6]. The process is simple and achieved by placing small volume samples held on vitrification carriers, such as cryoleaf, cryoloop, or ministraw directly into liquid nitrogen [62–64]. The method has become increasingly popular as a means of preserving cells that are particularly sensitive to cryoinjury such as embryonic cells [53]. However, current evidence suggests that the rapid freezing of whole teeth may be problematic due to progressive root resorption, and conditions will need to be further optimised if this method of cryopreservation is to be used for whole tooth banking [48]. A second potential caveat associated with the vitrification method is that rapid freezing may also generate an increased immunological response within the pulp [65]. Pragmatically, this method is also limited due to the large volumes of liquid nitrogen that would be required to scale up the technology to meet clinical demand.

Magnetic Freezing

Magnetic freezing applies a weak magnetic field to water or tissue that lowers the freezing point by approximately 6–7 °C. The first programmable device was developed by Hiroshima University and marketed as the cells alive system (CAS) [66]. It operates by producing an alternating magnetic field that prohibits water molecules from clustering during the freezing process, thereby preventing the formation of ice crystals [67]. This method is ostensibly cheaper than cryogenics and has been shown to provide comparatively high post-thaw viability for intact whole teeth and DPSCs alike [68, 69]. Furthermore, magnetic cryopreservation has been shown to reduce the concentration of DMSO required (5 %) to maintain cell viability and vitality [68, 69]. Magnetic cryopreservation has been successfully utilised for tooth banking with satisfactory implantation outcomes, indicating that viability of the dental pulp and periodontal ligament is adequately maintained [70]. Importantly, this method of cryopreservation may also eliminate the need for the addition of serum to the freezing medium, putting it in compliance with GMP.

Thawing Conditions

The standard method for thawing cryovials is to place them in a water bath at 37 °C. This facilitates a relatively quick thawing process that acts to remove and prevent the formation of ice crystals [71, 72]. However, studies examining the influence of temperature, time and method of thawing are limited. Of the few studies examining the effects of thawing on cell viability and vitality, some groups have examined the use of dry heat applied using 37 °C gel pads and compared this method with cells thawed using a water bath. It has been hypothesised that the application of dry heat provides a cleaner way of thawing cells that are to be used for clinical applications. These studies identified that samples thawed using dry heat had similar viabilities and clonogenic potential to those thawed using the standard procedure [73, 74]. Of the limited number of studies comparing the effects of thawing temperature, a report by Yang et al. identified no significant differences between cells that were thawed at 0, 20 and 37 °C for 20 min [75]. However, a comprehensive examination of the effects of thawing temperature and duration are required before any valid conclusions can be drawn. Lastly, upon thawing, removal of the cryoprotectant causes each cell first to shrink and then swell. Swelling can lead to damage as the cells expand. Such damage can be prevented by step-wise removal of the CPA using graded isotonic solutions [76]. An automated method designed to remove the CPA from haematopoietic grafts after banking has been shown to be feasible in preclinical models [77, 78] and the post-thaw removal of CPA is now standard for cells of bone marrow and peripheral blood. However, at present few studies on dental stem cell cryopreservation incorporate a step-wise removal of the CPA, instead opting to wash the thawed cells with a single isotonic solution to remove the CPA or often incorporating no washing step at all.

Post-Thaw Assessment

There are a number of possible negative cellular effects associated with cryopreservation, and at present it remains unclear whether post-thaw MSCs retain the same potential for therapeutic applications as their non-cryopreserved counterparts. Detrimental cellular effects associated with cryopreservation can include a reduction in cell viability due to cold-shock and/or the toxic effects of DMSO, changes in the expression of stem cell-related markers, cytoskeletal disassembly, delayed apoptosis, and osmotic and oxidative stresses [5, 79]. Consequently, these factors may have a downstream influence on the functionality of MSCs, thereby reducing their applicability for regenerative therapies. Additionally, ultrastructural changes in stem cell morphology remain largely unexamined during post-thaw assessment. However, such changes have been documented in cryopreserved adipose-derived stem cells (ADSCs) [80]. The subsequent downstream effects of these ultrastructural changes remain unknown. However, it is important that the consequences of these changes are fully determined before cell banking is applied on a national scale. At the most fundamental level it is crucial that stem cell viability, differentiation and other important characteristics required for effective tissue regeneration are extensively studied for each cell type, since these have been shown to vary for stem cells isolated from different tissue sources [55]. Here we highlight the importance of establishing protocols that are based on human studies and are tissue/cell specific. This is necessary since previous studies have shown that the viability and vitality of stem cells isolated from different species differs, with tissue-specific differences observed in post-thaw viability and gene expression [81].

Conclusions

Dental stem cell banking is becoming increasingly popular, with stem cells isolated from the dentition providing individuals with means of '*biological insurance*' to facilitate the regeneration of tissues lost to trauma or disease. In order to achieve this goal research and development is required to define optimum methods for long-term cryopreservation. The next phase of research will focus on the formulation of cryopreservation mediums that facilitate the safe and efficient transfer of cellular therapies within the clinic. This will mean eliminating the use of serum and toxic CPAs without compromising cell viability and vitality. Standardised protocols and procedures will also need to be developed that adhere to current good manufacturing practice (cGMP). Furthermore, pragmatic considerations relating to cost and scalability of cell banking procedures will need to be taken into consideration if cell therapies are to be applied on a national and international scale. Finally, perhaps one of the fundamental caveats concerning stem cell banking is that no studies have examined the effects of long-term cryopreservation (>10 years) on their viability and phenotype. This is a fundamental issue, and long-term studies will be required to demonstrate the efficacy of current cryopreservation procedures for future clinical applications.

References

1. Almansoori KA, Prasad V, Forbes JF et al (2012) Cryoprotective agent toxicity interactions in human articular chondrocytes. Cryobiology 64:185–191
2. Ock SA, Rho GJ (2011) Effect of dimethyl sulfoxide (DMSO) on cryopreservation of porcine mesenchymal stem cells (pMSCs). Cell Transplant 20:1231–1239
3. Schwartz O (1992) Cryopreservation of teeth before replantation or transplantation. In: Andreasen JO (ed) Atlas of replantation and transplantation of teeth. Mediglobe SA, Freiburg, pp 241–256
4. Orellana MD, De Santis GC, Abraham KJ et al (2015) Efficient recovery of undifferentiated human embryonic stem cell cryopreserved with hydroxyethyl starch, dimethyl sulfoxide and serum replacement. Cryobiology. doi:10.1016/j.cryobiol.2015.01.005
5. Li Y, Ma T (2012) Bioprocessing of cryopreservation for large-scale banking of human pluripotent stem cells. Biores Open Access 1(5):205–214
6. Kuleshova LL, Lopata A (2002) Vitrification can be more favourable than slow cooling. Fertil Steril 78(3):449–454
7. Mazur P (1981) Fundamental cryobiology and the preservation of organs by freezing. In: Karow AM, Pegg DE (eds) Organ preservation for transplantation. Marcel Dekker, New York, pp 143–175

8. Liu Y, Xu X, Ma X et al (2010) Cryopreservation of human bone marrow-derived mesenchymal stem cells with reduced dimethylsulfoxide and well-defined freezing solutions. Biotechnol Prog 26(6):1635–1643
9. Naaldijk Y, Staude M, Fedorova V et al (2012) Effect of different freezing rates during cryopreservation of rat mesenchymal stem cells using combinations of hydroxyethyl starch and dimethylsulfoxide. BMC Biotechnol 12:49
10. Gronthos S, Mankani M, Brahim J et al (2000) Postnatal human dental pulp stem cells (DPSCs) in vitro and in vivo. Proc Natl Acad Sci U S A 97(25):13625–13630
11. Kim SH, Kim YS, Lee SY et al (2011) Gene expression profile in mesenchymal stem cells derived from dental tissues and bone marrow. J Periodontal Implant Sci 41:192–200
12. Barlett PF, Reade PC (1972) Cryopreservation of developing teeth. Cryobiology 9:205–211
13. Temmerman L, De Pauw GA, Beele H et al (2006) Tooth transplantation and cryopreservation: state of the art. Am J Orthod Dentofacial Orthop 129(5):691–695
14. D'Aquino R, De Rosa A, Laino G et al (2009) Human dental pulp stem cells: from biology to clinical applications. J Exp Zool B Mol Dev Evol 312B:408–415
15. Shi S, Bartold PM, Miura M et al (2005) The efficacy of mesenchymal stem cells to regenerate and repair dental structures. Orthod Craniofac Res 8(3):191–199
16. Arova V, Arova P, Munshi AK et al (2009) Banking stem cells from human exfoliated deciduous teeth (SHED): saving for the future. J Clin Pediatr Dent 33:289–294
17. Zhang W, Walboomers XF, Shi S et al (2006) Multilineage differentiation potential of stem cells derived from human dental pulp after cryopreservation. Tissue Eng 12(10):2813–2823
18. Papaccio G, Graziano A, d'Aquino R et al (2006) Long-term cryopreservation of dental pulp stem cells (SBP-DPSCs) and their differentiated odontoblasts: a cell source for tissue repair. J Cell Physiol 208:319–325
19. Oh YH, Che ZM, Hong JC et al (2005) Cryopreservation of human teeth for future organization of a tooth bank—a preliminary study. Cryobiology 51(3):322–329
20. Schwartz O, Andreasen JO (1983) Cryopreservation of mature teeth before replantation in monkeys. I. Effect of different cryoprotective agents and freezing devices. Int J Oral Surg 12(6):425–436
21. Gioventu S, Andriolo G, Bonino F et al (2012) A novel method for banking dental pulp stem cells. Transfus Apher Sci 47(2):199–206
22. Lindemann D, Werle SB, Steffens D et al (2014) Effects of cryopreservation on the characteristics of dental pulp stem cells of intact deciduous teeth. Arch Oral Biol 59(9):970–976
23. Woods EJ, Perry BC, Hockema JJ et al (2009) Optimized cryopreservation method for human dental pulp-derived stem cells and their tissues of origin for banking and clinical use. Cryobiology 59(2):150–157
24. Ji EH, Song JS, Kim SO et al (2014) Viability of pulp stromal cells in cryopreserved deciduous teeth. Cell Tissue Bank 15(1):67–674
25. Abedini S, Kaku M, Kawata T et al (2011) Effects of cryopreservation with a newly-developed magnetic field programmed freezer on periodontal ligament cells and pulp tissues. Cryobiology 62(3):181–187
26. Tvinnereim HM, Lygre GB, Haug K et al (2014) A biobank of primary teeth within the Norwegian mother and child cohort study (MoBa) per 2014: a resource for the future. Norsk Epidemiologi 24(1–2):135–140
27. Woods EJ, Benson JD, Agca Y et al (2004) Fundamental cryobiology of reproductive cells and tissues. Cryobiology 48:146–156
28. Bansal R, Jain A (2015) Current overview on dental stem cells applications in regenerative dentistry. J Nat Sci Biol Med 6(1):29–34
29. Bernardi L, Luisi SB, Fernandes R et al (2011) The isolation of stem cells from human deciduous teeth pulp is related to the physiological process of resorption. J Endod 37(7):973–979
30. Gronthos S, Arthur A, Bartold PM et al (2011) A method to isolate and culture expand human dental pulp stem cells. Methods Mol Biol 698:107–121

31. Wang J, Liu X, Jin X et al (2010) The odontogenic differentiation of human dental pulp stem cells on nanofibrous poly(L-lactic acid) scaffolds in vitro and in vivo. Acta Biomater 6(10):3856–3863
32. Davies OG, Cooper PR, Shelton RM et al (2014) A comparison of the in vitro mineralisation and dentinogenic potential of mesenchymal stem cells derived from adipose tissue, bone marrow and dental pulp. J Bone Miner Metab 33(4):371–382
33. Takeyasu M, Nozaki T, Daito M (2006) Differentiation of dental pulp stem cells into a neural lineage. Pediatr Dent J 16(2):154–162
34. Huang CE, Hu FW, Yu CH et al (2014) Concurrent expression of Oct4 and Nanog maintains mesenchymal stem-like property of human dental pulp cells. Int J Mol Sci 15(10): 18623–18639
35. Ginis I, Grinblat B, Shirvan MH (2012) Evaluation of bone marrow-derived mesenchymal stem cells after cryopreservation and hypothermic storage in clinically safe medium. Tissue Eng Part C Methods 18(6):453–463
36. Zambelli A, Poggi D, Da Prada G et al (1998) Clinical toxicity of cryopreserved circulating progenitor cells infusion. Anticancer Res 18(6B):4705–4708
37. Iwatani M, Ikegami K, Kremenska Y et al (2006) Dimethyl sulfoxide has an impact on epigenetic profile in mouse embryoid body. Stem Cells 24(11):2549–2556
38. Galvao J, Davis B, Tilley M et al (2014) Unexpected low-dose toxicity of the universal solvent DMSO. FASEB J 28(3):1317–1330
39. Windrum P, Morris TC, Drake MB et al (2005) Variation in dimethyl sulfoxide use in stem cell transplantation: a survey of EBMT centres. Bone Marrow Transplant 36(7):601–603
40. Syme R, Bewick M, Stewart D et al (2004) The role of depletion of dimethyl sulfoxide before autografting: on hematologic recovery, side effects, and toxicity. Biol Blood Marrow Transplant 10:135–141
41. Martino M, Morabito F, Messina G et al (1996) Fractionated infusions of cryopreserved stem cells may prevent DMSO-induced major cardiac complications in graft recipients. Haematologica 81:59–61
42. De Rosa AF, De Francesco V, Tirino GA et al (2009) A new method for cryopreserving adipose-derived stem cells: an attractive and suitable large-scale and long-term cell-banking technology. Tissue Eng Part C Methods 15:659–667
43. Eroglu A, Russo MJ, Bieganski R et al (2000) Intracellular trehalose improves the survival of cryopreserved mammalian cells. Nat Biotechnol 18:163–167
44. Thirumala S, Gimble JM, Devireddy RV (2010) Cryopreservation of stromal vascular fraction of adipose tissue in serum-free freezing medium. J Tissue Eng Regen Med 4(3):224–234
45. Thirumala S, Wu X, Gimble JM et al (2010) Evaluation of polyvinylpyrrolidone as a cryoprotectant for adipose tissue-derived adult stem cells. Tissue Eng Part C Methods 16(4):783–792
46. Lee YA, Kim YH, Kim BJ et al (2013) Cryopreservation of mouse spermatogonial stem cells in dimethylsulfoxide and polyethylene glycol. Biol Reprod 89(5):109
47. Grein TA, Freimark D, Weber C et al (2010) Alternatives to dimethylsulfoxide for serum-free cryopreservation of human mesenchymal stem cells. Int J Artif Organs 33(6):370–380
48. Kawasaki N, Hamamoto Y, Nakajima T et al (2004) Periodontal regeneration of transplanted rat molars after cryopreservation. Arch Oral Biol 49(1):59–69
49. Thirumala S, Zvonic S, Floyd E et al (2005) Effect of various freezing parameters on the immediate post-thaw membrane integrity of adipose tissue derived adult stem cells. Biotechnol Prog 21(5):1511–1524
50. Carvalho KA, Cury CC, Oliviera L et al (2008) Evaluation of bone marrow mesenchymal stem cell standard cryopreservation procedure efficiency. Transplant Proc 40:839–841
51. Pal R, Hanwate M, Totey SM et al (2008) Effect of holding time, temperature and different parenteral solutions on viability and functionality of adult bone marrow-derived mesenchymal stem cells before transplantation. J Tissue Eng Regen Med 2:436–444
52. Thirumala S, Goebel WS, Woods EJ (2013) Manufacturing and banking of mesenchymal stem cells. Expert Opin Biol Ther 13:673–691

53. Li T, Mai Q, Gao J et al (2010) Cryopreservation of human embryonic stem cells with a new bulk vitrification method. Biol Reprod 82(5):848–853
54. Ha SY, Jee BC, Suh CS et al (2005) Cryopreservation of human embryonic stem cells without the use of a programmable freezer. Hum Reprod 20(7):1779–1785
55. Davies OG, Smith AJ, Cooper PR et al (2014) The effects of cryopreservation on cells isolated from adipose, bone marrow and dental pulp tissues. Cryobiology 69(2):342–347
56. Rosenfeld CS, Gremba C, Shadduck RK et al (1994) Engraftment with peripheral blood stem cells using noncontrolled-rate cryopreservation: comparison with autologous bone marrow transplantation. Exp Hematol 22(3):290–294
57. Detry G, Calvet L, Straetmans N et al (2014) Impact of uncontrolled freezing and long-term storage of peripheral blood stem cells at −80 °C on haematopoietic recovery after autologous transplantation. Report from two centres. Bone Marrow Transplant 49(6):780–785
58. Kumar A, Bhattacharyya S, Rattan V (2015) Effect of uncontrolled freezing on biological characteristics of human dental pulp stem cells. Cell Tissue Bank. doi:10.1007/s10561-9498-5
59. Keros V, Xella S, Hultenby K et al (2009) Vitrification versus controlled-rate freezing in cryopreservation of human ovarian tissue. Hum Reprod 24(7):1670–1683
60. Matsumara K, Bae JY, Kim HH et al (2011) Effective vitrification of human induced pluripotent stem cells using carboxylated ε-poly-L-lysine. Cryobiology 63(2):76–83
61. Luyet B (1937) The vitrification of organic colloids and protoplasm. Biodynamica 1(29):1–14
62. Chian RC, Gilbert L, Huang JY et al (2009) Live birth after vitrification of in vitro matured human oocytes. Fertil Steril 91:372–376
63. Lane M, Schoolcraft WB, Gardner DK et al (1999) Vitrification of mouse and human blastocysts using a novel cryoloop containerless technique. Fertil Steril 72:1073–1078
64. Liebermann J, Tucker MJ (2002) Effect of carrier system on the yield of human oocytes and embryos as assessed by survival and developmental potential after vitrification. Reproduction 124:483–489
65. Ahn SJ, Jang JH, Seo JS et al (2013) Influence of 2 cryopreservation methods to induce CCL-13 from dental pulp cells. J Endod 39(12):1562–1566
66. Morono Y, Terada T, Yamamoto Y et al (2015) Intact preservation of environmental samples by freezing under an alternating magnetic field. Environ Microbiol Rep 7(2):243–251
67. Kawata T, Abedini S, Kaku M et al (2012) Effects of DMSO (dimethyl sulfoxide) free cryopreservation with program freezing using a magnetic field on periodontal ligament cells and dental pulp tissues. Biomed Res 23(3):438–443
68. Lee SY, Sun CH, Kuo TF et al (2012) Determination of cryoprotectant for magnetic cryopreservation of dental pulp tissue. Tissue Eng Part C Methods 18(6):397–407
69. Lin SL, Chang WJ, Lin CY et al (2014) Static magnetic field increases survival rate of dental pulp stem cells during DMSO-free cryopreservation. Electromagn Biol Med 23:1–7
70. Lee SY, Chiang PC, Tsai YH et al (2010) Effects of cryopreservation of intact teeth on the isolated dental pulp stem cells. J Endod 36(8):1336–1340
71. Katayama Y, Yano T, Bessho A et al (1997) The effects of a simplified method for cryopreservation and thawing procedures on peripheral blood stem cells. Bone Marrow Transplant 19(3):283–287
72. Perry BC, Zhou D, Wu X et al (2008) Collection, cryopreservation, and characterization of human dental pulp-derived mesenchymal stem cells for banking and clinical use. Tissue Eng Part C Methods 14(2):149–156
73. Rollig C, Babatz J, Wagner I et al (2002) Thawing of cryopreserved mobilized peripheral blood—comparison between waterbath and dry warming device. Cytotherapy 4(6):551–555
74. Triana E, Ortega S, Azgueta C et al (2013) Thawing of cryopreserved hematopoietic progenitor cells from apheresis with a new dry-warming device. Transfusion 53(1):85–90
75. Yang H, Acker JP, Hannon J et al (2001) Damage and protection of UC blood cells during cryopreservation. Cytotherapy 3(5):377–386

76. Mukherjee IN, Song YC, Sambanis A (2007) Cryoprotectant delivery and removal from murine insulinomas at vitrification-relevant concentrations. Cryobiology 55(1):10–18
77. Calmels B, Houze P, Hengesse JC et al (2003) Preclinical evaluation of an automated closed fluid management device: Cytomate, for washing out DMSO from hematopoietic stem cell grafts after thawing. Bone Marrow Transplant 31(9):823–828
78. Decot V, Houze P, Stoltz JF et al (2009) Quantification of residual dimethylsulfoxide after washing cryopreserved stem cells and thawing tissue grafts. Biomed Mater Eng 19(4–5): 293–300
79. Pal R, Mamidi MK, Das AK et al (2012) Diverse effects of dimethyl sulfoxide (DMSO) on the differentiation potential of human embryonic stem cells. Arch Toxicol 86(4):651–661
80. James AW, Levi B, Nelson ER et al (2011) Deleterious effects of freezing on osteogenic differentiation of human adipose-derived stromal cells in vitro and in vivo. Stem Cells Dev 20(3):427–439
81. Liu Y, Xu X, Ma XH et al (2011) Effect of various freezing solutions on cryopreservation of mesenchymal stem cells from different animal species. Cryo Letters 32(5):425–435

Epigenetics of Dental Stem Cells

Henry F. Duncan and Paul R. Cooper

Introduction

Stem cells (SCs) are critical to tissue regeneration [1] and they are either embryonic (embryonic stem cells [ESCs]), or post-natal/adult in origin (SCs) [2, 3]. ESCs are in an epigenetic-state which enables them to self-renew or differentiate into any bodily cell type (pluripotent). ASCs, however, have more restricted lineage potential (multipotent) [4, 5]. Current research is aimed at understanding cellular regulators that control this lineage commitment to enable novel stem cell-based regenerative strategies to be developed in medicine and dentistry.

Epigenetic modification of histone proteins by acetylation or via DNA methylation, are key regulator processes involved in SC self-renewal and pluripotency [6, 7] and occur in many tissues including dental ones [8]. Epigenetic modifications occur due to environmental factors and result in gene expression modulation, independent of DNA sequence [9] (Fig. 1). In healthy tissues, synchronous genetic patterns and epigenetic modifications regulate transcription, however, these processes are often disrupted in disease [11]. Whilst epigenetic changes are heritable, and accumulate throughout life, they are however reversible and therefore provide exciting opportunities for clinical application via pharmacological inhibition and modulation [12]. Key epigenetic modifications include direct DNA methylation [13], the post-translational modifications of acetylation, methylation and phosphorylation of histone proteins [14] and nucleosome positioning [15]. Therapeutically, a range of inhibitors targeting these molecular events have been developed with selected

H.F. Duncan (✉)
Division of Restorative Dentistry and Periodontology, Dublin Dental University Hospital, Trinity College Dublin, Lincoln Place, Dublin 2, Ireland
e-mail: Hal.Duncan@dental.tcd.ie

P.R. Cooper
Oral Biology, School of Dentistry, University of Birmingham, Pebble Mill Road, Birmingham B5 7SA, UK

B. Zavan, E. Bressan (eds.), *Dental Stem Cells: Regenerative Potential*, Stem Cell Biology and Regenerative Medicine, DOI 10.1007/978-3-319-33299-4_4

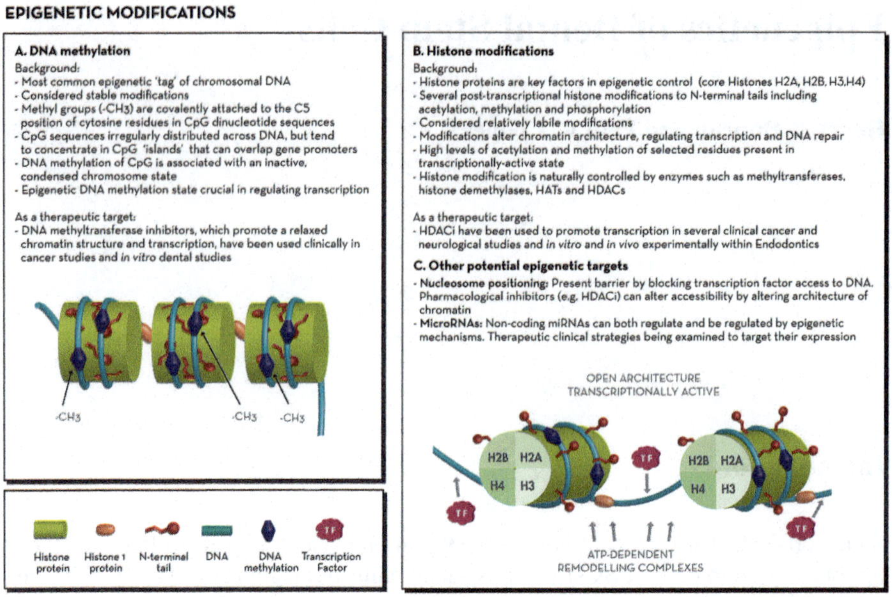

Fig. 1 Didactic classification of epigenetic modifications (reproduced with permission from [10]). (**a**) DNA methylation schematic illustrating the repeating nucleosome, which a single strand of DNA wound round the histone core. Methyl groups are covalently attached in CpG islands. (**b**) Histone modifications: the nucleosome comprises eight core histones. Linker DNA, connects the nucleosome structures and an additional histone protein (H1), enables further DNA compaction. The nucleosome core has a relatively long N-terminal amino acid extension, which is subject to several covalent epigenetic modifications (methylation, acetylation, phosphorylation, ubiquitination and SUMOylation). Histone modifications control both the conformation of the chromatin structure leading to an open or closed chromatin state and also nucleosome positioning. (**c**) The varying architecture of chromatin and position of the nucleosome regulates transcription, with an open conformation being transcriptionally active and a closed structure being transcriptionally repressive

therapeutics including Suberoylanilide hydroxamic acid (SAHA), Romidepsin and 5-azacytidine, already having United States Food and Drug Administration (FDA) clinical approval [16, 17].

Pharmacological agents, which epigenetically modify histone acetylation, include histone deacetylase inhibitors (HDACis). These molecules can reversibly change gene expression resulting in altered SC fate [18, 19]. Several HDACis, including Trichostatin A (TSA), Valproic acid (VPA) and Butyric acid, can induce SC proliferation, differentiation and also have anti-inflammatory activity [18, 20–22]. From a dental perspective, HDACis have been shown to epigenetically-modulate postnatal dental pulp SC (DPSC) differentiation and self-renewal potential due to their ability to alter the balance between histone deacetylases (HDACs) and histone acetyl transferases (HATs) [23]. Application of HDACis prevents deacetylation of the N-terminal histone tail within the nucleosome, leading to changes in chromatin architecture and transcription activity [24]. Eighteen human HDAC enzymes have been identified which differ in their expression profiles and cellular localisations [25]. While the entire functional role of HDACs is still to be fully elucidated [26], pan-HDACis and isoform specific inhibitors have been developed. Notably application of pan-HDACi have been shown to accelerate differentiation and mineralisation events in dental pulp cell populations [18, 27], although de-differentiation induction has also been shown in some immature SC populations [28].

In endodontics, tissue-resident and local SC populations including DPSCs and Stem Cells from the Apical Papilla (SCAP) may be key to regenerative responses [4]. These SCs are relatively easily accessible and have therapeutic application at other oral and bodily locations. For repair processes SC migration, proliferation and differentiation are necessary [29] and reportedly these cellular regenerative-associated events are epigenetically regulated with the role of HDACis in modifying these processes being recently highlighted [30, 31]. 'Regenerative endodontics' utilising SC technology continues to gather significant momentum and involves concerted academic and clinical effort to develop novel tissue engineering approaches which aim to utilise cell-free homing, platelet–derived plasma and triple antibiotic paste [32–35]. Targeted restorative dental materials, which harness innate biological reparative processes utilising epigenetic therapeutics may also have clinical utility in the more immediate future [36].

Furthermore for clinical application of dental SC technologies a detailed working knowledge of their cell isolation, expansion, and biobanking combined with knowledge of genotoxicity will be essential for clinical application. Training of the clinical team in these areas will be important to realising patient benefit with this technology. The aim of this chapter therefore is to review current knowledge of dental SCs and how epigenetics may play a role in their responses during health, disease and dental clinical application.

Review

Dental Stem Cells

Postnatal SCs, in particular ones derived from dental sites, provide a readily available source for tissue engineering and repair strategies throughout the body [2]. DPSCs and stem cells from exfoliated deciduous teeth (SHED), reportedly account for 1–5 % of total permanent and 2–9 % deciduous pulp cell populations, respectively [4, 37, 38]. Other dentally relevant SC sources have also been identified [2] and include stem cells of the apical papilla (SCAP), dental follicle progenitor cells (DFPCs) and periodontal ligament stem cells (PDLSCs) [38–40].

DPSCs are one of the most readily available SCs as they can be isolated by the dental practitioner from extracted third molar teeth which is regarded clinically as 'waste tissue'. Notably harvesting from this site avoids the donor site morbidity typically observed at other surgical locations. While the therapeutic application of DPSCs, such as dentine-pulp regeneration, appears quite promising, their overall utility may be limited due to the availability of autologous molar teeth for extraction along with the need for their in vitro expansion to obtain sufficient numbers for use [41]. Indeed during culture expansion, phenotypic change [42] may occur which also limits their clinical application. An understanding of the mediators, which are often epigenetic in nature, controlling self-renewal and lineage commitment, is therefore important. For regenerative dentistry other dental SC sources, such as DFPC and PDLSC may either be limited in their application due to diminished differentiation potential or in their availability [2, 8, 38]. SHED, however, are considered a rich source of dental SCs which can be harvested from waste tissue from extracted or exfoliated deciduous teeth. These dental SCs can be commercially biobanked as intact pulp tissue and subsequently SHED are isolated and expanded in vitro following enzymatic digestion of pulp tissue combined with fluorescence-activated cell sorting [43]. Although SHED are relatively easily obtained [43], currently their routine harvesting and clinical application is reportedly hampered by infection, lack of viable pulp tissue as well as lack of a standardised medicinal good manufacturing practice [37, 38].

Epigenetic Modifications and Dental Stem Cells

Classically ESCs exhibit an open chromatin structure characterised by relatively low levels of DNA methylation and high levels of histone acetylation [44]. After cellular lineage commitment and differentiation, the chromatin changes to a more tightly-bound conformation with reducing levels of acetylation and increasing levels of DNA methylation [45]. It is now evident that ESCs are maintained in a self-renewing state by a complex network of transcription factors [46], while differentiation and developmental potential are under epigenetic control [47]. During differentiation the pluripotency genes, Oct4 and Nanog, are repressed [48], an effect which is mediated initially by histone modifications, with the down-regulation maintained and cell reprogramming prevented by DNA methylation processes [49].

DNA methylation is cell type dependent and was traditionally thought to occur exclusively at CpG islands (Fig. 1), however, recently it was noted that up to 25 % of all ESC methylation was in a non-CG context [50]. Indeed the observed pattern of non-CG methylation disappeared when differentiation processes were induced in ESCs supporting a crucial role for methylation in determining cell fate [50]. Further characterization of epigenetic modifications in ESCs identified hypermethylation at 3 % of promoter regions in CgG islands, and notably the regions included developmental genes such as Rhox2 [51]. It has been suggested that although global methylation levels of CpG may be similar between cell types the distributions of methylation marks in ESCs are unlike any other cell type [52]. Other epigenetic modifications are also important regulators of SC function and self-renewal with

HDACs being vital to maintaining the self-renewal capabilities of mesenchymal SCs [7, 53] by maintaining the expression of key pluripotent transcription factors [6]. Amongst the best characterised cellular mediators of acetylation and methylation processes in SCs are the polycomb and trithorax protein complexes, which occupy ESC gene promoter sites and assist in catalysing specific histone modifications [54, 55]. Over the last 10 years knowledge of non-coding (nc) RNA, such as microRNA, has developed in relation to their epigenetic control due to their dysregulation in diseases such as cancer [56], their regulation of epigenetic mechanisms such as DNA methylation [57] and conversely their expression is regulated by epigenetic modifications [58]. Other ncRNAs, such as long-ncRNAs, also epigenetically modulate chromatin architecture by recruiting and binding chromatin-modifying proteins to specific genomic sites [59], while also guiding proteins, such as histone methyltransferases to those sites [60, 61].

Recently, interest in the epigenetic control of DPSC behaviour has led to suggestions that dental developmental anomalies, such as dentine dysplasia and dentinogenesis imperfecta, may be related to epigenetic modifications during odontoblast differentiation [62]. However, current research in this area is preliminary and has not, to date, analysed pure SC populations [63]. In a recent SC study, epigenetic states and related differentiation profiles were analysed in DPSCs and DFPCs via their expression of odontogenic genes, such as dentine sialophosphoprotein (DSPP) and dentine matrix protein-1 (DMP-1). Transcript levels were epigenetically-suppressed in DPFCs, while osteogenic stimulation in vitro demonstrated significant mineralisation increases only in DPSCs [8]. Interestingly, a highly dynamic histone modification response was demonstrated in mineralising DPFCs, but not in DPSCs, and the latter cells also expressed relatively high levels of the pluripotency-associated transcripts, Oct4 and Nanog. It was concluded that these two neural crest-derived SC populations were distinguished by epigenetic repression of dentinogenic genes with dynamic histone enrichment in DPFCs during mineralisation. These data demonstrated the essential role of epigenetic mechanisms in DPSC terminal differentiation and in controlling cell phenotype and lineage commitment [8].

Histone Acetylation Related Stem Cell Effects

Histone acetylation controls cellular transcription by altering the conformation of chromatin, a process balanced homeostatically by two cellular enzymes histone deacetylases (HDACs) and histone acetyl transferases (HATs). HDACs in particular have been a focus of attention as potential therapeutic targets [21], with 18 human HDAC family members characterised and divided into four classes [64]. Class I members (HDAC-1, -2, -3, -8) are ubiquitously expressed predominately in the cell nucleus [25], while the expression of class II enzymes (HDAC-4, -5, -6, -7, -9, -10) are tissue-restricted and dynamically shuttle from nucleus to cytoplasm [21, 65]. A recent analysis of selected HDAC expression in extracted third molar teeth demonstrated that HDACs -2 and -9 were expressed in selected pulp cell populations, while HDACs -1, -3 and -4 appeared only relatively weakly expressed [66]. The reported variations in tissue expression suggest different cellular roles for HDACs and indications are that class I HDACs regulate cell cycle progression and maintain ESC pluripotency, while class II HDACs are central to stem cell differentiation and mineralisation processes [6, 67, 68]. However, it is likely that their roles inter-link, with both Class I and II HDAC enzymes playing crucial roles in maintaining the pluripotency of stem cells and in governing cell fate [6, 53]. The labile effects of histone acetylation can be therapeutically targeted by HDACis, which alter the homeostatic HAT/HDAC balance, leading to changes in transcription and subsequently, stimulating pleiotropic cellular effects involving growth, differentiation and regeneration [31, 69, 70].

HDACi-Induced Stem Cell Effects

Several HDACi including TSA, VPA and SAHA, have been used already in clinical cancer trials, and their promise in treating neurodegenerative. conditions, as well as in bone engineering and regenerative medicine has also been highlighted [16, 71, 72]. Indeed, several HDACi, at relatively low concentrations, have been demonstrated to induce differentiation and increase mineralisation dose-dependently in primary dental pulp cell (DPC) populations containing stem cells [18, 23, 27]. Furthermore, HDACis and VPA in particular has also been shown to improve both the efficiency of somatic cell reprogramming to iPSCs, as well as increasing the expansion capacity of selected ASC populations [19, 73].

Certain HDAC members, -1 and -2, have been identified as playing crucial roles in controlling cellular proliferation and stem cell self-renewal by maintaining the expression of key pluripotent transcription factors such as Oct4, Nanog, Esrrb, and Rex1 [6]. In vitro HDACi stimulate the expansion of hematopoietic SCs, while decreasing the relative number of differentiated cells [19, 74], whereas in vivo HDACi-induced an increase in SC markers following digit amputation in mice [75] and during kidney organogenesis in zebrafish [30]. Conversely, several studies using various cell types have demonstrated that HDACi can reduce stem cell proliferation and promote terminal differentiation [76, 77, 78, 53]. These apparently divergent effects of HDACi were attributed to the level of baseline cell differentiation with stem cells tending to de-differentiate, while precursor blast cells differentiated [74]. Other researchers have further speculated that in addition to cell status, HDACi-dose may all be critical to the cell response, with undifferentiated cells and high doses of HDACi tending to promote differentiation, while low doses applied to cells undergoing self-renewal are more likely to stimulate a de-differentiation effect [28, 79].

Within dentistry several recent in vitro studies using mixed primary cell cultures [27], and pure DPSC populations [18, 23] have demonstrated an HDACi-induced stimulation of odontoblast-like cell differentiation with an associated increase in matrix mineral deposition. The effect of epigenetic modulation on regenerative processes highlighted differences between normal and immortalised cells with lower concentrations of HDACi and shorter durations of exposure being required to promote reparative effects in primary cells [27]. The effect of HDACis on human DPSC populations in vitro demonstrated increases in mineralisation, associated mineralisation-marker gene expression and down-regulation of HDAC-3 [18] and HDAC-2 [23], respectively. In vivo the response to a pre-natal systemic injection of TSA in dentine-pulp development was studied by histological analysis of tooth development at 7-days post-natal [23]. The volume of dentine deposited was thicker (1.64-fold) and the number of odontoblasts in the measured area was higher (1.74-fold) in the TSA group compared with the control [18]; this was attributed to an accelerated HDACi-induced differentiation effect. This reported ability of HDACi to manipulate stem cell fate renewal, although complex, is interesting from a regenerative medicine perspective as it potentially enables the scientist and clinician to promote differentiation, for example in tissue engineering, or electively maintain pluripotency if expanding SCs for cell-based therapies.

Induced Pluripotent Stem Cells (iPSCs)

The translational potential of pluripotent ESCs use has been somewhat limited due to expense, storage issues, ethics, as well as issues relating to immunological rejection [80]. In an attempt to address these issues, patient's own somatic cells have been reprogrammed into induced pluripotent stem cells (iPSCs) using a combination of four transcription factors (Oct4, Sox2, Klf4, c-Myc) and a viral induction protocol [81, 82]. However, as safety concerns were raised over the use of the proto-oncogenes, Klf4 and c-Myc, and low reprogramming rates (<0.001 %), the epigenetic modifiers, HDAC and DNA methyltransferase inhibitors, were used with and without these transcription factors [73]. The use of HDACi, in particular VPA, combined with all four transcription factors improved reprogramming efficiency up to 100-fold [73]. The presence of VPA, but absence of c-Myc, also increased reprogramming efficiency levels; however, when klf4 was eliminated the efficiency was greatly reduced indicating the importance of this molecule in the process [83]. It subsequently appears that VPA acts by inducing Oct4 predominately through the PI3K/Akt/mTOR pathway [84]. Further optimisation of iPSC techniques have been developed focusing on viral-free methods and avoidance of genomic modification approaches by using synthetic modified mRNAs for Oct4, Sox2, Klf4, and c-Myc, [85]. Although, VPA supplementation has also been combined with the synthetic mRNA methodology, only minor increases in reprogramming efficiency were observed [85]. The role of HDACi-supplementation in reprogramming demonstrates the pivotal role of epigenetic modification in the de-differentiation process and indicates the importance of the open chromatin architecture characteristic of ESCs [52].

Conclusions

A picture of the influence of epigenetic modifications on the mechanisms controlling stem cell fate and differentiation is developing, which critically involves DNMT and HDAC enzyme function, cell maturity and environmental signals. Further elucidation of the role of these enzymes is required to understand the influence of epigenetic modifications on pluripotency, their influence on stem cell fate and role in regenerative responses prior to any proposed clinical application. Regenerative tissue engineering approaches require SCs, signalling mediator molecules and a scaffold within which the cells can operate [86]. Therapeutically, HDAC and DNMT inhibitors could epigenetically manipulate the self-renewal capacity of SCs in vitro or indeed in vivo in the implanted stem cells. Additionally, there are opportunities for pharmacological epigenetic inhibitors to improve the efficiency of somatic cell reprogramming to iPSCs for subsequent craniofacial tissue regeneration, as availability of sufficient SCs may be limited. Finally, the availability of DSCs, their isolation, expansion, and epigenetic manipulation with HDACi could, at the very least, lead to a better understanding of DSC biology, which will enable their future clinical use within regenerative medical strategies.

References

1. Dupont KM, Sharma K, Stevens HY, Boerckel JD, Garcia AJ, Guldberg RE (2010) Human stem cell delivery for treatment of large segmental bone defects. Proc Natl Acad Sci U S A 107:3305–3310
2. Huang GT, Gronthos S, Shi S (2009) Mesenchymal stem cells derived from dental tissues vs. those from other sources: their biology and role in regenerative medicine. J Dent Res 9:792–806
3. Shukla V, Vaissière T, Herceg Z (2008) Histone acetylation and chromatin signature in stem cell identity and cancer. Mutat Res 637:1–15
4. Gronthos S, Mankani M, Brahim J, Robey PG, Shi S (2000) Postnatal human dental pulp cells (DPSCs) in vitro and in vivo. Proc Natl Acad Sci U S A 97:13625–13630
5. Ware CB, Wang L, Mecham BH et al (2009) Histone deacetylase inhibition elicits an evolutionarily conserved self-renewal program in embryonic stem cells. Cell Stem Cell 4:359–369
6. Jamaladdin S, Kelly RD, O'Regan L et al (2014) Histone deacetylase (HDAC) 1 and 2 are essential for accurate cell division and the pluripotency of embryonic stem cells. Proc Natl Acad Sci U S A 111:9840–9845
7. Romagnani P, Lasagni L, Mazzinghi B, Lazzeri E, Romagnani S (2007) Pharmacological modulation of stem cell function. Curr Med Chem 14:1129–1139
8. Gopinathan G, Kolokythas A, Luan X, Diekwisch TG (2013) Epigenetic marks define the lineage and differentiation potential of two distinct neural crest-derived intermediate odontogenic progenitor populations. Stem Cells Dev 22:1763–1778
9. Arnsdorf EJ, Tummala P, Castillo AB, Zhang F, Jacobs CR (2010) The epigenetic mechanism of mechanically induced osteogenic differentiation. J Biomech 43:2881–2886
10. Duncan HF, Smith AJ, Fleming GJ, Cooper PR (2015) Epigenetic modulation of dental pulp stem cells: implications for regenerative endodontics. Int Endod J. doi:10.1111/iej.12475 [Epub ahead of print]

11. Kelly TK, De Carvalho DD, Jones PA (2010) Epigenetic modifications as therapeutic targets. Nat Biotechnol 28:1069–1078
12. Wright J (2013) Epigenetics: reversible tags. Nature 498:S10–S11
13. Nagase H, Ghosh S (2008) Epigenetics: differential DNA methylation in mammalian somatic tissues. FEBS J 275:1617–1623
14. Vaissière T, Sawan C, Herceg Z (2008) Epigenetic interplay between histone modifications and DNA methylation in gene silencing. Mutat Res, Rev Mutat Res 659:40–48
15. Portela A, Esteller M (2010) Epigenetic modifications and human disease. Nat Biotechnol 28:1057–1068
16. Grant S, Easley C, Kirkpatrick P (2007) Vorinostat. Nat Rev Drug Discov 6:21–22
17. Griffiths EA, Gore SD (2008) DNA methyltransferase and histone deacetylase inhibitors in the treatment of myelodysplastic syndrome. Semin Hematol 45:23–30
18. Jin H, Park JY, Choi H, Choung PH (2013) HDAC inhibitor trichostatin A promotes proliferation and odontoblast differentiation of human dental pulp stem cells. Tissue Eng Part A 19:613–624
19. Mahmud N, Petro B, Baluchamy S et al (2014) Differential effects of epigenetic modifiers on the expansion and maintenance of human cord blood stem/progenitor cells. Biol Blood Marrow Transplant 20:480–489
20. Candido EP, Reeves R, Davie JR (1978) Sodium butyrate inhibits histone deacetylation in cultured cells. Cell 14:105–113
21. Marks PA (2010) Histone deacetylase inhibitors: a chemical genetics approach to understanding cellular functions. Biochim Biophys Acta 1799:717–725
22. Shuttleworth SJ, Bailey SG, Townsend PA (2010) Histone deacetylase inhibitors: new promise in the treatment of immune and inflammatory diseases. Curr Drug Targets 11:1430–1438
23. Paino F, La Noce M, Tirino V et al (2014) Histone deacetylase inhibition with valproic acid downregulates osteocalcin gene expression in human dental pulp stem cells and osteoblasts: evidence for HDAC2 involvement. Stem Cells 32:279–289
24. Bolden JE, Peart MJ, Johnstone RW (2006) Anticancer activities of histone deacetylase inhibitors. Nat Rev Drug Discov 5:769–784
25. Montgomery RL, Davis CA, Potthoff MJ et al (2007) Histone deacetylases 1 and 2 redundantly regulate cardiac morphogenesis, growth, and contractility. Genes Dev 21:1790–1802
26. Khan N, Jeffers M, Kumar S et al (2008) Determination of the class and isoform selectivity of small-molecule histone deacetylase inhibitors. Biochem J 409:581–589
27. Duncan HF, Smith AJ, Fleming GJ, Cooper PR (2013) Histone deacetylase inhibitors epigenetically promote reparative events in primary dental pulp cells. Exp Cell Res 319:1534–1543
28. McCool KW, Xu X, Singer DB, Murdoch FE, Fritsch MK (2007) The role of histone acetylation in regulating early gene expression patterns during early embryonic stem cell differentiation. J Biol Chem 282:6696–6706
29. Schröder U (1985) Effects of calcium hydroxide-containing pulp-capping agents on pulp cell migration, proliferation, and differentiation. J Dent Res 64:541–548
30. De Groh ED, Swanhart LM, Cosentino CC et al (2010) Inhibition of histone deacetylase expands the renal progenitor cell population. J Am Soc Nephrol 21:794–802
31. Duncan HF, Smith AJ, Fleming GJ, Cooper PR (2011) HDACi: cellular effects, opportunities for restorative dentistry. J Dent Res 90:1377–1388
32. Albuquerque MT, Valera MC, Nakashima M, Nör JE, Bottino MC (2014) Tissue-engineering-based strategies for regenerative endodontics. J Dent Res 93:1222–1231
33. Keswani D, Pandey RK (2013) Revascularization of an immature tooth with a necrotic pulp using platelet-rich fibrin: a case report. Int Endod J 46:1096–1104
34. Kontakiotis EG, Filippatos CG, Tzanetakis GN, Agrafioti A (2015) Regenerative endodontic therapy: a data analysis of clinical protocols. J Endod 41:146–154
35. Murray PE, Garcia-Godoy F, Hargreaves KM (2007) Regenerative endodontics: a review of current status and a call for action. J Endod 33:377–390

36. Ferracane JL, Copper PR, Smith AJ (2010) Can interaction of materials with the dentin-pulp complex contribute to dentin regeneration? Odontology 98:2–14
37. Coppe C, Zhang Y, Den Besten PK (2009) Characterization of primary dental pulp cells in vitro. Pediatr Dent 31:467–471
38. Miura M, Gronthos S, Zhao M et al (2003) SHED: stem cells from human exfoliated deciduous teeth. Proc Natl Acad Sci U S A 100:5807–5812
39. Seo BM, Miura M, Gronthos S et al (2004) Investigation of multipotent postnatal stem cells from human periodontal ligament. Lancet 364:149–155
40. Sonoyama W, Liu Y, Yamaza T et al (2008) Characterization of the apical papilla and its residing stem cells from human immature permanent teeth: a pilot study. J Endod 34:166–171
41. Volponi AA, Sharpe PT (2013) The tooth – a treasure chest of stem cells. Br Dent J 215: 353–358
42. Patel M, Smith AJ, Sloan AJ, Smith G, Cooper PR (2009) Phenotype and behaviour of dental pulp cells during expansion culture. Arch Oral Biol 54:898–908
43. Arora V, Arora P, Munshi AK (2009) Banking stem cells human exfoliated deciduous teeth (SHED): saving for the future. J Clin Pediatr Dent 33:289–294
44. Mohammad HP, Baylin SB (2010) Linking cell signaling and the epigenetic machinery. Nat Biotechnol 28:1033–1038
45. Surani MA, Hayashi K, Hajkova P (2007) Genetic and epigenetic regulators of pluripotency. Cell 128:747–762
46. Jaenisch R, Young R (2008) Stem cells, the molecular circuitry of pluripotency and nuclear reprogramming. Cell 132:567–582
47. Meissner A (2010) Epigenetic modifications in pluripotent and differentiated cells. Nat Biotechnol 28:1079–1088
48. Yamanaka S, Blau HM (2010) Nuclear reprogramming to a pluripotent state by three approaches. Nature 465:704–712
49. Feldman N, Gerson A, Fang J, Li E, Zhang Y, Shinkai Y, Cedar H, Bergman Y (2006) G9a-mediated irreversible epigenetic inactivation of Oct-3/4 during early embryogenesis. Nat Cell Biol 8:188–194
50. Lister R, Pelizzola M, Dowen RH et al (2009) Human DNA methylomes at base resolution show widespread epigenomic differences. Nature 462:315–322
51. Fouse SD, Shen Y, Pellegrini M, Cole S, Meissner A, Van Neste L, Jaenisch R, Fan G (2008) Promoter CpG methylation contributes to ES cell gene regulation in parallel with Oct 4/Nanog, PcG complex, and histone H3 K4/K27 trimethylation. Cell Stem Cell 2:160–169
52. Meissner A, Mikkelsen TS, Gu H et al (2008) Genome-scale DNA methylation maps of pluripotent and differentiated cells. Nature 454:766–770
53. Lee S, Park JR, Seo MS et al (2009) Histone deacetylase inhibitors decrease proliferation potential and multilineage differentiation capability of human mesenchymal stem cells. Cell Prolif 42:711–720
54. Brien GL, Gambero G, O'Connell DJ et al (2012) Polycomb PHF19 binds H3K36me3 and recruits PRC2 and demethylase NO66 to embryonic stem cell genes during differentiation. Nat Struct Mol Biol 19:1273–1281
55. Guenther MG, Young RA (2010) Transcription. Repressive transcription. Science 329: 150–151
56. Chawla JP, Iyer N, Soodan KS, Sharma A, Khurana SK, Priyadarshni P (2015) Role of miRNA in cancer diagnosis, prognosis, therapy and regulation of its expression by Epstein-Barr virus and human papillomaviruses: with special reference to oral cancer. Oral Oncol 51:731–737
57. Fabbri M, Garzon R, Cimmino A et al (2007) MicroRNA-29 family reverts aberrant methylation in lung cancer by targeting DNA methyltransferases 3A and 3B. Proc Natl Acad Sci U S A 104:15805–15810
58. Saito Y, Liang G, Egger G, Friedman JM, Chuang JC, Coetzee GA, Jones PA (2006) Specific activation of microRNA-127 with downregulation of the proto-oncogene BCL6 by chromatin-modifying drugs in human cancer cells. Cancer Cell 9:435–443

59. Khalil AM, Guttman M, Huarte M et al (2009) Many human large intergenic noncoding RNAs associate with chromatin-modifying complexes and affect gene expression. Proc Natl Acad Sci U S A 106:11667–11672
60. Guttman M, Donaghey J, Carey BW et al (2011) lincRNAs act in the circuitry controlling pluripotency and differentiation. Nature 477:295–300
61. Nagano T, Mitchell JA, Sanz LA, Pauler FM, Ferguson-Smith AC, Feil R, Fraser P (2008) The air noncoding RNA epigenetically silences transcription by targeting G9a to chromatin. Science 322:1717–1720
62. Sun Q, Liu H, Chen Z (2015) The fine tuning role of microRNA-RNA interaction in odontoblast differentiation and disease. Oral Dis 21:142–148
63. Cardoso FP, de Faria Amormino SA, Dutra WO, Ribeiro Sobrinho AP, Moreira PR (2014) Methylation pattern of the CD14 and TLR2 genes in human dental pulp. J Endod 40:384–386
64. Gregoretti IV, Lee YM, Goodson HV (2004) Molecular evolution of the histone deacetylase family: functional implications of phylogenetic analysis. J Mol Biol 338:17–31
65. Verdin E, Dequiedt F, Kasler HG (2003) Class II histone deacetylases: versatile regulators. Trends Genet 19:286–293
66. Klinz FJ, Korkmaz Y, Bloch W, Raab WH, Addicks K (2012) Histone deacetylases 2 and 9 are coexpressed and nuclear localized in human molar odontoblasts in vivo. Histochem Cell Biol 137:697–702
67. De Ruijter AJ, Van Gennip AH, Caron HN, Kemp S, Van Kuilenberg AB (2003) Histone deacetylases (HDACs): characterization of the classical HDAC family. Biochem J 370: 737–749
68. Witt O, Deubzer HE, Milde T, Oehme I (2009) HDAC family: what are the cancer relevant targets? Cancer Lett 277:8–21
69. Hochedlinger K, Plath K (2009) Epigenetic reprogramming and induced pluripotency. Development 136:509–523
70. Williams PJ, Nishu K, Rahman MM (2011) HDAC inhibitor trichostatin A suppresses osteoclastogenesis by upregulating the expression of C/EBP-β and MKP-1. Ann N Y Acad Sci 1240:18–25
71. Nuutinen T, Suuronen T, Kauppinen A, Salminen A (2010) Valproic acid stimulates clusterin expression in human astrocytes: implications for Alzheimer's disease. Neurosci Lett 475: 64–68
72. Schroeder TM, Westendorf JJ (2005) Histone deacetylase inhibitors promote osteoblast maturation. J Bone Miner Res 20:2254–2263
73. Huangfu D, Maehr R, Guo W et al (2008) Induction of pluripotent stem cells by defined factors is greatly improved by small-molecule compounds. Nat Biotechnol 26:795–797
74. Bug G, Gül H, Schwarz K et al (2005) Valproic acid stimulates proliferation and self-renewal of hematopoietic stem cells. Cancer Res 65:2537–2541
75. Wang G, Badylak SF, Heber-Katz E, Braunhut SJ, Gudas LJ (2010) The effects of DNA methyltransferase inhibitors and histone deacetylase inhibitors on digit regeneration in mice. Regen Med 5:201–220
76. De Boer J, Licht R, Bongers M, Der Klundert V, Arends R, Van Blitterswijk C (2006) Inhibition of histone acetylation as a tool in bone tissue engineering. Tissue Eng 12:2927–2937
77. Hsieh J, Nakashima K, Kuwabara T, Mejia E, Gage FH (2004) Histone deacetylase inhibition-mediated neuronal differentiation of multipotent adult neural progenitor cells. Proc Natl Acad Sci U S A 101:16659–16664
78. Humphrey GW, Wang YH, Hirai T, et al (2008) Complimentary roles for histone deacetylases 1, 2 an 3 in differentiation of pluripotent stem cells. Differentiation 76:348–356
79. Karantzali E, Schulz H, Hummel O, Hubner H, Hatzopoulos AK, Kretsovali A (2008) Histone deacetylase inhibition accelerates the early events of stem cell differentiation:transcriptomic and epigenetic analysis. Genome Biol 9:R65

80. Jiang Z, Han Y, Cao X (2014) Induced pluripotent stem cell (iPSCs) and their application in immunotherapy. Cell Mol Immunol 11:17–24
81. Takahashi K, Yamanaka S (2006) Induction of pluripotent stem cells from mouse embryonic and adult fibroblast cultures by defined factors. Cell 126:663–676
82. Takahashi K, Tanabe K, Ohnuki M et al (2007) Induction of pluripotent stem cells from adult human fibroblasts by defined factors. Cell 131:861–872
83. Huangfu D, Osafune K, Maehr R et al (2008) Induction of pluripotent stem cells from primary human fibroblasts with only Oct4 and Sox2. Nat Biotechnol 26:1269–1275
84. Teng HF, Li PN, Hou DR et al (2014) Valproic acid enhances Oct4 promoter activity through PI3K/Akt/mTOR pathway activated nuclear receptors. Mol Cell Endocrinol 383:147–158
85. Warren L, Manos PD, Ahfeldt T, Loh Y-H, Li H (2010) Highly efficient reprogramming to pluripotency and directed differentiation of human cells with synthetic modified mRNA. Cell 7:618–630
86. Hargreaves KM, Diogenes A, Teixeira FB (2013) Treatment options: biological basis of regenerative endodontic procedures. J Endod 39:S30–S43

Dental Stem Cells and Growth Factors

Paolo Ghensi

Stem Cells, Scaffolds and Growth Factors

The aim of the regenerative medicine and tissue engineering is to regenerate and repair the damaged cells and tissues in order to establish the normal functions [1]. The regenerative medicine involves the use of scaffolds, growth factors and stem cells [2]. Regeneration of the tissues exists naturally due to the presence of stem cells with the potential to self-regenerate and differentiate into one of more specialized cell types [3].

Stem cells are immature and unspecialized cells with the ability to renew and divide themselves indefinitely through "self-renewal" and able to differentiate into multiple cell lineages [4]. In the last decade, several improvements have been produced in the comprehension of stem cells properties in view of the fact that these cells have an important role in the repair of every organ and tissue [3].

In general, the stem cells are divided into three main types that can be utilized for tissue repair and regeneration: (1) the embryonic stem cells derived from embryos (ES) [5, 6]; (2) the adult stem cells that are derived from adult tissue [7]; and (3) the induced pluripotent stem (iPS) cells that have been produced artificially via genetic manipulation of the somatic cells [8]. ES and iPS cells are considered pluripotent stem cells because they can develop into all types of cells from all three germinal layers. On the contrary, adult stem cells are multipotent because they can only differentiate into a restricted number of cell types. Adult stem cells, also termed postnatal stem cells or somatic stem cells, are discovered in a particular area of each tissue named "stem cell niche" [3].

P. Ghensi (✉)
Centre for Integrative Biology (CIBIO), University of Trento, Trento, Italy

Department of Neurosciences, University of Padova, Padova, Italy
e-mail: dr.ghensi@gmail.com

B. Zavan, E. Bressan (eds.), *Dental Stem Cells: Regenerative Potential*, Stem Cell Biology and Regenerative Medicine, DOI 10.1007/978-3-319-33299-4_5

Different type of postnatal stem cells resides in numerous mesenchymal tissues and these cells are at the same time referred to as mesenchymal stem cells (MSCs) [7, 9]. Different studies have showed that MSCs can be isolated from different tissues, such as peripheral blood, umbilical cord blood, amniotic membrane, adult connective, adipose and dental tissues [10–14].

Today, every cell population which has the following characteristics independently of its tissue source, is usually referred as MSCs: (1) they adhere to plastic and have a fibroblast-like morphology; (2) they have the capacity of self-renewal and could differentiate into cells of the mesenchymal lineage such as osteocytes, chondrocytes and adipocytes. In addition, MSCs also can also differentiate, under appropriate conditions, into cells of the endoderm and ectoderm lineages such as hepatocytes and neurons, respectively [15, 16].

Research related to MSC from oral origin began in 2000 [17] and every year numerous investigations have demonstrated that oral tissues, which are simply available for dentists, are a rich source for mesenchymal stem cells [18, 19].

Today numerous types of MSCs have been isolated from teeth: in 2000 MSCs were first isolated by Gronthos et al. from dental pulp (DPSCs) [17, 19]. Subsequently, MSCs have been also isolated from dental pulp of human exfoliated deciduous teeth (SHEDs) [20]. The periodontal ligament is another adult MSCs source in dental tissue, and periodontal ligament stem cells (PDLSCs) were isolated from extracted teeth [21]. Moreover, MSCs have been also isolated from developing dental tissues such as the dental follicle (DFPCs) [22] and apical papilla (SCAPs) [23].

These cells all share a number of features with the well-studied bone marrow-derived mesenchymal stem cells, including their ability to differentiate, albeit with varying efficacy, into cells of the major mesenchymal lineages: osteoblasts, chondrocytes, and adipocytes. However, these cells seem to differ in their propensity to differentiate into functional odontoblasts [24]. Further, some of these cells, such as the SCAPs, are isolated from specific developmental stages (in this case the apex of a developing tooth), and thus, do not serve as practical sources of cells for autologous cell-based transplantation [25].

Dental pulp stem cells (DPSCs), stem cells from human exfoliated deciduous teeth (SHEDs), SCAPs, and PDLSCs have all been successfully isolated and characterized both in vitro and in vivo. Because of the nature of the deciduous teeth, guiding the development of permanent teeth and eventual replacement by the permanent teeth, deciduous teeth may act as a very practical and easily accessibly reservoir for autologous stem cells and hold the most value in stem cell therapy [20].

Much excitement has recently surrounded the identification and application of these stem cells with therapeutic value, but equally important is the development of delivery methods for transplantation of these cells and the promotion of their efficient engraftment. Traditionally, stem cells are introduced into the body via injection into the site of interest or into the circulation. However, the efficacy of this methodology is questionable, since studies have shown poor cell survival, engraftment, and unpredictable differentiation in vivo [25].

The potential efficacy of stem cell delivery and differentiation may be improved with the adoption of tissue-engineered scaffolds for cell delivery and structural support. Delivery of dental stem cells can potentially be supported by scaffolds that provide both mechanical and molecular (growth Factors, integrin receptor engagement) cues for differentiation [26].

Properties of scaffolds include providing cell adhesion, enabling cell proliferation and differentiation, and mimicking microenvironment observed in natural tissues and organs. Unlike injection-based delivery, scaffolds allow for superior control for stem cell delivery and allow for impregnation with time-release growth factors, modulation of stiffness, pore size, and cell–substrate interaction. Some evidence suggests that in addition to directly participating in tissue regeneration, transplanted stem cells can also regulate tissue regeneration by secreting trophic factors [27]. These two properties of stem cells suggest that the proper local distribution of stem cells is critical, making bioscaffold-based delivery one of the most promising strategies in cell-based regenerative dentistry [25].

Apart from stem cells and an appropriate scaffold, regenerative medicine and tissue engineering also require the use of growth factors and ECM molecules that induce specific differentiation pathways and maintain the cellular phenotype.

Incorporation of growth factors and/or cytokines has been shown to increase the efficacy of cell-based regeneration, both in the context of exogenous cell delivery, and recruitment of patients' endogenous stem cells from the local environment in vivo [25]. Growth factors enhance dental stem cell activities by promoting the migration of endogenous cells and the subsequent proliferation, differentiation, angiogenesis and neuronal growth. Growth factors and cytokines are either autocrine or paracrine in nature and modulate cellular behavior by mediating intracellular communication [25]. Kim et al. recently discussed the effects of some growth factors in pulp-dentin regeneration [28]. These growth factors include platelet-derived growth factors (PDGF), transforming growth factor β (TGFb) family, vascular endothelial growth factor (VEGF), bone morphogenic proteins (BMPs), fibroblast growth factor (FGF) and etc. When delivered alone or with stem cells to sites of interest, they share common properties of promoting dentinogenesis, cellular proliferation and odontoblastic differentiation. PDGF, for example, normally released by platelets, induces chemotaxis and proliferation of stem cells by promoting angiogenesis and cell proliferation [29]. Rizk and Rabie demonstrated chondrogenesis of DPSCs transplanted into nude mice on poly-L-lactic acid/polyethylene glycol electrospun fiber scaffolds. These DPSC transplants were more efficient at chondrogenesis when transduced with TGFb3 virus, further supporting the notion that codelivery of growth factors can stimulate basal levels of stem cell differentiation upon transplantation in vivo [30].

Definition and Mechanism of Action of Growth Factors (GFs)

The term growth factors denominates a group of polypeptides of approx. 6–45 kD which have a role in controlling biological processes, such as cell growth, proliferation, differentiation and morphogenesis of tissues and organs during embryogenesis, postnatal growth and adulthood. GFs cannot pass through a cell's membrane; they must bind to high-affinity cell receptors in order to take effect. Many GFs stimulate several cell populations, while others are less versatile and specific to a particular cell line [31].

Growth factors can act as mitogenes in that they enhance proliferation of certain cell types. Some growth factors are also morphogenic in that they change the phenotype of their target cells. Growth factors act in an autocrine manner by directing the secreting cell itself and in a paracrine fashion by affecting neighbouring cells. For some growth factors, an endocrine effect is assumed or even established because of elevated serum levels. Many growth factors are deposited in the extracellular matrix where they are released during matrix degradation and act as part of a complex network of signals with mutual effects during tissue remodelling and regeneration. The effect of growth factors is mediated through surface receptors on the target cells by activating intracellular phosphorylating enzymes, which in turn induce an intracellular signalling pathway by aggregation of co-factors and other proteins, which migrate to the nucleus. Together with other transcription factors they activate a set of genes, which then exert the specific changes in cellular activity or phenotype. In vivo, the effect of growth factors is regulated through a complex system of feed back loops, which involve other growth factors, enzymes and binding proteins [31].

Postnatal therapeutic application of growth factors for repair of damaged tissue or an organ is used to accomplish regeneration or generation of tissue by reinducing the developmental process that had created this organ or body part during foetal or postnatal growth [31]. Several authors have described the importance of growth factors in tissue repair processes, in fact, they are important elements for new tissue production, moreover, they perform feedback controls on inflammatory processes within the tissue graft, in cases of regenerative surgery [32–35].

As growth and development are subject to a complex regulation that is also essential for the suppression of malignant growth, there are concerns that unreflected enhancement of isolated signals in this complex system might initiate a process of uncontrolled growth. This is particularly true for factors that are also associated with the development of malignant tumours. Consideration of the use of growth factors, therefore, also has to include aspects of growth control and interactions with suppressing and activating co-factors [31].

Dental Stem Cells and Growth Factors

Human Recombinant and Autogenous Growth Factors

The identification of the genetic code of growth factors has enabled the production of human recombinant proteins more than 20 years ago. This has repeatedly raised hopes that effective clinical application is just around the corner. Unfortunately, despite extensive research and successful preclinical testing, the clinical use of human recombinant growth factors has not yet reached a level of application that would reflect the numerous successful experimental reports that have been published for more than 15 years [36]. Costs and regulatory issues as well as the complexity of mimicking physiological dosage levels at the site of delivery have rendered the introduction of human recombinant growth factors into clinical use more difficult than it has been anticipated [36, 37].

An alternative strategy that has been applied more frequently than the application of recombinant proteins is the use of patients' own growth factors. Autogenous growth factors are present in rather large quantities in platelets and are easily available from platelet enriched centrifugation products of whole blood. Different concepts and preparation methods have been developed, that vary in the content of platelets, leukocytes and fibrin and hence the condition of application (liquid vs. gel) [38].

Platelet concentrates (PCs) contain a combination of growth factors. The concentration may vary between individual patients and according to the method of production [39–41]. The growth factors are contained in the alpha granules of the platelets. They can be released by addition of calcium chloride, allogeneic thrombin or by autologous fibrin depending on the preparation technique [38, 39, 42–45]. Other than individual recombinant growth factors, PCs convey a number of signals that may vary in their relative strength according to the dose levels present in the individual preparations. According to the nature of the growth factors contained in PCs, they are supposed to enhance mesenchymal and epithelial proliferation as well as angiogenesis.

Whitman [46] and Marx [47] published the first studies on the use of growth factors contained in platelet gel, called Platelet-Rich Plasma (PRP). Thanks to Marx's studies, it was possible to verify that the platelet concentrate is a very effective tool for the modulation of wound healing and tissue regeneration [3].

However, the PRP showed a number of disadvantages, such as the need of having to run a complex and expensive protocol for its production. To overcome some of these problems, the PRGF (Plasma Rich in Growth Factors) was introduced in the list of platelet concentrates. The PRGF is considered an evolution of the PRP and it allows a higher concentration of growth factors in platelet preparation [41, 48].

Among the advantages of the PRGF, we can cite the lesser amount of blood taken for the preparation and a procedure relatively faster, while, among the disadvantages we can mention the rapid clot formation, which require speed in its surgical use [3].

In 2001, Choukroun et coll. have instead proposed an alternative technique: the PRF (Platelet Rich Fibrin). PRF is derived from a simple preparation protocol that does not require alteration of the blood; it is a platelet concentrate rich in GFs that contains a three-dimensional matrix of autologous, elastic and flexible fibrin [36].

Dohan et al. have shown that platelet cytokines (PDGF, TGFbeta1 and IGF-1) are present in three-dimensional fibrin matrix derived from these platelet concentrates; moreover, PRF matrix traps glycosaminoglycans such as heparin and hyaluronic acid, which have considerable affinity with some peptides present in the bloodstream and therefore show strong ability of chemotaxis and diapedesis, useful for the healing of tissue damaged, for example, by trauma [3, 49].

Nowadays use of autogenous proteins is hardly impaired by regulatory issues and is available at low costs. The ease of production and the almost unrestrained use has fuelled a widespread clinical application of PCs in regenerative procedures [32, 36, 50].

Effect of Growth Factors on Dental Stem Cells

Fundamental to our understanding of regenerative medicine is the knowledge of growth factors that effect on a broad range of cellular activities including migration, proliferation, differentiation and apoptosis of cells, including stem/progenitor cells. Growth factors and cytokines may act as signaling molecules that modulate cell behavior by mediating intracellular communication. Growth factors are polypeptides or proteins that bind to specific receptors on the surface of target cells [51, 52]. They can initiate a cascade of intracellular signaling and act in either an autocrine or paracrine manner [53]. Cytokines are typically referred to as immunomodulatory proteins or polypeptides [54]. Cytokines are often used interchangeably with growth factors because many cytokines share similar actions as growth factors. As opposed to systemic effects by hormones on target cells, growth factors or cytokines typically act locally on target cells [28]. The effects of various growth factors on dental stem cells and how they may participate in dental pulp-dentin regeneration will now follow.

Platelet-Derived Growth Factor (PDGF)

Platelet derived growth factor (PDGF) is released by platelets, and has potency in promoting angiogenesis and cell proliferation [28, 55–58]. PDGF is a highly basic dimeric glycoprotein of 30 kD consisting of two disulphide bonded polypeptides encoded by different genes [30]. There are three isoforms characterized by the combination of A- and B-chains featuring two homodimeric (PDGF-AA and PDGF-BB) and one heterodimeric isoform (PDGF-AB) [59–61]. PDGF-BB and PDGF-AB are systemically circulating isoforms contained in alpha granules of platelets from where they are released after adhesion of platelets to injured sites of vessel walls, whereas PDGF-AA is secreted by unstimulated cells of the osteoblastic lineage [60–62]. PDGF stimulates specific target cells by binding to cell surface receptors with tyrosine kinase activity. PDGF plays an important role in the development of many tissues and organs in the growing embryo. It is essential not only for mesoderm induction and mesenchymal-epithelial interactions during organ development but also for development of the neural crest and patterning of limbs and myotomes in early embryogenesis [30, 63, 64].

The chemotaxis and proliferation of mesenchymal stem cells can be induced by PDGF in the injury site. In trauma, hemorrhage is followed by blood clot formation in dental pulp. Platelets in the blood clot release α-granules containing PDGFs and attract neutrophils and macrophages [56]. These cells play key roles in early wound healing by producing other signaling molecules for the formation of granulation tissues. However, PDGFs appear to have little effects on the formation of the dentin-like nodule in dental pulp stem cells isolated from rat lower incisors although PDGF-AB and -BB isoforms stimulate the expression of dentin sialoprotein (DSP) [65]. PDGFs stimulate cell proliferation and dentin matrix protein synthesis but appear to inhibit alkaline phosphatase (ALP) activity in dental pulp stem cells in culture [66–68]. DSP expression is inhibited by PDGF-AA, but enhanced by PDGF-AB and PDGF-BB although the mineralized tissue formation is inhibited, suggesting diverging effects of PDGFs on odontoblastic differentiation depending on dimeric form [65]. PDGF senhance the proliferation of fibroblasts in human dental pulp [66]. PDGF-BB may increase the expression of VEGF in osteoblasts and promotes angiogenesis at the site of dental pulp injury [69]. In vivo, PDGF promotes de novo formation of dental-pulp-like tissues in endodontically treated human teeth that are implanted in rats [28, 70].

Transforming Growth Factor-β (TGFβ)

Transforming growth factor β (TGFβ) is among the most widespread and versatile cytokine and plays an important role in the formation and development of many types of tissue [30]. There are more than 30 proteins, which belong to the TGFβ-superfamily such as TGFβ itself, bone morphogenetic proteins (BMPs), growth and differentiation factors (GDFs), Anti-Mullerian hormone (AMH), activins and nodal [28, 71, 72].

TGFβ is composed of ~390 amino acids, which are released mainly from platelets, macrophages, and bone [73]. This inactive polypeptide undergoes proteolytic cleavage to create the active C-terminal 112-amino acid form. The active form of TGFβ dimerizes to form 25 kD homodimers [74]. The name transforming growth factor is derived from the fact, that TGFβ belong to a group of growth regulators, that were originally detected in tumour extracts and thought to induce or maintain the neoplastic or transformed cell type [75]. Although TGFβ suppresses proliferation of epithelial cells, it is overexpressed in a number of epithelial malignancies.

The three isoforms present in mammals, which are TGFβ1, TGFβ2, TGFβ3, are detected also in human dentin [76].

The effect of TGFβ is highly variable and dependent on the type of cells and tissues. TGFβ1 regulates a wide range of cellular activities, such as cell migration, cell proliferation, cell differentiation, and extracellular matrix synthesis [28, 77–81]. TGFβ1 has been shown to increase cell proliferation and production of the extracellular matrix in dental pulp tissue culture [81], and promotes odontoblastic differentiation of dental pulp stem cells [82]. The effect of TGFβ1 can be synergistically upregulated by fibroblast growth fatocr-2 (FGF2), as evidenced by the increased ALP activity, the formation of mineralized nodule, and the expression of DSP and dentin matrix protein-1 [82]. The dentinogenic ability of dental pulp stem cells in the mechanically exposed dental pulp of dog teeth is shown to be induced by exogenous TGFβ1 [83]. TGFβ is chemotactic on dental pulp stem cells in vitro [84] TGFβ1 also plays an important role in the immune response during the dental pulp injury [28, 85, 86].

Bone Morphogenetic Protein (BMP)

Bone morphogenic proteins (BMPs) are homodimeric proteins of approx. 30 kD with the two identical strands linked by a cysteine binding group [30]. Nearly 20 modifications of BMPs with slightly different modifications in small secondary structure elements [87] have been identified so far [88], among which growth differentiation factors (GDFs) were included. The content of BMPs in bone has been estimated to be 1 μg/g bone tissue[30]. BMP2–BMP9 belong to the TGFβ superfamily with a high degree of homology with the TGFβs [30]. TGF β and BMP have a common scaffold with the cysteine knot motif and two double-stranded beta sheets [87]. BMPs activity is mediated through a transmembraneous heteromeric receptor complex on the cell surface that is identical to that of TGFβ. The difference in signalling pathway is characterized by the SMAD proteins that are activated by BMPs [30].

BMPs play important roles in the development and remodeling of the bone, and they can promote the chemotaxis and aggregation of cells into osteogenic site in different ways and facilitate the differentiation into osteoblasts. In addition, these proteins can also promote the angiogenesis, regulate the activity of some growth factors and affect the production of these growth factors, which is helpful for the osteogenesis. BMPs have been considered as the most potent growth factors that can promote the bone regeneration. To date, more than 20 BMPs have been identified and BMP-2, -4, -6 and -7 [osteogenic protein 1 (OP1)] have found to the osteogenic potential [89–93]. BMP9 is also known as growth differentiation factor 2 (GDF-2) and mainly expressed in the liver [94]. BMP9 can induce and maintain the cholinergic differentiation of embryonic neurons, regulate the metabolism of glucose and fatty acid, modulate the dynamic balance of iron and exert other important biological functions [95–97]. However, the role of BMP9 in the osteogenesis and bone regeneration is poorly understood.

Human recombinant BMP2 stimulates the differentiation of dental pulp cells into odontoblasts [98], inducing mRNA expression of dentin sailophosphoproteins (DSPPs) and higher ALP activity upon BMP2 application, but no effect on cell proliferation. DSPP expression and odontoblastic differentiation are regulated likely via BMP2-induced nuclear transcription factor Y signaling [99]. BMP2 also stimulates the differentiation of dental pulp stem cells into odontoblasts in vivo and in vitro [100]. Bovine dental pulp cells treated with BMP2 and BMP4 differentiate into preodontoblasts [101]. BMP7, also known as osteogenic protein-1, promotes dentin formation when placed over amputated dental pulp in macaque teeth [102, 103]. The dentinogenic effect of BMP7 on amputated dental pulp has been shown in several animal models including rats [104], ferrets [105], and miniature swine [106]. Dental pulp cells transfected with BMP11, also known as GDF11, yields mineralization [107]. Dentin matrix protein 1, ALP, DSPP, enamelysin, and phosphate-regulating gene are highly expressed in BMP11-tranfected cells [107]. Transplantation of BMP11-transfected cell pellets induces formation of dentin-like tissue on amputated dental pulp in dogs [107]. Ultrasound-mediated gene delivery of BMP11 stimulates odontoblastic differentiation of dental pulp stem/progenitor cells in vitro and reparative dentin formation in vivo [108].

Detection of alkaline phosphatase (ALP) and calcium deposition showed dental follicle stem cells transfected with BMP9 gene could significantly promote the osteogenesis. In addition, SB203580 and PD98059 were employed to block the p38 mitogen-activated protein kinase (p38MAPK) and extracellular signal-regulated kinase (ERK1/2), respectively. Detection of ALP and calcium deposition revealed the BMP9 induced osteogenic differentiation of dental follicle stem cells depended on MAPK signaling pathway [109].

BMP9 can effectively induce odontogenic differentiation of the stem cells from mouse apical papilla (SCAPs). Using a reversible immortalization system expressing SV40 T flanked with Cre/loxP sites, it was demonstrated that the SCAPs can be immortalized, resulting in immortalized SCAPs (iSCAPs) that express mesenchymal stem cell markers. BMP9 upregulates Runx2, Sox9, and PPARc2 and odontoblastic markers, and induces alkaline phosphatase activity and matrix mineralization in the iSCAPs. The in vivo stem cell implantation studies indicate that iSCAPs can differentiate into bone, cartilage, and, to lesser extent, adipocytes upon BMP9 stimulation. Results demonstrate that the conditionally iSCAPs not only maintain long-term cell proliferation but also retain the ability to differentiate into multiple lineages, including osteo/odontoblastic differentiation [110].

Fibroblast Growth Factor (FGF)

Fibroblast Growth Factor (FGF) plays key roles in cell migration, proliferation and differentiation during embryonic development [111] and wound healing [112]. Currently 22 members have been identified in humans [113], of which FGF2 appears to be significant in regeneration of the pulp-dentin complex [28]. FGF2 is a basic FGF (bFGF), while FGF1 is acidic. Basic FGF (bFGF) is known to be a critically important factor for maintaining the self-renewal ability of human embryonic stem (hES) cells in cultures and is a potent mitogen for cultured neural progenitors [114, 115].

bFGF dramatically induces the mRNA expression of dentin sialophosphoprotein (DSPP) and bone sialoprotein (BSP) in immature adult rat incisor dental pulp cells [116], suppresses alkaline phosphatase (ALP) activity and osteogenesis differentiation in human dental pulp cells (hDPCs), and inhibits osteogenesis in mouse adipose tissue-derived stromal cells, whereas bFGF sustains their proliferative and osteogenic potential state [117–119]. bFGF induces the migration of dental pulp stem cells [120]. Using a transwell migration assay, significantly more dental pulp cells are recruited by bFGF (FGF2) into a 3D collagen gel than controls without cytokines and BMP7 [120]. bFGF also stimulates the proliferation of dental pulp stem cells without differentiation, whereas bFGF combined with TGFβ1 induces differentiation of dental pulp stem cells into odotoblast-like cells, and synergistically upregulates the effect of TGFβ1 on odontoblast differentiation [82].

bFGF stimulation in the osteogenic differentiation period decreased the in vitro osteogenic differentiation ability of DPSCs. One week pre-treatment with bFGF increased the in vitro osteogenic differentiation ability of DPSCs, whereas 2 weeks pre-treatment with bFGF decreased the in vitro osteogenic differentiation ability of DPSCs. The pre-treatment period was vital for the osteogenic differentiation of DPSCs in vitro. The in vivo results were similar to the in vitro results. bFGF affected the osteogenic differentiation of DPSCs in a treatment-dependent manner both in vitro and in vivo [121].

In vitro studies showed that different timing of the presence of bFGF might either suppress osteogenic/dentinogenic differentiation of SCAP or enhance osteogenic/dentinogenc differentiation [122]. These findings indicate that the effects of bFGF are differentiation stage specific and that bFGF might modulate cell differentiation by acting at distinct stages of cell maturation [123]. Under certain conditions, therefore, bFGF enhances SCAP stemness by up-regulating stem cell gene expression, increasing proliferation ability, and potentiating differentiation potency [122]. Results showed that short-term bFGF treatment enhanced OCT4, REX1, and NANOG mRNA expression as well as colony forming unit ability in stem cells isolated from human exfoliated deciduous teeth (SHEDs). The FGFR inhibitor pre-treatment was able to attenuate the influence of bFGF on pluripotent stem cell marker expression, confirming bFGF function. In addition, cells cultured in high passage number had decreased in cell proliferation, colony forming unit capacity, and pluripotent stem cell maker mRNA expression. However, bFGF supplementation in culture medium enhanced both pluripotent stem cell marker expression and colony forming unit capacity in later passage, though the effect was not robust. Together, these results indicate that high passage number may attenuate pluripotent properties of SHEDs and bFGF supplementation could be the beneficial approach to maintain SHEDs' stemness properties [124].

Insulin-Like Growth Factor (IGF)

Insulin-like growth factors (IGFs) are single chain peptides that exist in two isoforms (IGF-I (70 amino acids) and IGF-II (67 amino acids)). IGFs have approximately 40–50 % homology between themselves and with insulin. Despite this significant homology between insulin and IGFs, all three have unique binding sites to their receptors [125]. In contrast to many other growth factors, which act merely on a local or regional level, IGFs are also factors with a general both metabolic and growth promoting activity in many cell and tissue types. IGFs, comprising IGF-1 and IGF-2, contribute to odontogenesis and dental tissue repair by cell proliferation and differentiation. [28, 126]. Of the two isoforms, IGF-1, also known as somatomedin C, has potency in growth and differentiation of dental pulp cells [127]. IGF-1 induces proliferation and differentiation of dog dental pulp cells into odontoblast-like cells in serum-free medium [127]. IGF-1 with PDGF-BB has a synergistic effect on the proliferation of dental pulp cells in vitro [128]. IGF-1 and IGF-1R have a higher level of expression in dental pulp tissue from teeth with complete root development than teeth with incomplete root formation, suggesting that IGF-1 stimulates mineralization and cell differentiation [128–130]. IGF-1 could promote proliferation and osteogenic differentiation of human dental pulp stem cells (DPSCs) via mTOR pathways, which might have clinical implications for osteoporosis [131].

IGF-1 treated stem cells from apical papilla (SCAPs) present the morphological and ultrastructural changes. Cell proliferation, alkaline phosphatase (ALP) activity and mineralization capacity of SCAPs are increased by IGF-1. Western blot and quantitative RT-PCR analyses further demonstrate that the expression of osteogenic-related proteins and genes (e.g., alkaline phosphatase, runt-related transcription factor 2, osterix, and osteocalcin) is significantly up-regulated in IGF-1 treated SCAPs, whereas the expression of odontoblast-specific markers (e.g., dentin sialoprotein and dentin sialophosphoprotein) is downregulated by IGF-1. In vivo results reveal that IGF-1 treated SCAPs mostly give birth to bone-like tissues while untreated SCAPs mainly generate dentin-pulp complex-like structures after transplantation. IGF-1 can promote the osteogenic differentiation and osteogenesis capacity of SCAPs, but weaken their odontogenic differentiation and dentinogenesis capability, indicating that IGF-1 treated SCAPs can be used as a potential candidate for bone tissue engineering [132].

Vascular Endothelial Growth Factor (VEGF)

Vascular Endothelial Growth Factor (VEGF) is a heparin-binding protein with specific affinity to endothelial cells and plays a keyrole in angiogenesis [133]. The functions of VEGF involve the proliferation of endothelial cells and their enhanced survival [134], stimulating neovascularization in the area of injury [28].

VEGF, belonging to a family of homodimeric proteins consisting of six members (VEGF-A, -B, -C, -D, -E; placenta growth factor). In particular, VEGF-A exists in five different isoforms among which VEGF-A like other isoforms, binding specific co-receptor proteins, named Neuropilin-1 and -2, activates two different tyrosine kinase receptors VEGFR-1 (named also Flt-1) and VEGFR-2 (KDR/Flk-1), the stimulation of which triggers different signals in a large number of cells. For a long time, VEGF has mainly been considered as a potent mitogenic factor for vascular endothelial cells, involved in the modulation of physiological angiogenesis, vascular permeability and also the occurrence/of tissue inflammation. More recently, VEGF has been recognized as a positive regulator of bone development, skeletal growth and fracture repair, also stimulating proliferation and differentation of bone-derived osteoblasts [135, 136].

VEGF is produced by human dental pulp cells, in physiological and even more in pathological conditions [137, 138], causing in these cells a potent induction of chemotaxis, a marginal activation of proliferation and an increase in the alkaline phosphatase (ALP) expression [137]. Recently, D'Aquino et al. [139] showed that human DPSC comprise at least two types of cells expressing different immunophenotypic profiles, which spontaneously diffrentiate, both in vitro and in vivo, into osteoblasts and, to a lesser extent, into endothelial cells over a long period. The spontaneous commitment of DPSC towards vascular phenotype cells may be accelerated by plating DPSC on an appropriate substrate (Matrigel) and exposing them to VEGF for 8 days [140].

VEGF appears to induce the differentiation of human dental pulp stem cells into endothelial cells [140]. Dental pulp cells become positive for CD29, CD44, CD73, CD105, CD166 but negative for CD14, CD34, CD45 after VEGF treatment [140]. VEGF increases the expression of VEGFR1 (fms-like tyrosine kinase, Flt-1) and VEGFR2 (kinase-insert domain containing receptor, KDR) and microvessel formation in a three-dimensional fibrin mesh seeded with dental pulp stem cells [28, 140]. MSC obtained from human dental pulp (DPSC) of normal impacted third molars, when cultured in lineage-specific inducing media, differentiate into osteoblasts and adipocytes (evaluated by Alizarin Red S and Red Oil O stainings, respectively), thus showing a multipotency. DPSC, grown under undifferentiating conditions, are negative for hematopoietic (CD45, CD31, CD34, CD144) and positive for mesenchymal (CD29, CD90, CD105, CD166, CD146, STRO-1) markers, that underwent down-regulation when cells were grown in osteogenic medium for 3 weeks. In this condition, they also exhibit an increase in the expression of osteogenic markers (RUNX-2, alkaline phosphatase) and extracellular calcium deposition, whereas the expression of receptors (VEGFR-1 and -2) for vascular endothelial growth factors (VEGF) and related VEGF binding proteins was similar to that found in undifferentiated DPSC [136]. Exposure of DPSC growing under undifferentiating or osteogenic conditions to VEGF-A165 peptide (10–40 ng/ml) for 8 days dose- and time-dependently increased the number of proliferating cells without inducing differentiation towards endothelial lineage, as evaluated by the lack of expression of specific markers (CD31, CD34, CD144). Additionally, exposure of DPSC cultured in osteogenic medium to VEGF-A165 for a similar period enhanced cell differentiation towards osteoblasts as evaluated after 14 and 21 days by Alizarin Red S staining and alkaline phosphatase activity quantification. These findings may have clinical implications possibly facilitating tissue repair and remodeling [136].

References

1. Mason C, Dunnill P (2008) A brief definition of regenerative medicine. Regen Med 3:1–5
2. Sundelacruz S, Kaplan DL (2009) Stem cell- and scaffold-based tissue engineering approaches to osteochondral regenerative medicine. Semin Cell Dev Biol 20:646–655
3. Tatullo M, Marrelli M, Paduano F (2015) The regenerative medicine in oral and maxillofacial surgery: the most important innovations in the clinical application of mesenchymal stem cells. Int J Med Sci 12(1):72–77
4. Weissman IL (2000) Stem cells: units of development, units of regeneration, and units in evolution. Cell 100:157–168

5. Evans MJ, Kaufman MH (1981) Establishment in culture of pluripotential cells from mouse embryos. Nature 292:154–156
6. Martin GR (1981) Isolation of a pluripotent cell line from early mouse embryos cultured in medium conditioned by teratocarcinoma stem cells. Proc Natl Acad Sci U S A 78: 7634–7638
7. Korbling M, Estrov Z (2003) Adult stem cells for tissue repair – a new therapeutic concept? N Engl J Med 349:570–582
8. Yu J, Vodyanik MA, Smuga-Otto K, Antosiewicz-Bourget J, Frane JL, Tian S, Nie J, Jonsdottir GA, Ruotti V, Stewart R, Slukvin II, Thomson JA (2007) Induced pluripotent stem cell lines derived from human somatic cells. Science 318:1917–1920
9. Keating A (2012) Mesenchymal stromal cells: new directions. Cell Stem Cell 10:709–716
10. Zhang Y, Huang B (2012) Peripheral blood stem cells: phenotypic diversity and potential clinical applications. Stem Cell Rev 8:917–925
11. Zarrabi M, Mousavi SH, Abroun S, Sadeghi B (2014) Potential uses for cord blood mesenchymal stem cells. Cell 15:274–281
12. Bongso A, Fong CY (2013) The therapeutic potential, challenges and future clinical directions of stem cells from the Wharton's jelly of the human umbilical cord. Stem Cell Rev 9:226–240
13. Romagnoli C, Brandi ML (2014) Adipose mesenchymal stem cells in the field of bone tissue engineering. World J Stem Cells 6:144–152
14. Egusa H, Sonoyama W, Nishimura M, Atsuta I, Akiyama K (2012) Stem cells in dentistry—part I: Stem cell sources. J Prosthodont Res 56:151–165
15. Wu XB, Tao R (2012) Hepatocyte differentiation of mesenchymal stem cells. Hepatobiliary Pancreat Dis Int: HBPD INT 11:360–371
16. Ferroni L, Gardin C, Tocco I, Epis R, Casadei A, Vindigni V, Mucci G, Zavan B (2013) Potential for neural differentiation of mesenchymal stem cells. Adv Biochem Eng Biotechnol 129:89–115
17. Gronthos S, Mankani M, Brahim J, Robey PG, Shi S (2000) Postnatal human dental pulp stem cells (DPSCs) in vitro and in vivo. Proc Natl Acad Sci U S A 97:13625–13630
18. Mao JJ, Prockop DJ (2012) Stem cells in the face: tooth regeneration and beyond. Cell Stem Cell 11:291–301
19. Tatullo M, Marrelli M, Shakesheff KM, White LJ (2015) Dental pulp stem cells: function, isolation and applications in regenerative medicine. J Tissue Eng Regen Med 9(11): 1205–1216
20. Miura M, Gronthos S, Zhao M, Lu B, Fisher LW, Robey PG, Shi S (2003) Shed: stem cells from human exfoliated deciduous teeth. Proc Natl Acad Sci U S A 100:5807–5812
21. Seo BM, Miura M, Gronthos S, Bartold PM, Batouli S, Brahim J, Young M, Robey PG, Wang CY, Shi S (2004) Investigation of multipotent postnatal stem cells from human periodontal ligament. Lancet 364:149–155
22. Morsczeck C, Gotz W, Schierholz J, Zeilhofer F, Kuhn U, Mohl C, Sippel C, Hoffmann KH (2005) Isolation of precursor cells (PCs) from human dental follicle of wisdom teeth. Matrix Biol 24:155–165
23. Sonoyama W, Liu Y, Yamaza T, Tuan RS, Wang S, Shi S, Huang GT (2008) Characterization of the apical papilla and its residing stem cells from human immature permanent teeth: a pilot study. J Endod 34:166–171
24. Huang GT, Gronthos S, Shi S (2009) Mesenchymal stem cells derived from dental tissues vs. those from other sources: their biology and role in regenerative medicine. J Dent Res 88:792
25. Feng R, Lengner C (2013) Application of stem cell technology in dental regenerative medicine. Adv Wound Care (New Rochelle) 2(6):296–305
26. Mooney DJ, Vandenburgh H (2008) Cell delivery mechanisms for tissue repair. Cell Stem Cell 2:205
27. Capone C, Frigerio S, Fumagalli S, Gelati M, Principato MC, Storini C, Montinaro M, Kraftsik R, De Curtis M, Parati E, De Simoni MG (2007) Neurosphere-derived cells exert a neuroprotective action by changing the ischemic microenvironment. PloS One 2, e373

28. Kim SG, Zhou J, Solomon C, Zheng Y, Suzuki T, Chen M, Song S, Jiang N, Cho S, Mao JJ (2012) Effects of growth factors on dental stem/progenitor cells. Dent Clin North Am 56(3):563–575
29. Aubin JE, Liu F, Malaval L, Gupta AK (1995) Osteoblast and chondroblast differentiation. Bone 17:77S
30. Rizk A, Rabie AB (2013) Human dental pulp stem cells expressing transforming growth factor beta3 transgene for cartilage-like tissue engineering. Cytotherapy 15:712
31. Schliephake H (2002) Bone growth factors in maxillofacial skeletal reconstruction. Int J Oral Maxillofac Surg 31(5):469–484, Review
32. Gibble JW, Ness PM (1990) Fibrin glue: the perfect operative sealant? Transfusion 30:741–747
33. Saltz R, Sierra D, Feldman D, Saltz MB, Dimick A, Vasconez LO (1991) Experimental and clinical applications of fibrin glue. Plast Reconstr Surg 88:1005–1015, Discussion 1016–7
34. Hotz G (1991) Alveolar ridge augmentation with hydroxylapatite using fibrin sealant for fixation. Part I: An experimental study. Int J Oral Maxillofac Surg 20:204–207
35. Hotz G (1991) Alveolar ridge augmentation with hydroxylapatite using fibrin sealant for fixation. Part II: Clinical application. Int J Oral Maxillofac Surg 20:208–213
36. Schliephake H (2015) Clinical efficacy of growth factors to enhance tissue repair in oral and maxillofacial reconstruction: a systematic review. Clin Implant Dent Relat Res 17(2): 247–273
37. Boyne PJ (2001) Application of bone morphogenetic proteins in the treatment of clinical oral and maxillofacial osseous defects. J Bone Joint Surg Am 83A(Suppl 1):S146–S150
38. Dohan Ehrenfest DM, Rasmusson L, Albrektsson T (2009) Classification of platelet concentrates: from pure platelet-rich plasma (P-PRP) to leucocyte- and platelet-rich fibrin (L-PRF). Trends Biotechnol 27:158–167
39. Mazzucco L, Balbo V, Cattana E, Guaschino R, Borzini P (2009) Not every PRP-gel is born equal. Evaluation of growth factor availability for tissues through four PRP-gel preparations: Fibrinet, RegenPRP-Kit, Plateltex and one manual procedure. Vox Sang 97:110–118
40. Fréchette JP, Martineau I, Gagnon G (2005) Platelet-rich plasmas: growth factor content and roles in wound healing. J Dent Res 84:434–439
41. Weibrich G, Kleis WK, Hafner G, Hitzler WE, Wagner W (2003) Comparison of platelet, leukocyte, and growth factor levels in point-of-care platelet-enriched plasma, prepared using a modified Curasan kit, with preparations received from a local blood bank. Clin Oral Implants Res 14:357–362
42. Kumar RV, Shubhashini N (2013) Platelet rich fibrin: a new paradigm in periodontal regeneration. Cell Tissue Bank 14:453–463
43. Gulliksson H (2012) Platelets from platelet-rich-plasma versus buffy-coat-derived platelets: what is the difference? Rev Bras Hematol Hemoter 34:76–77
44. Mazzocca AD, McCarthy MB, Chowaniec DM et al (2012) Platelet-rich plasma differs according to preparation method and human variability. J Bone Joint Surg Am 94: 308–316
45. Dohan Ehrenfest DM et al (2012) Do the fibrin architecture and eukocyte content influence the growth factor release of platelet concentrates? An evidence-based answer comparing a pure platelet-rich plasma (P-PRP) gel and a leukocyte and platelet-rich fibrin (L-PRF). Curr Pharm Biotechnol 13:1145–1152
46. Castillo TN, Pouliot MA, Kim HJ, Dragoo JL (2011) Comparison of growth factor and platelet concentration from commercial platelet-rich plasma separation systems. Am J Sports Med 39:266–271
47. Whitman DH, Berry R, Green D (1997) Platelet gel: an autologous alternative to fibrin glue with applications in oral and maxillofacial surgery. J Oral Maxillofac Surg 55:1294–1299
48. Marx RE, Carlson ER, Eichstaedt RM et al (1998) Platelet-rich plasma: growth factor enhancement for bone grafts. Oral Surg Oral Med Oral Pathol Oral Radiol Endod 85: 638–646

49. Weibrich G, Kleis WKG, Hitzler WE, Hafner G (2005) Comparison of the platelet concentrate collection system with the plasma-rich-in-growth-factors kit to produce platelet-rich plasma: a technical report. Int J Oral Maxillofac Implants 20:118–123
50. Del Fabbro M, Bortolin M, Taschieri S, Weinstein R (2011) Is platelet concentrate advantageous for the surgical treatment of periodontal diseases? A systematic review and meta-analysis. J Periodontol 82:1100–1111
51. Plachokova AS, Nikolidakis D, Mulder J, Jansen JA, Creugers NH (2008) Effect of platelet-rich plasma on bone regeneration in dentistry: a systematic review. Clin Oral Implants Res 19:539–545
52. Mao JJ, Kim SG, Zhou J, Ye L, Cho S, Suzuki T, Fu SY, Yang R, Zhou X (2012) Regenerative endodontics: barriers and strategies for clinical translation. Dent Clin North Am 56(3):639–649
53. Lind M (1996) Growth factors: possible new clinical tools. A review. Acta Orthop Scand 67(4):407–417
54. Lázár-Molnár E, Hegyesi H, Tóth S et al (2000) Autocrine and paracrine regulation by cytokines and growth factors in melanoma. Cytokine 12(6):547–554
55. Nicholas C, Lesinski GB (2011) Immunomodulatory cytokines as therapeutic agents for melanoma. Immunotherapy 3(5):673–690
56. Seppä H, Grotendorst G, Seppä S et al (1982) Platelet-derived growth factor in chemotactic for fibroblasts. J Cell Biol 92(2):584–588
57. Deuel TF, Senior RM, Huang JS et al (1982) Chemotaxis of monocytes and neutrophils to platelet-derived growth factor. J Clin Invest 69(4):1046–1049
58. Hellberg C, Ostman A, Heldin CH (2010) PDGF and vessel maturation. Recent Results Cancer Res 180:103–114
59. Bouletreau PJ, Warren SM, Spector JA et al (2002) Factors in the fracture microenvironment induce primary osteoblast angiogenic cytokine production. Plast Reconstr Surg 110(1): 139–148
60. Hock JM, Canalis E (1994) Platelet-derived growth factor enhances bone cell replication but not differentiated function of osteoblasts. Endocrinology 134:1423–1428
61. Rydziel S, Shaikh S, Canalis E (1994) Platelet-derived growth factor-AA and -BB (PDGF-AA and -BB) enhance the synthesis of PDGF-AA in bone cell cultures. Endocrinology 134:2441–2546
62. Canalis E, Varghese S, McCarthy TL, Centrella M (1992) Role of platelet derived growth factor in bone cell function. Growth Regul 2:151–155
63. Shinbrot E, Peters KG, Williams LT (1994) Expression of the platelet-derived growth factor beta receptor during organogenesis and tissue differentiation in the mouse embryo. Dev Dyn 199:169–175
64. Soriano P (1997) The PDGF alpha receptor is required for neural crest cell development and for normal patterning of the somites. Development 124:2691–2700
65. Yokose S, Kadokura H, Tajima N et al (2004) Platelet-derived growth factor exerts disparate effects on odontoblast differentiation depending on the dimers in rat dental pulp cells. Cell Tissue Res 315(3):375–384
66. Rutherford RB, Trailsmith MD, Ryan ME et al (1992) Synergistic effects of dexamethasone on platelet-derived growth factor mitogenesis in vitro. Arch Oral Biol 37(2):139–145
67. Denholm IA, Moule AJ, Bartold PM (1998) The behaviour and proliferation of human dental pulp cell strains in vitro, and their response to the application of platelet-derived growth factor-BB and insulin-like growth factor-1. Int Endod J 31(4):251–258
68. Nakashima M (1992) The effects of growth factors on DNA synthesis, proteoglycan synthesis and alkaline phosphatase activity in bovine dental pulp cells. Arch Oral Biol 37(3):231–236
69. Mullane EM, Dong Z, Sedgley CM et al (2008) Effects of VEGF and FGF2 on the revascularization of severed human dental pulps. J Dent Res 87(12):1144–1148
70. Kim JY, Xin X, Moioli EK et al (2010) Regeneration of dental-pulp-like tissue by chemotaxis-induced cell homing. Tissue Eng Part A 16(10):3023–3031

71. Watabe T, Miyazono K (2009) Roles of TGF-beta family signaling in stem cell renewal and differentiation. Cell Res 19(1):103–115
72. Burt DW, Law AS (1994) Evolution of the transforming growth factor-beta superfamily. Prog Growth Factor Res 5(1):99–118
73. Wahl SM (1992) Transforming growth factor beta (TGF-beta) in inflammation: a cause and a cure. J Clin Immunol 12(2):61–74
74. Derynck R, Jarrett JA, Chen EY et al (1985) Human transforming growth factor-beta complementary DNA sequence and expression in normal and transformed cells. Nature 316(6030): 701–705
75. Centrella M, Canalis E (1985) Transforming and non transforming growth factors are present in medium conditioned by fetal rat calvariae. Proc Natl Acad Sci U S A 82:7335–7339
76. Cassidy N, Fahey M, Prime SS et al (1997) Comparative analysis of transforming growth factor-beta isoforms 1–3 in human and rabbit dentine matrices. Arch Oral Biol 42(3): 219–223
77. Lambert KE, Huang H, Mythreye K et al (2011) The type III transforming growth factor-β receptor inhibits proliferation, migration, and adhesion in human myeloma cells. Mol Biol Cell 22(9):1463–1472
78. Verrecchia F, Mauviel A (2002) Transforming growth factor-beta signaling through the Smad pathway: role in extracellular matrix gene expression and regulation. J Invest Dermatol 118(2):211–215
79. Massagué J, Blain SW, Lo RS (2000) TGFbeta signaling in growth control, cancer, and heritable disorders. Cell 103(2):295–309
80. Moses HL, Serra R (1996) Regulation of differentiation by TGF-beta. Curr Opin Genet Dev 6(5):581–586
81. Melin M, Joffre-Romeas A, Farges JC et al (2000) Effects of TGFbeta1 on dental pulp cells in cultured human tooth slices. J Dent Res 79(9):1689–1696
82. He H, Yu J, Liu Y et al (2008) Effects of FGF2 and TGFbeta1 on the differentiation of human dental pulp stem cells in vitro. Cell Biol Int 32(7):827–834
83. Tziafas D, Papadimitriou S (1998) Role of exogenous TGF-beta in induction of reparative dentinogenesis in vivo. Eur J Oral Sci 106(Suppl 1):192–196
84. Howard C, Murray PE, Namerow KN (2010) Dental pulp stem cell migration. J Endod 36(12):1963–1966
85. Wahl SM (1999) TGF-beta in the evolution and resolution of inflammatory and immune processes. Introduction. Microbes Infect 1(15):1247–1249
86. Farges JC, Romeas A, Melin M et al (2003) TGF-beta1 induces accumulation of dendritic cells in the odontoblast layer. J Dent Res 82(8):652–656
87. Scheufler C, Sebald W, Hülsmeyer M (1999) Crystal structure of human bone morphogenetic protein-2 at 2.7 A resolution. J Mol Biol 287:103–105
88. Miyazono K (2000) Positive and negative regulation of TGF-β signalling. J Cell Sci 113:1101–1109
89. McIntosh CJ, Lawrence S, Smith P, Juengel JL, McNatty KP (2012) Active immunization against the proregions of GDF9 or BMP15 alters ovulation rate and litter size in mice. Reproduction 143(2):195–201
90. Valentin-Opran A, Wozney J, Csimma C, Lilly L, Riedel GE (2002) Clinical evaluation of recombinant human bone morphogenetic protein-2. Clin Orthop Relat Res 395:110–120
91. Kaneda A, Fujita T, Anai M, Yamamoto S, Nagae G, Morikawa M, Tsuji S, Oshima M, Miyazono K, Aburatani H (2011) Activation of Bmp2-Smad1 signal and its regulation by coordinated alteration of H3K27 tri-methylation in Ras-induced senescence. PLoS Genet 7(11), e1002359
92. Açil Y, Springer IN, Broek V, Terheyden H, Jepsen S (2002) Effects of bone morphogenetic protein-7 stimulation on osteoblasts cultured on different biomaterials. J Cell Biochem 86(1):90–98
93. Kanayama M, Hashimoto T, Shigenobu K, Yamane S, Bauer TW, Togawa D (2006) A prospective randomized study of posterolateral lumbar fusion using osteogenic protein-1 (OP-1)

versus local autograft with ceramic bone substitute: emphasis of surgical exploration and histologic assessment. Spine (Phila Pa 1976) 31(10):1067–1074

94. Miller AF, Harvey SA, Thies RS, Olson MS (2000) Bone morphogenetic protein-9. An autocrine/paracrine cytokine in the liver. J Biol Chem 275(24):17937–17945
95. Chen C, Grzegorzewski KJ, Barash S, Zhao Q, Schneider H, Wang Q, Singh M, Pukac L, Bell AC, Duan R, Coleman T, Duttaroy A, Cheng S, Hirsch J, Zhang L, Lazard Y, Fischer C, Barber MC, Ma ZD, Zhang YQ, Reavey P, Zhong L, Teng B, Sanyal I, Ruben SM, Blondel O, Birse CE (2003) An integrated functional genomics screening program reveals a role for BMP-9 in glucose homeostasis. Nat Biotechnol 21(3):294–301
96. Truksa J, Peng H, Lee P, Beutler E (2006) Bone morphogenetic proteins 2, 4, and 9 stimulate murine hepcidin 1 expression independently of Hfe, transferrin receptor 2 (Tfr2), and IL-6. Proc Natl Acad Sci U S A 103(27):10289–10293
97. López-Coviella I, Berse B, Krauss R, Thies RS, Blusztajn JK (2000) Induction and maintenance of the neuronal cholinergic phenotype in the central nervous system by BMP-9. Science 289(5477):313–316
98. Saito T, Ogawa M, Hata Y et al (2004) Acceleration effect of human recombinant bone morphogenetic protein-2 on differentiation of human pulp cells into odontoblasts. J Endod 30(4):205–208
99. Chen S, Gluhak-Heinrich J, Martinez M et al (2008) Bone morphogenetic protein 2 mediates dentin sialophosphoprotein expression and odontoblast differentiation via NF-Y signaling. J Biol Chem 283(28):19359–19370
100. Iohara K, Nakashima M, Ito M et al (2004) Dentin regeneration by dental pulp stem cell therapy with recombinant human bone morphogenetic protein 2. J Dent Res 83(8):590–595
101. Nakashima M, Nagasawa H, Yamada Y et al (1994) Regulatory role of transforming growth factor-beta, bone morphogenetic protein-2, and protein-4 on gene expression of extracellular matrix proteins and differentiation of dental pulp cells. Dev Biol 162(1):18–28
102. Rutherford RB, Spångberg L, Tucker M et al (1994) The time-course of the induction of reparative dentine formation in monkeys by recombinant human osteogenic protein-1. Arch Oral Biol 39(10):833–838
103. Rutherford RB, Wahle J, Tucker M et al (1993) Induction of reparative dentine formation in monkeys by recombinant human osteogenic protein-1. Arch Oral Biol 38(7):571–576
104. Six N, Lasfargues JJ, Goldberg M (2002) Differential repair responses in the coronal and radicular areas of the exposed rat molar pulp induced by recombinant human bone morphogenetic protein 7 (osteogenic protein 1). Arch Oral Biol 47(3):177–187
105. Rutherford RB, Gu K (2000) Treatment of inflamed ferret dental pulps with recombinant bone morphogenetic protein-7. Eur J Oral Sci 108(3):202–206
106. Jepsen S, Albers HK, Fleiner B et al (1997) Recombinant human osteogenic protein-1 induces dentin formation: an experimental study in miniature swine. J Endod 23(6): 378–382
107. Nakashima M, Iohara K, Ishikawa M et al (2004) Stimulation of reparative dentin formation by ex vivo gene therapy using dental pulp stem cells electrotransfected with growth/differentiation factor 11 (Gdf11). Hum Gene Ther 15(11):1045–1053
108. Nakashima M, Tachibana K, Iohara K et al (2003) Induction of reparative dentin formation by ultrasound-mediated gene delivery of growth/differentiation factor 11. Hum Gene Ther 14(6):591–597
109. Li C, Yang X, He Y, Ye G, Li X, Zhang X, Zhou L, Deng F (2012) Bone morphogenetic protein-9 induces osteogenic differentiation of rat dental follicle stem cells in P38 and ERK1/2 MAPK dependent manner. Int J Med Sci 9(10):862–871
110. Zhang H, Wang J, Deng F, Huang E, Yan Z, Wang Z, Deng Y, Zhang Q, Zhang Z, Ye J, Qiao M, Li R, Wang J, Wei Q, Zhou G, Luu HH, Haydon RC, He TC, Deng F (2015) Bone morphogenetic protein-9 effectively induces osteo/odontoblastic differentiation of the reversibly immortalized stem cells of dental apical papilla. Biomaterials 39:145–154
111. Metzger RJ, Krasnow MA (1999) Genetic control of branching morphogenesis. Science 284(5420):1635–1639

112. DiMario J, Buffinger N, Yamada S et al (1989) Fibroblast growth factor in the extracellular matrix of dystrophic (mdx) mouse muscle. Science 244(4905):688–690
113. Itoh N, Ornitz DM (2011) Fibroblast growth factors: from molecular evolution to roles in development, metabolism and disease. J Biochem 149(2):121–130
114. Kitchens DL, Snwyder EY, Gottlieb DI (1994) FGF and EGF are mitogens for immortalized neural progenitors. J Neurobiol 25:797–807
115. Amit M, Carpenter MK, Inokuma MS et al (2000) Clonally derived human embryonic stem cell lines maintain pluripotency and proliferative potential for prolonged periods of culture. Dev Biol 227:271–278
116. Nakao K, Itoh M, Tomita Y, Tomooka Y, Tsuji T (2004) FGF-2 potently induces both proliferation and DSP expression in collagen type I gel cultures of adult incisor immature pulp cells. Biochem Biophys Res Commun 325:1052–1059
117. Solchaga LA, Penick K, Porter JD, Goldberg VM, Caplan AI, Welter JF (2005) FGF-2 enhances the mitotic and chondrogenic potentials human adult bone marrow-derived mesenchymal stem cells. J Cell Physiol 203:398–409
118. Shimabukuro Y, Ueda M, Ozasa M et al (2009) Fibroblast growth factor-2 regulates the cell function of human dental pulp cells. J Endod 35:1529–1535
119. Quarto N, Longaker MT (2006) FGF-2 inhibits osteogenesis in mouse adipose tissue-derived stromal cells and sustains their proliferative and osteogenic potential state. Tissue Eng 12:1405–1418
120. Suzuki T, Lee CH, Chen M et al (2011) Induced migration of dental pulp stem cells for in vivo pulp regeneration. J Dent Res 90(8):1013–1018
121. Qian J, Jiayuan W, Wenkai J, Peina W, Ansheng Z, Shukai S, Shafei Z, Jun L, Longxing N (2015) Basic fibroblastic growth factor affects the osteogenic differentiation of dental pulp stem cells in a treatment-dependent manner. Int Endod J 48(7):690–700
122. Wu J, Huang GT, He W, Wang P, Tong Z, Jia Q, Dong L, Niu Z, Ni L (2012) Basic fibroblast growth factor enhances stemness of human stem cells from the apical papilla. J Endod 38(5):614–622
123. Debiais F, Hott M, Graulet AM, Marie PJ (1998) The effects of fibroblast growth factor-2 on human neonatal calvaria osteoblastic cells are differentiation stage specific. J Bone Miner Res 13:645–654
124. Sukarawan W, Nowwarote N, Kerdpon P, Pavasant P, Osathanon T (2014) Effect of basic fibroblast growth factor on pluripotent marker expression and colony forming unit capacity of stem cells isolated from human exfoliated deciduous teeth. Odontology 102(2):160–166
125. O'Connor R (1998) Survival factors and apoptosis. Adv Biochem Eng Biotechnol 62: 137–166
126. Joseph BK, Savage NW, Young WG et al (1993) Expression and regulation of insulin-like growth factor-I in the rat incisor. Growth Factors 8(4):267–275
127. Onishi T, Kinoshita S, Shintani S et al (1999) Stimulation of proliferation and differentiation of dog dental pulp cells in serum-free culture medium by insulin-like growth factor. Arch Oral Biol 44(4):361–371
128. Caviedes-Bucheli J, Canales-Sánchez P, Castrillón-Sarria N et al (2009) Expression of insulin-like growth factor-1 and proliferating cell nuclear antigen in human pulp cells of teeth with complete and incomplete root development. Int Endod J 42(8):686–693
129. Caviedes-Bucheli J, Angel-Londoño P, Díaz-Perez A et al (2007) Variation in the expression of insulinlike growth factor-1 in human pulp tissue according to the root-development stage. J Endod 33(11):1293–1295
130. Caviedes-Bucheli J, Muñoz HR, Rodríguez CE et al (2004) Expression of insulin-like growth factor-1 receptor in human pulp tissue. J Endod 30(11):767–769
131. Feng X, Huang D, Lu X, Feng G, Xing J, Lu J, Xu K, Xia W, Meng Y, Tao T, Li L, Gu Z (2014) Insulin-like growth factor 1 can promote proliferation and osteogenic differentiation of human dental pulp stem cells via mTOR pathway. Dev Growth Differ 56(9):615–624

132. Wang S, Mu J, Fan Z, Yu Y, Yan M, Lei G, Tang C, Wang Z, Zheng Y, Yu J, Zhang G (2012) Insulin-like growth factor 1 can promote the osteogenic differentiation and osteogenesis of stem cells from apical papilla. Stem Cell Res 8(3):346–356
133. Leung DW, Cachianes G, Kuang WJ et al (1989) Vascular endothelial growth factor is a secreted angiogenic mitogen. Science 246(4935):1306–1309
134. Nör JE, Christensen J, Mooney DJ et al (1999) Vascular endothelial growth factor (VEGF)-mediated angiogenesis is associated with enhanced endothelial cell survival and induction of Bcl-2 expression. Am J Pathol 154(2):375–384
135. Ferrara N (2004) Vascular endothelial growth factor: basic science and clinical progress. Endocr Rev 25(4):581–611
136. D'Alimonte I, Nargi E, Mastrangelo F et al (2011) Vascular endothelial growth factor enhances in vitro proliferation and osteogenic differentiation of human dental pulp stem cells. J Biol Regul Homeost Agents 25(1):57–69
137. Matsushita K, Motani R, Sakuta T, Yamaguchi N, Koga T, Matsuo K, Nagaoka S, Abeyama K, Maruyama I, Torii M (2000) The role of vascular endothelial growth factor in human dental pulp cells: induction of chemotaxis, proliferation, and differentiation and activation of the AP-1-dependent signaling pathway. J Dent Res 79(8):1596–1603
138. Artese L, Rubini C, Ferrero G, Fioroni M, Santinelli A, Piattelli A (2002) Vascular endothelial growth factor (VEGF) expression in healthy and inflamed human dental pulps. J Endod 28(1):20–23
139. d'Aquino R, Graziano A, Sampaolesi M, Laino G, Pirozzi G, De Rosa A, Papaccio G (2007) Human postnatal dental pulp cells co-differentiate into osteoblasts and endotheliocytes: a pivotal synergy leading to adult bone tissue formation. Cell Death Differ 14(6):1162–1171
140. Marchionni C, Bonsi L, Alviano F et al (2009) Angiogenic potential of human dental pulp stromal (stem) cells. Int J Immunopathol Pharmacol 22(3):699–706

Nano Surface and Stem Cells for Implants

G.K. Thakral

Introduction

Nanotechnology can be defined as the science and engineering involved in the design, synthesis, characterization and application of materials and devices whose smallest functional organization, in at least one dimension is on the nanometer (nm) scale or one billionth of a meter [1].

The technology involves increasing the complexity of the surface topography with the addition of nanoscale molecules (Fig. 1) [2].

Nanotechnology therefore, involves materials that have a nano-sized topography or are composed of nano-sized materials ranging between 1 and 100 nm [3]. Nano-biotechnology can be described as a hybrid science that has evolved by blending two predominant technologies—nanotechnology and biotechnology. It essentially involves the further improvement and enhancement of the physical and chemical properties of biomaterials for their better and advantageous use in the medical science [4]. Since, bio-molecules are present in nanometer dimensions, nano-sized biomaterials are expected to integrate and assimilate well in biomedical devices and procedures [5].

An evaluation of nano-mechanical properties of the surrounding bone by nano-indentation revealed that while both implants exhibited similar bone-to-implant contact, the nano-indentation demonstrated that the tissue quality was significantly enhanced around the hydroxyapatite (HA)-coated implants [6]. Since osteoblasts readily adhere to this novel surface, dental implants coated with TiO_2 nanotubes could significantly improve healing following dental implant surgery (Fig. 2) [2].

Nanotechnology may involve one-dimensional concepts (nanodots and nanowires) or the self-assembly of more complex structures (nanotubes). Materials are

G.K. Thakral (✉)
Department of Dentistry, SGRR Institute of Medical & Health Sciences, Dehradun, India
e-mail: gkthakral@yahoo.com

B. Zavan, E. Bressan (eds.), *Dental Stem Cells: Regenerative Potential*, Stem Cell Biology and Regenerative Medicine, DOI 10.1007/978-3-319-33299-4_6

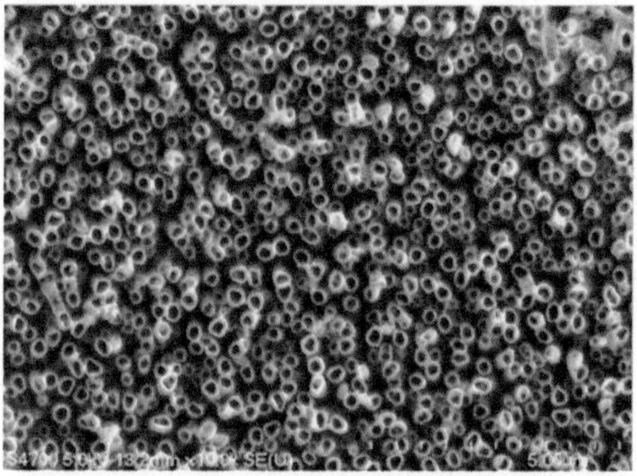

Fig. 1 A cluster of titanium dioxide nano-tubes [2]

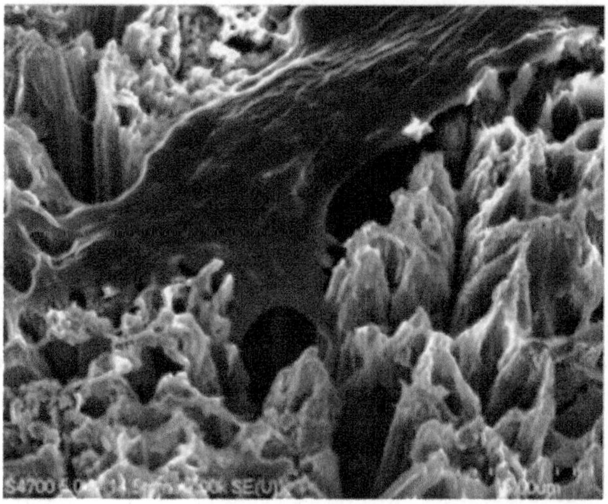

Fig. 2 Bone cell anchoring to a surface of titanium dioxide nano-tubes [2]

also classified according to their form and structure as nano-structures, nano-crystals, nano-coatings, nano-particles, and nano-fibers [3]. Nano-scale modification of the titanium endosseous implant surface may lead to alteration in the topography as well as chemistry of the surface. Therefore, the goal of nanoscale modification should be a specific chemical modification of commercially pure (cp) titanium (grade 4 and 5). A distinct implication associated with nano-scale manipulation of any material is that it also leads to inherent chemical changes on the material surface.

The topographic changes invariably create an increased surface area and nano-scale surface roughness leading to better biological responses of osteogenic cells and effective tissue-implant mechanical interlocking, apart from making the implant surface high wear resistant. These qualities of the nano-biomaterials make it more favourable for implant procedures as compared to other materials. Nano modifications in biomaterials can be undertaken as under:

- Biomaterial Surface Modifications
  - Alteration of Surface Topography
  - Alteration of Surface Chemistry
- Scaffolds involving Nano Materials
  - Nano-ceramic Composites
  - Nano-fibre Composites

These surface modifications enhance solubility and stability of nano materials in aqueous media and make them biologically more compatible. The surface modifications can be achieved by employing physical, chemical and biological methods. Physical modifications can be achieved through molecular coating, molecular adsorption, surface entrapment, plasma spraying and ozone ablation etc. Whereas, chemical modifications may involves procedures like silanisation, surface oxidation and self-assembled monolayers etc.

The advantages of nano-topography based implants or nano-material based implants are significant and may include improved mechanical properties, wear resistance, bone-to-implant contact, protein adsorption, cell response and osseointegration [7, 8].

Dental implants closely mimic natural teeth, but lack in tissue dynamics associated with periodontal tissues. Stem cells offer a promising source for tissue regeneration due to their peculiar quality of proliferation, differentiation and natural plasticity. A positive and desirable interaction of stem cells with various biomaterials offers an enormous scope for their use in various therapeutic applications. Dental stem cells derived from exfoliated deciduous teeth as well as from wisdom teeth are an easy and rich resource for tissue regeneration including osseointegration of implants.

Biomaterials

Biomaterials can be defined as the materials that are well tolerated and accepted when placed inside or adjacent to the living tissues in a body. Biocompatibility therefore, refers to a positive response a material generates in close proximity to the living tissues. The placement of a material in close proximity to the living tissues creates an interface. The material and the tissues come face to face at this interface leading to some reaction or response from both sides. This response is not merely a static reaction but a dynamic one. This reaction or interaction between the material and the living tissues may lead to certain specifically outcomes like:

- The degeneration of tissues affecting the properties of the material
- The degeneration of the material affecting health of tissues
- Regeneration of the tissues and survival of the material

Table 1 Classification of implant biomaterials

Classification	Response	Effect
Bio-tolerant	Formation of thin connective tissue capsules that does not adhere to the implant surface	Rejection of the implant, leading to implant failure
Bio-active	Formation of bony tissue around the implant material and strongly integrates with the implant surface	Acceptance of the implant leading to success of implant
Bio-reabsorbable	Replaced by the autologous tissue	Acceptance of the implant leading to success of implant

It is this dynamic interaction that determines both—the biological response of the tissues to the material and survival of the material in the biological environment against corrosion and degeneration. That is the basic concept of biocompatibility of a material. Biocompatibility therefore, is not merely a property of the material. Biocompatible activity of a material at the interface of material and the tissue depends on factors like—site of the material placement, its duration in the biological environment, properties of the material and health of the host. Therefore, a material that is biocompatible as a crown and bridge material may not necessarily be compatible as an implant material and vice-versa. Health conditions of the host also make an impact on the biocompatibility response of a material. For that matter, biocompatibility of a young person may be different from that of an older person. The biological response of the same material may vary from a healthy young adult to a young adult affected by some debilitating disease or conditions that hinder or inhibit tissue response.

Therefore, it is important to evaluate and understand the physical, chemical and biological properties of basic components of a material before any biological applications, since these properties determine the success or the failure of an implant material in a long term use.

Broadly speaking, the implant biomaterials can be classified into three groups based on their interaction with the tissues (Table 1).

Bio-tolerant: The term refers to the behavior of a material that depicts minimal interaction with its surrounding tissue when placed in the body. The tissue response may be limited to formation of some connective (scar) tissue at the bone implant interface without any integration to the implant surface.

Bio-active: It refers to a material that actively interacts with the surrounding bone and soft tissues. The bio-active materials may either initiate or promote some regenerative activity once placed adjacent to the living tissues. The tissue response leads to formation of bony tissue at the bone implant interface that strongly integrates with the implant surface. This is the most widely used bio-material category for dental implants.

Bio-resorbable: It refers to that property of a material that upon placement within the human body makes it to gradually dissolve and get replaced by adjacent tissues. The tissue response creates formation of autologous tissue leading to implant success. These types of bio-materials are more frequently used in orthopedic implants.

Desired Properties of Biomaterials

Development of biomaterials is a highly skilled job requiring interdisciplinary collective involvement of professionals like pathologists, clinicians, material scientists, material engineers, biomedical scientists etc. The basic properties that ensure success of an implant adjacent to human tissues can be summarized as—

Mechanical Properties

The mechanical properties of a material play a decisive role in its selection for fabrication of implants. Some of the important properties include hardness, tensile strength, modulus of elasticity and elongation. The response of the material to repeated occlusal loads is dependent on the fatigue strength of the material and determines the long term success of implant under recurring occlusal stresses. The modulus of elasticity of bone varies in magnitude from 4 to 30 Gpa depending on the type of bone and the direction of measurement [9, 10].

The material used for implants should have modulus of elasticity equivalent to that of the bone. Inadequate strength or any mismatch in mechanical properties between implant material and the bone is bound to lead to fracture of implant making it biomechanically incompatible. At the same time, an implant material that has a strength much higher than the bone will prevent transfer of the stress to the adjacent bone leading to bone resorption around the implant and implant failure. The biomechanical incompatibility that leads to death of the bone cells is known as 'Stress Shielding Effect [11]. Therefore, the material of choice for dental implants has to have a combination of high strength and low modulus of elasticity closer to the bone.

Biocompatibility

The materials used for implants are expected to be well accepted and tolerated by the human tissues without causing any adverse effect or reactions in the body. This quality of a material is known as biocompatibility [12]. Interaction and some reaction start between implant surface and adjacent tissues immediately upon placement of implant in the human body. The reaction involves blood coagulation and adhesion of blood platelets to biomaterial surface and encapsulation of the biomaterial by fibrous tissues.

The biological compatibility of implant materials is directly dependent upon their chemical composition, design of the implant, topographic mechanics and wettability of the implant material. Use of nanotechnology may lead to the development of dental implants with controlled topography and chemistry leading to the development of implant surfaces with predictable and favourable tissue response during osseointegration [13, 14].

High Corrosion and Wear Resistance

The implants should be developed using materials with high wear and corrosion resistance in order to ensure longevity of the dental implant under biological conditions. The longevity of the implant is usually determined by its resistance to abrasion and wear in the mouth. The implant materials having low wear and corrosion resistance release non-compatible metal ions in the body fluids. The released ions cause allergic and toxic reactions in the body [15].

A low resistance to wear leads to formation and deposition of wear debris at the implant-tissue interface causing several chemical and biological reactions in the tissues with ultimate failure of the implant due to its loosening [16].

Osseointegration

Osseointegration is a term founded by Professor Per-Ingvar Brånemark after his important breakthrough in the 1950s when he discovered that bone can integrate with titanium components. Osseointegration has been described as a direct structural and functional bone to implant contact under load [17]. Professor Brånemark named his discovery after Latin word 'os' that means bone, and 'integrate' that means make whole. When conjoined together it points out to interactive coexistence. It was observed that when a screw-shaped implant fixture made of titanium was carefully placed in the bone, the genetic code that usually makes bone reject a foreign material was not activated. The bone cells were able to attach to the titanium surface resulting in a firm and permanent anchorage for a prosthetic reconstruction. It actually promotes bone regeneration that fills the micro gaps between the implant surface and the adjacent bone. The term has been used to explain the success or failure of the implants in medical and dental science ever since. In the absence of osseointegration, fibrous tissue is formed between the bone and the implant surface. The implant therefore, is not well integrated into the bone resulting in implant loosening and subsequent implant failure [18]. Surface topography, surface roughness, surface material and surface chemistry, all play a significant role for a successful osseointegration.

Impact of Nano Topography in Implants

Titanium and Titanium alloys Ti_6Al_4V have over the years emerged as the most biocompatible material for dental implants due to their excellent corrosion resistance property. Resistance to corrosion in Titanium occurs due to the formation of biologically inert oxide layer on the surface of the implant body. This oxide layer spontaneously converts into tenacious surface oxide on exposure to air or physiologic saline. This surface oxide transformation leads to the formation of three types of oxides on the implant surface; namely TiO (Anastase), TiO_2 (Rutile) and Ti_2O_3 (Brookite). Of the three oxides, TiO_2 (Rutile) is the most stable and the most frequently formed oxide layer that has the potential to regenerate instantly. For that matter, if implant surface gets abraded or scratched during implant placement procedure, this oxide layer can recreate itself spontaneously.

Majority of dental implants today are made of cp titanium (1–4 grades) or titanium alloys like Ti_6Al_4V. Factors that contribute to the ultimate success of an implant include—

1. Physiological conditions of the patient
2. Implant placement procedure
3. Implant material
4. Implant design, and
5. Implant surface

Albrektsson and Wennerberg [19] divided implant surface quality into three categories—

1. Mechanical properties,
2. Topographic properties, and
3. Physico-chemical properties.

They pointed out that these characteristics are inter-related and a change in any of these groups affects the others as well. This significant observation is quite relevant to the study of nano-surface modifications of the endosseous cp titanium implant surface.

The success of dental implantology to a great extent depends on two basic eventualities. The first is to achieve osseointegration and maintain it through the life of implants. This biological bonding of the implant with the surrounding bone ensures a sound mechanical anchoring in situ. The second is an excellent adaptation and blending of the gingival tissues at the neck of dental implant. This ensures a good sealing of the implant in the soft tissue and isolation of implant body from oral environment and thereby preventing any bacterial lodgment in the region and subsequent peri-implantitis.

This initial tissue response to osseointegration renders primary stability to the dental implant that gradually decreases to pave the way for a secondary tissue response of biological union at the implant tissue interface. The implant material, implant design and implant surface play a very significant and crucial role in the success of secondary tissue response for a lasting osseointegration of the dental implant and may be preceded by a transient phase of decreased implant stability. Surface topography of the dental implants plays a significant role on the quality and quantity of osseointegration post implantation. In order to have favourable osseointegration, the implant surface should promote adsorption of proteins, the initial adhesion followed by differentiation of cells and homogenous tissue integration.

It has been established through studies that Calcium Phosphate (CaP) coatings provide titanium implants with an osteo-conductive surface. The CaP coatings get dissolved in the peri-implant region after implant placement leading to increased ionic strength and blood saturation that in turn initiates precipitation of biological apatite nano-crystals on the implant surface. The inherent proteins in the biological apatite layer promote adhesion of osteo-progenitor cells that generate extracellular matrix of bony tissues [20]. Studies have also revealed that osteoclasts that are capable to resorb the bone cells cause degradation of the CaP coating through enzymatic reactions and create resorption pits on the CaP coated implant surface [21].

CaP particles on the implant surface serve as a catalyst and promote almost immediate osseointegration of implants into the adjacent bone. It is significant to create CaP coating on the implant surface that would disintegrate at a rate similar to apposition of bone so that a direct bone contact may develop on the surface of implant.

A favourable biological interactions between implant and bone interface are greatly dependent upon the topography, chemical compositions and wettability of the implant surface. An implant surface that allows predictable, controlled and guided tissue regeneration is more likely to promote contact osteogenesis [22, 23].

Moreover, nano-surface modifications substantially increase the total surface areas at the interface for osteoapposition leading to better osteointegration of the implant. Introduction of Ti nanotubes (300 nm) resulted in significant increase of up to 3.1 times in the strength of bone-titanium interface in rat femur [22].

Methods of Imparting Nano-Features

General modification in implant materials may be broadly categorized into—physical modifications, chemical modifications and biological modifications and can be summarized as under:

1. Physical Modification
   (a) Molecular Coating
   (b) Surface Entrapment
   (c) Plasma Spraying
   (d) Ozon Ablation
2. Chemical Modification
   (a) Surface Oxidation
   (b) PEG (Poly-ethylene-Glycol) Chemistry
   (c) Silane (Silicone-Based) Chemistry
   (d) Self-assembled Monolayers
3. Biological Modification
   (a) Antibody–Antigen
   (b) Receptor–Ligand (Ligand is a molecule like an antibody, hormone or drug that binds to a receptor)
   (c) DNA–DNA Hybridization

Table 2 Methods for creating nano-features on cp titanium implants [24]

Methods	Characteristics
Self-assembly of monolayers	The exposed functional end group could be a molecule with different functions (an osteoinductive or cell adhesive molecule)
Compaction of nanoparticles	Conserves the chemistry of the surface among different topographies. Not readily applied over implant surfaces
Ion beam deposition	Can impart nanofeatures to the surface based on the material used
Acid etching	Combined with other methods (sandblasting and/or peroxidation) can impart nanofeatures to the surface and remove contaminants
Peroxidation	Produces a titania gel layer. Both chemical and topography changes are imparted
Alkali treatment (NaOH)	Produces a sodium titanate gel layer allowing hydroxyapatite deposition. Both chemical and topographic changes are imparted
Anodization	Can impart nanofeatures to the surface creating a new oxide layer (based on the material used)
Sol–gel (colloidal particle adsorption)	Creates a thin-film of controlled chemical characteristics. Atomic-scale interactions display strong physical interactions
Discrete crystalline deposition	Superimposes a nanoscale surface topographical complexity on the surface
Lithography and contact printing technique	Many different shapes and materials can be applied over the surface. Approaches are labor intensive and require considerable development prior to clinical translation and application on implant surface

There are various methods to create nanoscale features at the implant surface (Table 2) [24].

These methods include:

1. Physical methods, like self-assembly of mono-layers, compaction of nano particles and ion beam deposition;
2. Chemical methods, like acid etching, peroxidation, alkali treatment (NaOH) and anodization;
3. Nano particle deposition like sol–gel (colloidal particle deposition) and discrete crystalline deposition; and
4. Lithography and contact printing technique

Influence of Implant Surface on Osseointegration

One of the main concerns related to coating the implant surface is the risk of coating detachment and resultant toxicity of related debris. An evaluation of relationship of particle size and cell viability and proliferation compared to micron-particles revealed that nano-particles of titanium and alumina had less negative impact in cell viability and proliferation as compared to conventional particles. There may be an advantage to nano-scale modification of surfaces using sol–gel coating methods. The quantum interaction of high electron density at the atomic level can enforce high bond strength between the substrate and nano-scale coating [25].

Studies reveal that the addition of a nanometer-scale calcium phosphate treatment to a dual acid-etched implant surface appeared to increase the extent of bone development after 4 and 8 weeks of healing. It was observed that this rapid accrual of bone at the implant surface expedites the implant healing period and supports early loading protocols [26].

Nanotechnology and Cellular Activity

The observation that a micron-scale rough surface prepared by grit blasting and subsequent acid etching was capable of rapid and increased bone accumulation further strengthened an earlier report that a TiO_2 grit blasted surface also supported more rapid and increased bone accrual at cp titanium implants [27].

The study also pointed to a significant fact that the cp titanium surface could be modified to enhance bone accumulation and suggested that cp titanium was not only "bioinert" or "biocompatible", but was also capable to influence cellular activity or tissue responses leading to better and greater osteogenesis and thereby promoting better osseointegration. Nano-topography seems to influence cell interactions at surface of the material being used. It also leads to changed cell behavior when compared to conventional sized topography. The cellular protein adsorption is altered by nanoscale modification of bulk material. Depending on the nano-architecture, cell spreading may be increased or decreased. The present undefined mechanisms indicate that cell proliferation appears to be enhanced by nanoscale topography. Several investigators have shown that nanoscale topography enhances osteoblast differentiation [28–30].

Protein Adsorption (Surface Wettability) in Nano-Surface

Alteration in initial protein surface interaction is believed to control osteoblast adhesion, a critical aspect of the osseointegration process [31]. When implant comes in contact with a biological environment, the initiation of protein adsorption (e.g. plasma fibronectin) promotes subsequent cell attachment and proliferation. Change in surface energy or wettability of a biomaterial corresponds to a typical way of altering cell interactions with the surface. Nano-scale topography is now an established way of altering protein interactions within a surface. An increased vitronectin adsorption has been observed on nano-structured surfaces when compared to conventional surfaces [32, 33]. The study also suggested an increased osteoblast adhesion when compared to other cell types, such as fibroblasts, on the nanosurfaces [33].

Cell Adhesion, Spreading and Motility in Nanosurface

Irrespective of the surface-adsorbed proteins, cells are remarkable in their ability to sense nanostructure. Nano-features of a surface affect both cell adhesion and cell motility. Both of these cell traits are attributed in part, to the function of integrins

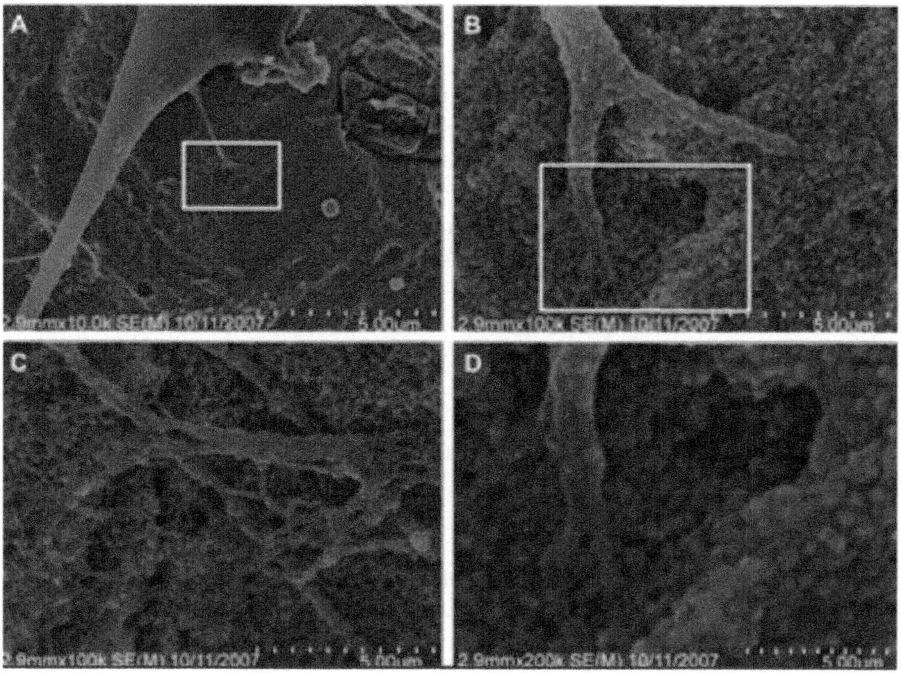

Fig. 3 Nanoscale topography-cell interactions on a nano-surface produced by H_2O_2/H_2SO_4. Treatment (**a**) 10,000 image of adherent cell, (**b**, **c**) represent 100,000 images of the same adherent cell and (**d**) 200,000 magnification of the cell with nano-features. (**b**) Higher magnification of the *rectangle* in (**a**). (**d**) Higher magnification of the *rectangle* in (**b**) [33]

[33, 34]. Underlying substratum topography influences cell behaviors by both direct and indirectinteractions. In 20–40 nm features produced by H_2O_2/H_2SO_4 treatment there were definitive interactive points for lamellipodia of spreading cells (Fig. 3) [33]. Cell behavior is affected by both, the dimension and the density of the nano structures.

Proliferation in Nano-Surface

Nanosurface modifications promote adherent cell proliferation. Zhao et al., utilizing three distinct methods—electrochemical machining, anodization and chemical etching to produce reproducible submicron-scale structures on titanium surfaces reported an opposing relationship between cell proliferation and cell differentiation with increase in micro scale of surface features.

It has further been supported by studies that observed an increase in the osteoblast proliferation on nano scale materials like alumina, titanium and hydroxyapatite [35, 36]. However, the mechanism involved is still not clear as to how nano-structured surfaces modulate the adherent osteoblast response.

Selectivity of Adhesion in Nano-Surface

Selectivity of cell adhesion is an interesting feature attributed to nanoscale topographic surfaces. Several studies have revealed a relative lowering of fibroblast adhesion compared to osteoblast adhesion on evaluation of nano and micron-structured surfaces [37, 38]. The affinity ratio between osteoblasts and fibroblasts was 3 to 1 on nanosized materials as compared to the conventional materials depicting a ratio of 1 to 1 [33]. Studies on other cell types such as smooth muscle cells and chondrocytes have also reported similar results [39].

All these observations may lead to some major implications in specification of tissue response at bone and mucosal surface of the dental implant/abutment assembly.

Differentiation

A rapid differentiation of adherent mesenchymal cells along the osteoblast lineage is as significant for the process of osseointegration as supporting osteoblast-specific adhesion and adherent cellular proliferation. Studies have revealed that alkaline phosphatase synthesis and calcium mineral content increased in cell layers formed on nano sized materials after 21 and 28 days [40, 41] (Table 3).

Nano-Surface and Bacterial Proliferation

Another significant finding with nano-surfaces has been found to be a diminished bacterial adhesion and proliferation [70]. There was a marked decrease in bacterial colonization on nano structured TiO_2 and ZnO_2 irrespective of the fact that these surfaces promote osteoblast adhesion and differentiation. These initial observations may suggest the need for further exploration of the implant abutment surface with focus on biofilm accumulation and peri-implantitis.

Nano-Surface and Surface Reactivity

Endosseous implant surface reactivity may get influenced by nano-surface modifications. Insignificant bone bonding occurs at endosseous titanium implants, especially during the initial phases of bone formation [71]. Nano-scale topographic modifications tend to change the chemical reactivity of materials and presence of bone on implant surface during early stages [72].

Bone bonding seems to be an advantage associated with titanium implants with nano-scale surface modifications. Advantages of nano-topography on biomaterials have been demonstrated as early as 1999 [30].

Table 3 Summary of the osteochondro-inductive nanostructures presented and the cell behavior outcomes observed [42]

Nanostructured materials	Fabrication techniques	Cell types	Cell behaviors
TiO_2 nanotubes with Ta coating [43]	Two electrode setup anodization and vacuum-deposited	Osteoblasts	Improvement of viability and faster Mineralization
TiO_2 nanotubes [44]	Two electrode setup anodization	Chondrocytes	Promotion of chondrogenesis
Carbon nanotubes-nanocomposite of CHI fibers + HA crystals [45]	Arc discharged method, freeze-drying and lyophilization	Osteoblasts	Improvement of adhesion and proliferation
Graphene oxide (GO) + CHI + HA [46]	Covalent liaison of CHI and GO in solution	MC3T3-E1 fibroblastic cells	Improvement of adhesion, proliferation and osteogenic differentiation
Graphene oxide with PMMA [47]	Chemical vapor deposition	Mesenchymal stem cells	Improvement of osteogenic differentiation
Collagen-GAG scaffolds with biomolecular pattern (fibronectin) [48]	Direct photolithography	MC3T3-E1 fibroblastic cells	Improvement of the speed of cell attachment
Nanostructured alumina surfaces [49]	Chemical vapor deposition	Osteoblasls	Induction of osteogenic differentiation
Polycrystalline titanium nanostructured surface with conformal bioactive calcium phosphate thin films sputter [50]	RF magnetron sputter deposition	Bone marrow mesenchymal stem cells	Improvement of adherence, proliferation and osteogenic differentiation
Electrochemically grooved nanostructured stainless steel implant with pre-adsorption of Protein [51]	Phase reversion-induced nanograined structure	Osteoblasts	Improvement of osteoblastic function and activity
Micro- and nanopatterned transplantable poly(lactic-co-glycolic-acid) polymer [52]	Capillary force lithography with a surface micro-wrinkling	Mesenchymal stem cells	Improvement of adhesion, osteogenic differentiation and pattern-controlled bone regeneration
Patterned silicon topographically-patterned surface [53]	Nanolithography	Mesenchymal stem cells	Specific size scale of topographic cue promotes osteogenic differentiation wilh or without osteogenic agents
Carbon nanotubes-reintorced HA coating on titanium implants [54]	Plasma-spray	Osteoblasts	CNT addition improves osseo integration

(continued)

Table 3 (continued)

Nanostructured materials	Fabrication techniques	Cell types	Cell behaviors
Nanofibrous CHI-nanocrystalline HA scaffolds with single-walled carbon nanotubes [45]	Freeze-drying and lyophilization	Osteoblasts	Improvement of cytocompatibility for osteoblasts adhesion and proliferation
Nanofibrous PCL with BMP-2 nanoreservoirs [55]	Electrospinning and layer by layer deposition	Osteoblasts	Improvement of osteogenic gene expression and mineralization
Nanofibrous gelatin [56]	Thermally induced phase separation and porogene-leaching	Osteoblasts	Improvement of migration, proliferation and mineralization
Collagen scaffold and heparin-binding peptide amphiphiles with nanofiber-heparan sulfate [57]	Peptide synthesis	In vivo implantation without cells	Large volumes of regenerated bone
Nano-fibrous Poly(l-lactic)acid scaffolds [58]	Freeze-drying and lyophilization	Osteoblasts	Improvement of osteoblast phenotype, mineralization and earlier differentiation
Supra. molecular self-assembled nanofibers of peptide amphiphiles [59]	Standard solid phase methods and self assembly	Mesenchymal stem cells	Improvement of viability and chondrogenic differentiation
Peptide hydrogel KLD12 ($[KLDL]_3$) and RAD16-1 ($[RADA]_4$)	Self assembly	Chondrocytes [60] Bone marrow stromal cells [61]	Promotion of chondrogenesis Promotion of chondrogenesis
Poly (lactic acid-co-glycolic acid) nanocapsules with bone morphogenetic protein BMP-2 and poly(3-Hydroxybutyrate-co-3-hydraxyvalerate) nanocapsules with BMP-7 embedded in CHI scaffold [62]	Co-electrospinning	Bone marrow stromal cells	Improvement of osteogenic differentiation (ALP activity)
Nanofibrous CHI, silk Fibroin (SF) and CHI/SF [63]	Electrospinning	Bone marrow mesenchymal stem cells	Promotion of proliferation and osteogenic differentiation

Nanostructured materials	Fabrication techniques	Cell types	Cell behaviors
Nanofibrous PCL with BMP-2 nanoreservoirs [64]	Electrospinning and layer by layer deposition	Osteoblasts	Promotion of mineralization and proliferation
Nanofibrous PCL trilaminarcomposite scaffolds [62]	Electrospinning	Chondrocytes	Support chondrogenesis and higher mechanical properties
Aligned nanofibrous PCI [65]	Electrospinning	Chondrocytes	Higher resistance to damage
Bilayered nanofibrous PCL [66]	Electrospinning	Chondrocytes correct topology	Promotion of chondrogenesis
Nanofibrous Polyurethane (PU) and PU-HA composite [67]	Electrospinning	Osteoblasts, embryonic mesenchymal progenitor cells	Higher mechanical properties and improvement of bone matrix formation
Oriented and aligned nanofibers of biodegradable poly-Dl-lactide with embedded multi-walled carbon nanotubes [68]	Electrospinning	Osteoblasts	Improvement of osteoblast functions
Interspersed PLA and gelatin fibers [69]	Co-etectrospinning	Chondrocytes	Improvement of proliferation and differentiation

Biomechanics of Stem Cells and Nano Surface

Nano Topography Supports Stem Cell Differentiation

The quality of stem cells to have extensive expansion and differentiation capability has led to their use as a reliable source for tissue engineering. Nano-fibrous scaffolds used for stem cell culture have been found to support basic processing of stem cells like proliferation and multi-differentiation [73]. Since there are no ethical or legal issues involved with dental stem cells coupled with their ease of retrieval and availability from deciduous teeth as well as permanent third molar, they are emerging as one of the most suited sources of stem cells for tissue engineering.

Bone has a distinct structure and chemical composition that makes it suitable to support external loads and the capability to absorb shocks associated with external loading. The bone consists of two distinct layers of different density and biological quality—the outer layer composed of compact bone and the inner layer of cortical bone. The two different cell types in the bone render it the potential to constantly undergo a dynamic remodeling. The osteoblasts continuously keep on building new bone whereas osteoclasts keep on digesting the old bone. MSCs have the capability to differentiate into various connective tissue cells in order to replenish them. Bone tissue engineering employs the use of osteoblasts, osteoprogenitor cells and stem cells. The stem cells when seeded with electrospun nano-fibrous scaffolds promote and allow osteogenic differentiation in vitro as well as in vivo [74, 75]. The addition of calcium carbonate resulted in increased mineralization in the cellular assemblies, a prelude to bone formation.

The osteo-conductivity of calcium carbonate and composite scaffold along with osteogenic promotion of nano-fibers is successfully being used in guided bone regeneration therapy to repair alveolar bone defects.

Interactions Between Implant Surface and Mesenchymal Stem Cells

Immediately following implantation procedure, the MSCs are attracted to the implant surface and play a specific role in the healing at the implant-tissue interface. The complete healing process encompasses certain stages of tissue reaction and transformation as under:

1. Tissue Migration,
2. Tissue Adhesion,
3. Tissue Proliferation, and
4. Tissue Differentiation

An instant favourable interaction between the implant surface and the adjacent osseous and gingival tissues defines the successful osseointegration of dental implant. The implant surface should have the potential to guide cells colonization

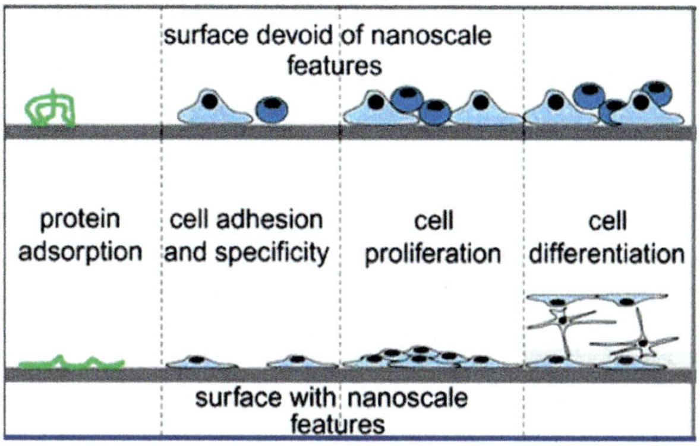

Fig. 4 Depiction of broad range of nano-scale topography effects observed in cellular protein adsorption in bulk materials. Both cell specificity and extent of cell adhesion are altered. Depending on the nano-architecture cell spreading may be increased or decreased [30]

and differentiation. Cell migration, adhesion, and proliferation on implant surfaces are a prerequisite to initiate the tissue regeneration (Fig. 4) [30].

Successful osteogenesis is directly proportional to factors like, growth of specific cells like progenitor cells and osteoblasts, development of a mineralized ECM, presence of soluble bioactive molecules like growth factors, ions, vitamins and hormones etc. and mechanical stimulus. These factors play a significant role right from initiation to the complete bone formation. All these factors contribute to differentiation of MSCs towards osteogenic lineage, a pre-requisite to oseointegration of dental implants.

Studies have revealed that some biological factors present in the tissues, released during the inflammatory phase have the capability to attract MSCs to the site of trauma [76, 77]. It is an established fact that the platelet factors initiate and stimulate the proliferation of Mesenchymal stem cells [78]. In addition, the plasma clot adjacent to the implant surface serves as a three-dimensional microporous structure, thus allowing the diffusion of regulatory factors [79, 80].

The migration and adhesion of the MSCs to the implant surface and the extracellular matrix follows a substantial proliferation so as to generate new tissues. Needless to say, nano-level topography of the implant surface is a strong determinant parameter to the biological response. The plasma clot at the site of injury acts as a storehouse of fibrin molecules that in turn release various bioactive factors including growth factors that initiate migration and subsequent differentiation of MSCs into specific lineages [81–83]. These factors stimulate MSCs to differentiate into osteoblastic cells adjacent to bone and into fibroblastic cells adjacent to the gingival tissues. The fibroblastic adhesion and subsequent differentiation in the gingival portion of the dental implant is highly desirable. Studies have shown this fibroblastic adhesion and proliferation to be considerably lower on the nano-scale surface as compared to the conventional surface [84]. The findings point out to the fact that nano-scale topography is neither required nor advisable at the neck of dental implants. The nano-surface of dental implants restricts/decreases fibrous tissue adhesion and proliferation at the implant body and favours adhesion and proliferation of osteoblastic cells, leading to success of the implant [85, 86]. In fact, it serves as a deterrent to fibrous tissue proliferation on the main implant body.

Interaction of Stem Cells to Nano Surface Dental Implants

The size, shape and surface texture offers three-dimensional surface topography to the dental implants and serves as one of the most significant parameters to influence cellular reactions at the implant-bone interface. Osteoblasts from the mesenchymal progenitor cells are the main architect of osseointegration of nano-structured implants with the adjacent bone leading to osteoid calcification and formation of hydroxyapatite. Osteoblasts further differentiate into osteocytes upon being encircled by mineralized bone. However, as on date the topography of the various implant surfaces has been categorized on the basis of their manufacturing procedure and not on the basis of the precise measurements of the topographic modifications [87, 88].

Cellular Attachment and Adhesion

The very first response of a biomaterial in the biological environment is the rapid adsorption of proteins to its surface [89]. The amount, type and composition of the adsorbed proteins guides and determines the subsequent process of cell adhesion, protein exchange [90–92], cell migration, proliferation and cell differentiation. The capability of a biomaterial to adsorb proteins is dependent upon its physiochemical qualities like surface energy or hydrophobicity and also on the surrounding environment like *p*H, concentration of ions, composition and functional groups of proteins, strength of solution, temperature etc. (*Vroman Effect*) [93].

The protein adsorption on a biomaterial surface takes place in two phases. In the first phase small rapidly diffusing proteins get attached to the implant surface. The second phase comprises of a progressive replacement of the smaller proteins by larger proteins having high affinity to the substrate. In the later phase, proteins containing sequences such as fibronectin or vitronectin having chemotactic or adhesive properties to bone cells act as cell receptors. These peptides also have strong impact on matrix maturation and bio-mineralization [94–96]. The conditioning of the naked biomaterial by protein adsorption promotes a rapid cell attachment on the protein coated surface of the implant material [97]. Studies have revealed that the contact angle (CA), a significant parameter for wettability of cpTi surfaces increases linearly with the average roughness having angles more than 45°, whereas angles less than 45° resulted into a decrease in linearly [98].

Nano-tubes and nano-particle surface of Ti created by anodization has been found to promote osteoblast adhesion up to 3 times as compared to non-anodized surface [99]. This enhancement in osteoblastic adhesion could possibly be explained due to increase in the surface area created by nano topography of the biomaterial. Enhanced protein and cell binding properties have also been found in porous HA materials due to a larger surface area and increased degree of roughness of the implant surface [100].

Helical Rosette Nano-tubes (HRN), capable of building self-assembly surface structures explains the significance of surface topography for protein binding and osteoblastic adhesion. Significant change of HRN coverage upon heating has been found to enhance protein binding and osteoblastic adhesion quality of Ti surface [101, 102]. The initially enhanced protein deposition and cell adhesion due to surface topography favourably modifies the surface properties of the implant surface leading to more advantageous interaction between the biomaterial and the tissues.

Cellular Migration and Proliferation

Structured nano surfaces exhibit a predictable osteoblastic orientation and migration that eventually leads to well defined cell colonization during directed tissue formation [103, 104]. Titanium Oxide with nano topography has been found to promote cellular spreading and to induce osteoblastic differentiation [105, 106]. The rate of cell migration also depends on the type of cells and the stage of its differentiation. A lower level of osteoblastic differentiation leads to a higher migration rate. The cells with low motility have been found to have the potential to form stronger focal contacts (FC). Cellular adhesion through FC may strengthen the linkage between the cell and extracellular matrix (ECM) and may also influence the rate of migration [107].

Cellular Differentiation, Gene Expression and Protein Synthesis

Different cell phenotypes exhibit different levels of sensitivities upon adhesion to nano-topography biomaterials [108, 109]. Osteoblasts have been found to be present in the surfaces as low as up to 10 nm dimensions, the size of a single collagen fibre [110]. The topography and the physio-chemical surface of the biomaterials significantly influence the qualitative and quantitative kinetics in gene and protein expression [111].

It has also been observed that micro-level Discontinuous Edge Surface (DES), having square boxes with a depth of 10 μm led to alteration of osteoblast adhesion and migration but enhanced cell multi-layering, matrix deposition and mineralization in comparison to the smooth surface control groups [112].

In vitro experiments have shown that nano-surface biomaterials were able to substantially enhance not only calcium and phosphorus deposition by osteoblasts but also encouraged calcium and phosphorus precipitation from culture media without osteoblasts in contrast to micro-surface Ti6Al4V and CoCrMo [113].

A marked difference in proliferation of MSCs on various sizes of TiO_2 nanotubes has been demonstrated. The MSCs appeared to be more round, stationary and devoid of any noticeable filopodia extensions and cellular propagation. Whereas, on nano-tube surfaces of larger diameters a large number of prominent filopodia and

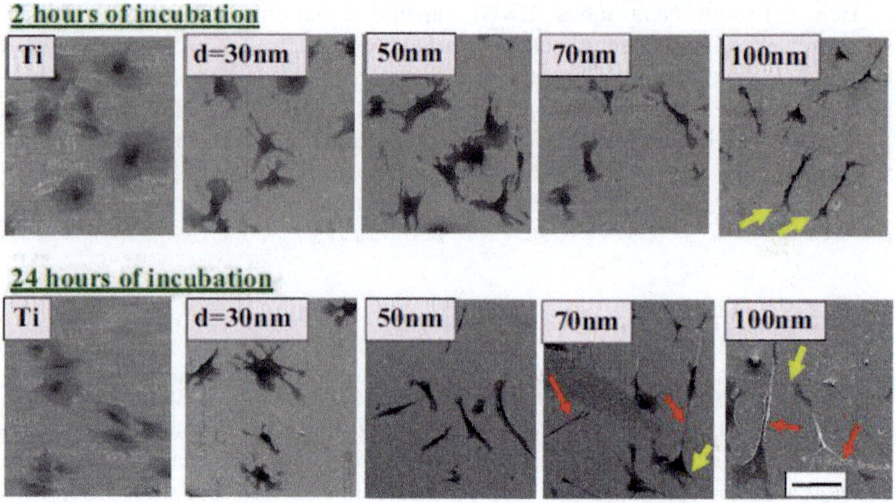

Fig. 5 SEM micrographs of human mesenchymal stem cells (hMSCs) on flat Ti and 30-, 50, 70-, and 100 nm diameter TiO_2 nanotube surfaces after 2 h of culture. (Scale bar, 100 m.) Cells are flat, spread out, and round-shaped on the flat Ti substrate, they are somewhat flat and rounded on 30-nm nanotubes, and they become progressively elongated as the nanotube diameter is increased to 50-nm diameter and beyond. Extraordinary cell elongation is induced on nanotubes with diameters of 70 and 100 nm (see *red arrows*), especially after the 24 h culture. More mobile morphologies are indicated by the presence of somewhat elongated leading edges of lamellipodia (*yellow arrows*) seen on 70 and 100 nm nanotubular surfaces. The cell shapes suggest that cells are more elongated on the bigger TiO_2 nanotubes [114]

unidirectional lamellipodia extensions were evident even after 2 h culture (Fig. 5, Upper). The proliferation became more visible after 24 h of culture and there was significantly pronounced when nano-tube diameter was increased (Fig. 5, Lower) [114].

Higher magnification SEM images of TiO_2 nano-tube surface after 2 h of culture revealed the adhesion of many round protein aggregates modifying the nano-surface with a protein coating for further adhesion and proliferation of MSCs. Comparatively much less protein aggregates with sparse distribution was observed on flat Ti samples (Fig. 6) [114].

Conclusion

Nano-surface modifications in biomaterials, tissue engineering and effective use of stem cells still requires further research and analysis in order to achieve the goal of successful osseointegration of dental implants. Further studies on stem cell behavior and interactions between ECM and various nano-surface scaffolds may pave the way for functionally advantageous tissue engineering and a more effective use of

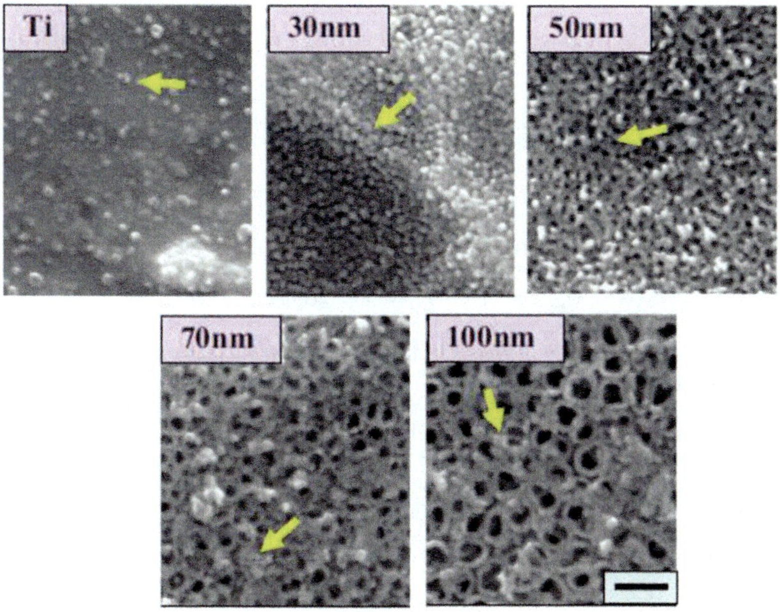

Fig. 6 SEM micrographs showing extracellular matrix aggregates on the surfaces of flat Ti and 30-, 50-, 70-, and 100-nm diameter TiO_2 nanotubes after 2 h of hMSC culture. (Scale bar, 200 nm.) Note that the presence of protein aggregates is infrequent on Ti, abundant on 300-nm nanotubes, and much less on the larger-diameter 70- and 100-nm nanotubes [114]

the bio-technology in dental implantology. Current information pointing to a favourable tissue growth of stem cells on nano-surface of various bio-materials may serve as a platform to discover a more conclusive physical, chemical and biological technology in the future for more assuring and lasting implants in the oral cavity. The outcome in the coming years may completely revolutionize the implant selection, implant procedure, their success and longevity of the oral implants.

Application of nano-topography in biomaterials and use of stem cells in bioengineering is passing though the phase of infancy and much needs to be explored and achieved in order to take full advantage of the technology.

A positive bone response on nano-surface biomaterials is already an established fact. Up to what extent the stem cell tissue engineering and nano-topography of biomaterials can be used to improve tissue and abutment interface bonding remain to be seen. Commercially available current implants using nano-topography may not have acquired the level of microstructure desired for the ultimate goal of perfection in the area. Further research and study in the field may open up new horizons of a more favorable use of stem cells and titanium with nano-surface in implant dentistry. The development may well make dental implants last for life in the oral cavity like natural teeth with practically no failures.

References

1. Silva GA (2004) Introduction to nanotechnology and its application to medicine. Surg Neurol 61:216–220
2. Shokuhfar T et al (2013) Intercalation of anti-inflammatory drug molecules within TiO_2 nanotubes. JRSC Adv 3(38):17380–17386
3. Christenson EM et al (2007) Nanobiomaterial applications in orthopedics. J Orthop Res 25:11–22
4. Shelton DC, Samuel AW, Gregory ML (2007) Nanotechnological applications in medicine. Curr Opin Biotechnol 18:26–30
5. Laval JM, Thomas D, Mazeran PE (1999) Nanobiotechnology and its role in the development of new analytical devices. Analyst 125:29–33
6. Jimbo R et al (2012) Nano hydroxyapatite-coated implants improve bone nanomechanical properties. J Dent Res 91:121172–121177
7. Xu T, Zhang N, Nichols HL, Shi D, Wen X (2007) Modification of nanostructured materials for biomedical applications. Mater Sci Eng C 27:579–594
8. Wang X, Liu LH, Ramstrom O, Yan M (2009) Engineering nanomaterial surfaces for biomedical applications. Exp Biol Med 234:1128–1139
9. Lawrence KJ (1980) Nature 283:106–107
10. Black J, Hastings GW (1998) Handbook of biomaterials properties. Chapman and Hall, London
11. Sumner DR, Turner TM, Igloria R, Urban RM, Galante JO (1998) J Biomech 31:909–917
12. Williams DF (2008) Biomaterials 29:2941–2953
13. Le Guéhennec L, Soueidan A, Layrolle P, Amouriq Y (2007) Surface treatments of titanium dental implants for rapid osseointegration. Dent Mater 23(7):844–854
14. Lavenus S, Ricquier J-C, Louarn G, Layrolle P (2010) Cell interaction with nanopatterned surface of implants. Nanomedicine 5(6):937–947
15. Hallab NJ, Anderson S, Stafford T, Glant T, Jacobs JJ (2005) J Orthop Res 23(2):384–391
16. Sargeant A, Goswami T (2006) Mater Des 27:287–307
17. Brånemark PI, Adell R, Albrektsson T, Lekholm U, Lundkvist S, Rockler B (1983) Osseointegrated titanium fixtures in the treatment of edentulousness. Biomaterials 4(1):25–28
18. Viceconti M, Muccini R, Bernakiewicz M, Baleani M, Cristofolini L (2000) J Biomech 33:1611–1618
19. Albrektsson T, Wennerberg A (2004) Oral implant surfaces: Part 1—Review focusing on topographic and chemical properties of different surfaces and in vivo responses to them. Int J Prosthodont 17:536–543
20. Geesink RGT (2002) Osteoconductive coatings for total joint arthroplasty. Clin Orthop Relat Res 395:53–65
21. Leeuwenburgh S, Layrolle P, Barrre F et al (2001) Osteoclastic resorption of biomimetic calcium phosphate coatings in vitro. J Biomed Mater Res 56(2):208–215
22. Kubo K, Tsukimura N, Iwasa F et al (2009) Cellular behavior on TiO_2 nanonodular structures in a micro-to-nanoscale hierarchy model. Biomaterials 30(29):5319–5329
23. LeGeros RZ, Craig RG (1993) Strategies to affect bone remodeling: osteointegration. J Bone Miner Res 8(2):S583–S596
24. Webster TJ, Ejiofor JU (2004) Increased osteoblast adhesion on nanophasemetals: Ti, Ti_6Al_4V, and CoCrMo. IJ Biomaterials 25:4731–4739
25. Gutwein LG, Webster TJ (2004) Increased viable osteoblast density in the presence of nanophase compared to conventional alumina and titania particles. IJ Biomaterials 25:4175–4183
26. Goené RJ et al (2007) Influence of a nanometer-scale surface enhancement on de novo bone formation on titanium implants: a histomorphometric study in human maxillae. Int J Periodontics Restorative Dent 27:211–219

27. Klabunde KJ et al (1996) Nanocrystals as stoichiometric reagents with unique surface chemistry. J Phys Chem 100:121
28. Wu SJ et al (1996) Sintering of nanophase g-Al_2O_3 powder. J Am Ceram Soc 79:2207
29. Baraton MI et al (1997) FTIR study of nanostructured alumina nitride powder surface: determination of the acidic/basic sites by CO, CO_2, and acetic acid adsorptions. J Nanomater 8:435
30. Balasundaram G et al (2006) Using hydroxyapatite nanoparticles and decreased crystallinity to promote osteoblast adhesion similar to functionalizing with RGD. IJ Biomaterials 27:2798–2805
31. Webster TJ et al (2001) Mechanisms of enhanced osteoblast adhesion on nanophase alumina involve vitronectin. J Tissue Eng 7:291–301
32. Webster TJ et al (2000) Specific proteins mediate enhanced osteoblast adhesion on nanophase ceramics. J Biomed Mater Res 51:475–483
33. Brunette DM (1988) The effects of implant surface topography on the behavior of cells. Int J Oral Maxillofac Implants 3:231–234
34. Cavalcanti Adam EA et al (2007) Cell spreading and focal adhesion dynamics are regulated by spacing of integrin ligands. Biophys J 92:2964–2974
35. Webster TJ et al (2000) Enhanced functions of osteoblasts on nanophase ceramics. IJ Biomaterials 21:1803–1810
36. Price RL et al (2003) Osteoblast function on nanophase alumina materials: influence of chemistry, phase, and topography. J Biomed Mater Res A 67:1284–1293
37. McManus AJ et al (2005) Evaluation of cytocompatibility and bending modulus of nanoceramic/polymer composites. J Biomed Mater Res A 72:98–106
38. Price RL et al (2004) Nanometer surface roughness increases select osteoblast adhesion on carbon nanofiber compacts. J Biomed Mater Res A 70:129–138
39. Webster TJ et al (1999) Osteoblast adhesion on nanophase ceramics. IJ Biomaterials 20:1221–1227
40. Meirelles L (2007) On nano size structures for enhanced early bone formation [Ph.D.]. J Gothenburg University, Gothenburg
41. Colon G et al (2006) Increased osteoblast and decreased Staphylococcus epidermidis functions on nanophase ZnO and TiO_2. J Biomed Mater Res A 78:595–604
42. Schwinté P, Keller L, Lemoine S, Gottenberg JE, Benkirane-Jessel N et al (2015) Nanoengineered scaffold for osteoarticular regenerative medicine. J Nanomed Nanotechnol 6:258
43. Frandsen CJ, Brammer KS, Noh K, Johnston G, Jin S (2014) Tantalum coating on TiO_2 nanotubes induces superior rate of matrix mineralization and osteofunctionality in human osteoblasts. Mater Sci Eng C 37:332–341
44. Brammer KS, Oh S, Frandsen CJ, Varghese S, Jin S (2010) Nanotube surface triggers increased chondrocyte extracellular matrix production. Mater Sci Eng C 30:518–525
45. Im O, Li J, Wang M, Zhang LG, Keidar M (2012) Biomimetic three-dimensional nanocrystalline hydroxyapatite and magnetically synthesized single-walled carbon nanotube chitosan nanocomposite for bone regeneration. Int J Nanomedicine 7:2087–2099
46. Depan D, Pesacreta TC, Misra RDK (2014) The synergistic effect of a hybrid graphene oxide-chitosan system and biomimetic mineralization on osteoblast functions. Biomater Sci 2:264–274
47. Nayak TR, Andersen H, Makam VS, Khaw C, Bae S et al (2011) Graphene for controlled and accelerated osteogenic differentiation of human mesenchymal stem cells. ACS Nano 5:4670–4678
48. Martin TA, Caliari SR, Williford PD, Harley BA, Bailey RC (2011) The generation of biomolecular patterns in highly porous collagen-GAG scaffolds using direct photolithography. Biomaterials 32:3949–3957
49. Damien E, Hing K, Saeed S, Revell PA (2003) A preliminary study on the enhancement of the osteointegration of a novel synthetic hydroxyapatite scaffold in vivo. J Biomed Mater Res A 66A:241–246

50. McCafferty MM, Burke GA, Meenan BJ (2014) Calcium phosphate thin films enhance the response of human mesenchymal stem cells to nanostructured titanium surfaces. J Tissue Eng 5
51. Nune KC, Misra RD (2014) Pre-adsorption of protein on electrochemically grooved nano-structured stainless steel implant and relationship to cellular activity. J Biomed Nanotechnol 10:1320–1335
52. Kim J, Bae WG, Choung HW, Lim KT, Seonwoo H et al (2014) Multiscale patterned trans-plantable stem cell patches for bone tissue regeneration. Biomaterials 35:9058–9067
53. Watari S, Hayashi K, Wood JA, Russell P, Nealey PF et al (2012) Modulation of osteogenic differentiation in hMSCs cells by submicron topographically-patterned ridges and grooves. Biomaterials 33:128–136
54. Facca S, Lahiri D, Fioretti F, Messadeq N, Mainard D et al (2011) In vivo osseointegration of nano-designed composite coatings on titanium implants. ACS Nano 5:4790–4799
55. Ferrand A, Eap S, Richert L, Lemoine S, Kalaskar D et al (2013) Osteogenetic properties of electrospun nanofibrous PCL scaffolds equipped with chitosan-based nanoreservoirs of growth factors. Macromol Biosci 14:45–55
56. Sachar A, Strom TA, Serrano MJ, Benson MD, Opperman LA et al (2012) Osteoblasts responses to three-dimensional nanofibrous gelatin scaffolds. J Biomed Mater Res A 100A:3029–3041
57. Lee SS, Huang BJ, Kaltz SR, Sur S, Newcomb CJ et al (2013) Bone regeneration with low dose BMP-2 amplified by biomimetic supramolecular nanofibers within collagen scaffolds. Biomaterials 34:452–459
58. Woo KM, Jun JH, Chen VJ, Seo J, Baek J et al (2007) Nano-fibrous scaffolding promotes osteoblast differentiation and biomineralization. Biomaterials 28:335–343
59. Shah RN, Shah NA, Del Rosario Lim MM, Hsieh C, Nuber G et al (2010) Supramolecular design of self-assembling nanofibers for cartilage regeneration. Proc Natl Acad Sci 107:3293–3298
60. Liu J, Song H, Zhang L, Xu H, Zhao X (2010) Self-assembly-peptide hydrogels as tissue-engineering scaffolds for three-dimensional culture of chondrocytes in vitro. Macromol Biosci 10:1164–1170
61. Kopesky PW, Vanderploeg EJ, Sandy JS, Kurz B, Grodzinsky AJ (2010) Self-assembling peptide hydrogels modulate in vitro chondrogenesis of bovine bone marrow stromal cells. Tissue Eng Part A 16:465–477
62. McCullen SD, Autefage H, Callahan A, Gentleman E (2012) Anisotropic fibrous scaffolds for articular cartilage regeneration. Tissue Eng Part A 18:2073–2083
63. Lai GJ, Shalumon KT, Chen SH, Chen J-P (2014) Composite chitosan/silk fibroin nanofibers for modulation of osteogenic differentiation and proliferation of human mesenchymal stem cells. Carbohydr Polym 111:288–297
64. Mendoza-Palomares C, Ferrand A, Facca S, Fioretti F, Ladam G et al (2011) Smart hybrid materials equipped by nanoreservoirs of therapeutics. ACS Nano 6:483–490
65. Accardi MA, Autefage H, McCullen SD, Callahan A, Chung S et al (2013) Effects of fiber orientation on the frictional properties and damage of regenerative articular cartilage surfaces. Tissue Eng Part A 19:2300–2310
66. Steele JA, McCullen SD, Callahan A, Autefage H, Accardi MA et al (2014) Combinatorial scaffold morphologies for zonal articular cartilage engineering. Acta Biomater 10:2065–2075
67. Tetteh G, Khan AS, Delaine-Smith RM, Reilly GC, Rehman IU (2014) Electrospun polyurethane/hydroxyapatite bioactive scaffolds for bone tissue engineering: the role of solvent and hydroxyapatite particles. J Mech Behav Biomed Mater 39:95–110
68. Shao S, Zhou S, Li L, Li J, Luo C et al (2011) Osteoblast function on electrically conductive electrospun PLA/MWCNTs nanofibers. Biomaterials 32:2821–2833
69. Torricelli P, Gioffrè M, Fiorani A, Panzavolta S, Gualandi C et al (2014) Coelectrospun gelatin-poly(l-lactic acid) scaffolds: modulation of mechanical properties and chondrocyte response as a function of composition. Mater Sci Eng C 36:130–138

70. Davies JE (2007) Bone bonding at natural and biomaterial surfaces. IJ Biomaterials 28:5058–5067
71. Tasker LH et al (2007) Applications of nanotechnology in orthopaedics. Clin Orthop Relat Res 456:243–249
72. Ellingsen JE et al (2004) Improved retention and bone-to-implant contact with fluoride-modified titanium implants. Int J Oral Maxillofac Implants 19:659–666
73. Li WJ, Tuli R, Huang X, Laquerriere P, Tuan RS (2005) Multilineage differentiation of human mesenchymal stem cells in a three-dimensional nanofibrous scaffold. Biomaterials 26:5158–5166
74. Yoshimoto H, Shin YM, Terai H, Vacanti JP (2003) A biodegradable nanofiber scaffold by electrospinning and its potential for bone tissue engineering. Biomaterials 24:2077–2082
75. Shin M, Yoshimoto H, Vacanti JP (2004) In vivo bone tissue engineering using mesenchymal stem cells on a novel electrospun nanofibrous scaffold. Tissue Eng 10:33–41
76. Agis H, Kandler B, Fischer MB, Watzek G, Gruber R (2009) Activated platelets increase fibrinolysis of mesenchymal progenitor cells. J Orthop Res 27(7):972–980
77. Vogel JP, Szalay K, Geiger F, Kramer M, Richter W, Kasten P (2006) Platelet-rich plasma improves expansion of human mesenchymal stem cells and retains differentiation capacity and in vivo bone formation in calcium phosphate ceramics. Platelets 17(7):462–469
78. Rock G, Neurath D, Lu M, Alharbi A, Freedman M (2006) The contribution of platelets in the production of cryoprecipitates for use in a fibrin glue. Vox Sang 91(3):252–255
79. Catelas I, Sese N, Wu BM, Dunn JCY, Helgerson S, Tawil B (2006) Human mesenchymal stem cell proliferation and osteogenic differentiation in fibrin gels in vitro. Tissue Eng 12(8):2385–2396
80. Schildhauer TA, Seybold D, Geßmann J, Muhr G, Köller M (2007) Fixation of porous calcium phosphate with expanded bone marrowcells using an autologous plasma clot. Mater Werkst 38(12):1012–1014
81. Catelas I, Dwyer JF, Helgerson S (2008) Controlled release of bioactive transforming growth factor beta-1 from fibrin gels in vitro. Tissue Eng C 14(2):119–128
82. Wong C, Inman E, Spaethe R, Helgerson S (2003) Fibrin-based biomaterials to deliver human growth factors. Thromb Haemost 89(3):573–582
83. Mosesson MW (2005) Fibrinogen and fibrin structure and functions. J Thromb Haemost 3(8):1894–1904
84. Eisenbarth E, Meyle J, Nachtigall W, Breme J (1996) Influence of the surface structure of titanium materials on the adhesion of fibroblasts. Biomaterials 17(14):1399–1403
85. Cohen A, Liu-Synder P, Storey D, Webster TJ (2007) Decreased fibroblast and increased osteoblast functions on ionic plasma deposited nanostructured Ti coatings. Nanoscale Res Lett 2(8):385–390
86. Miller D, Vance R, Thapa A, Webster T, Haberstroch K (2005) Comparison of fibroblast and vascular cell adhesion to nano structured poly(lactic co glycolic acid) films. Appl Bionics Biochem 2(1):1–7
87. Meyer U, Büchter A, Wiesmann HP, Joos U, Jones DB (2005) Basic reactions of osteoblasts on structured material surfaces. Eur Cell Mater 9:39–49
88. Macdonald W, Campbell P, Fisher J, Wennerberg A (2004) Variation in surface texture measurements. J Biomed Mater Res B 70(2):262–269
89. Andrade JD, Hlady V (1987) Plasma protein adsorption: the big twelve. Ann N Y Acad Sci 516:158–172
90. Hakansson M, Linse S (2002) Protein reconstitution and 3D domain swapping. Curr Protein Pept Sci 3(6):629–642
91. Iuliano DJ, Saavedra SS, Truskey GA (1993) Effect of the conformation and orientation of adsorbed fibronectin on endothelial cell spreading and the strength of adhesion. J Biomed Mater Res 27(8):1103–1113
92. Steele JG, Dalton BA, Johnson G, Underwood PA (1995) Adsorption of fibronectin and vitronectin onto Primaria(TM) and tissue culture polystyrene and relationship to the mechanism

of initial attachment of human vein endothelial cells and BHK-21 fibroblasts. Biomaterials 16(14):1057–1067
93. Vroman L (1987) The importance of surfaces in contact phase reactions. Semin Thromb Hemost 13(1):79–85
94. Meyer U, Meyer T, Jones DB (1998) Attachment kinetics, proliferation rates and vinculin assembly of bovine osteoblasts cultured on different pre-coated artificial substrates. J Mater Sci Mater Med 9(6):301–307
95. Moursi AM, Globus RK, Damsky CH (1997) Interactions between integrin receptors and fibronectin are required for calvarial osteoblast differentiation in vitro. J Cell Sci 110(18):2187–2196
96. Rezania A, Healy KE (2000) The effect of peptide surface density on mineralization of a matrix deposited by osteogenic cells. J Biomed Mater Res 52(4):595–600
97. Boyan BD, Hummert TW, Dean DD, Schwartz Z (1996) Role of material surfaces in regulating bone and cartilage cell response. Biomaterials 17(2):137–146
98. Lim YJ, Oshida Y (2001) Initial contact angle measurements on variously treated dental/medical titanium materials. Biomed Mater Eng 11(4):325–341
99. de Oliveira PT, Nanci A (2004) Nanotexturing of titanium-based surfaces upregulates expression of bone sialoprotein and osteopontin by cultured osteogenic cells. Biomaterials 25(3):403–413
100. Rouahi M, Gallet O, Champion E, Dentzer J, Hardouin P, Anselme K (2006) Influence of hydroxyapatite microstructure on human bone cell response. J Biomed Mater Res A 78(2):222–235
101. Chun AL, Moralez JG, Webster TJ, Fenniri H (2005) Helical rosette nanotubes: a biomimetic coating for orthopedics? Biomaterials 26(35):7304–7309
102. Fenniri H, Mathivanan P, Vidale KL et al (2001) Helical rosette nanotubes: design, self-assembly, and characterization. J Am Chem Soc 123(16):3854–3855
103. Gleiche M, Chi LF, Fuchs H (2000) Nanoscopic channel lattices with controlled anisotropic wetting. Nature 403(6766):173–175
104. Lenhert S, Meier M-B, Meyer U, Chi L, Wiesmann HP (2005) Osteoblast alignment, elongation and migration on grooved polystyrene surfaces patterned by Langmuir-Blodgett lithography. Biomaterials 26(5):563–570
105. Mustafa K, Wennerberg A, Wroblewski J, Hultenby K, Lopez BS, Arvidson K (2001) Determining optimal surface roughness of TiO_2 blasted titanium implant material for attachment, proliferation and differentiation of cells derived from human mandibular alveolar bone. Clin Oral Implants Res 12(5):515–525
106. Mustafa K, Wroblewski J, Hultenby K, Lopez BS, Arvidson K (2000) Effects of titanium surfaces blasted with TiO_2 particles on the initial attachment of cells derived from human mandibular bone: a scanning electron microscopic and histomorphometric analysis. Clin Oral Implants Res 11(2):116–128
107. Diener A, Nebe B, Lüthen F et al (2005) Control of focal adhesion dynamics by material surface characteristics. Biomaterials 26(4):383–392
108. Curtis ASG, Wilkinson C (1999) New depths in cell behaviour: reactions of cells to nanotopography. Biochem Soc Symp 65:15–26
109. Dalby MJ, Yarwood SJ, Riehle MO, Johnstone HJH, Affrossman S, Curtis ASG (2002) Increasing fibroblast response to materials using nanotopography: morphological and genetic measurements of cell response to 13-nmhigh polymer demixed islands. Exp Cell Res 276(1):1–9
110. Rice JM, Hunt JA, Gallagher JA, Hanarp P, Sutherland DS, Gold J (2003) Quantitative assessment of the response of primary derived human osteoblasts and macrophages to a range of nanotopography surfaces in a single culture model in vitro. Biomaterials 24(26): 4799–4818

111. Rouahi M, Champion E, Hardouin P, Anselme K (2006) Quantitative kinetic analysis of gene expression during human osteoblastic adhesion on orthopaedic materials. Biomaterials 27(14):2829–2844
112. Hamilton DW, Wong KS, Brunette DM (2006) Microfabricated discontinuous-edge surface topographies influence osteoblast adhesion, migration, cytoskeletal organization, and proliferation and enhance matrix and mineral deposition in vitro. Calcif Tissue Int 78(5):314–325
113. Ward BC, Webster TJ (2006) The effect of nanotopography on calcium and phosphorus deposition on metallic materials in vitro. Biomaterials 27(16):3064–3074
114. Oh S, Brammer KS, Julie Li YS, Teng D, Engler AJ, Chien S, Jin S (2009) Stem cell fate dictated solely by altered nanotube dimension. PNAS 106(7):2130–2135

Dental Pulp Stem Cells and Hydrogel in Pulp Regeneration

Waruna Lakmal Dissanayaka and Chengfei Zhang

Dental Pulp Stem Cells

The major functional cell population residing in the dental pulp is called odontoblasts, a subpopulation of mesenchymal cells which differentiated due to epithelial-dental papilla interactions during tooth morphogenesis. These cells deposit distinct mineralized matrix, dentin, during development of the tooth. Although, once formed, dental hard tissue do not undergo remodelling, it was shown that even after tooth eruption reactionary dentin formation continues in response to dentin destruction due to caries, erosion or mild chronic trauma [1]. Several studies have suggested that a precursor population that can proliferate and differentiate into odontoblasts reside somewhere in the adult dental pulp [2]. However, until recently only a little was known about the characteristics of such precursor cells.

The isolation and characterization of an odontogenic progenitor population named dental pulp stem cells (DPSCs) from adult dental pulp tissue was first reported by Gronthos et al. in 2000 [3]. In this study, a clonogenic and highly proliferative cell population was isolated and showed the capability of regenerating a dentin/pulp-like structure after xenogenic transplantation.

DPSCs can easily be isolated from discarded permanent teeth (3rd molars, supernumerary teeth, orthodontically unwanted teeth) with no serious ethical issues. This technique is minimally invasive and safe compared to that of other mesenchymal stem cell sources; and can be used to obtain large pools of autologous cells. For these reasons, DPSCs is considered as a promising population of stem cells in regenerative medicine. DPSCs share many similar characteristics with bone marrow mesenchymal stem cells (BMMSCs) including a similar gene expression profile [4]. Interestingly, DPSCs show a higher proliferation rate compared to that of BMMSCs [3].

W.L. Dissanayaka (✉) • C. Zhang
Comprehensive Dental Care, Faculty of Dentistry, The University of Hong Kong, Hong Kong, Hong Kong
e-mail: warunad@hku.hk

B. Zavan, E. Bressan (eds.), *Dental Stem Cells: Regenerative Potential*, Stem Cell Biology and Regenerative Medicine, DOI 10.1007/978-3-319-33299-4_7

There are mainly two methods to isolate DPSCs from the dental pulp; namely, enzymatic digestion method and the explant outgrowth method. It was shown that both methods give heterogenous populations of mesenchymal stem cells composed of more than one stem cell population [5]. Studies which aimed to characterize the differentiation capacity of DPSCs found that single colony derived DPSC strains vary from each other with regards to their odontogenic differentiation capacity [6]. These results further suggested that dental pulp contain a hierarchy of stem cells including a minor population of highly-proliferative, self-renewing, multipotent stem cells, among more committed population of progenitors.

Therefore, to select more defined clonal subset of stem/progenitor cells, several studies have used different surface markers for immunoselection by flow cytometry or magnetic-activated cell sorting. The markers used are STRO-1 [7–10], c-kit/CD117 [7, 11], CD34 [12], low-affinity nerve growth factor receptor (LANGFR/p75/CD271) [13], CD105/endoglin [14, 15], stage specific embryonic antigen 4 (SSEA-4) [16], and chemokine receptor type 4 (CXCR4) [17]. Alternatively, isolation and enrichment of a side population fraction based on the efflux of DNA binding fluorescent dye Hoechst 33342 has been used to isolate more potent stem cells [18–21]. However, one major challenge in utilizing DPSCs in clinical setting is to generate clinical grade human DPSCs. Immunoselection is not considered as a safe procedure to isolate cells for clinical application. Therefore, heterogenous DPSC populations still play a major role as the likely cell population which can be used clinically. Nevertheless, standardization of the isolation procedure is important to optimize the reproducibility and safety of clinical grade DPSC population in order to be used in regenerative endodontics.

Since DPSCs were isolated and reported, many studies have emerged to describe its multipotentiality and possible use under different conditions. These studies showed that DPSCs not only can differentiate into odontogenic lineage but also to osteogenic [11, 12], chondrogenic [22], adipogenic [6, 18, 22], endothelial [23], myogenic [24–26] and neurogenic lineages [27]. Many other studies subsequently showed that DPSCs share similar characteristics with mesenchymal stem cell populations of different origins, such as bone marrow and adipose tissue.

Over the past few years, there has been a significant revolution in the understanding of the functionality of mesenchymal stem cells (MSCs) as a source of cell replacement. More recently MSCs have been utilized for mediating tissue regeneration by release of "paracrine" factors and this in turn may forward investigations into more productive directions. Emerging data suggest that paracrine factors secreted by MSCs exert their effects to induce tissue recovery on many different ways, including activation of endogenous progenitor/stem cells, inhibition of apoptosis of susceptible cells, remodelling of extracellular matrix, and induction of neo-vasculogenesis.

Scaffolds

Engineering of three dimensional tissues requires cells to seed into an artificial structure capable of supporting their survival, proliferation and differentiation. These structures, generally known as scaffolds, should be able to recapitulate the in vivo milieu of the differentiating cells and should allow cells to manipulate their own microenvironments. Scaffolds often serve as carriers, to which the cells and/or growth factors can attach and migrate. They also act as vessels that enable diffusion of nutrients and metabolic products. Further, scaffolds provide mechanical and biological support to modulate the performance of the cells.

A scaffold of interest must meet certain requirements in order to accomplish the goal of tissue regeneration. Biodegradability is one of the crucial factors since scaffolds should be degraded allowing the regenerating tissue to unite with the in situ environment. The degradation rate has to be well-matched with the rate of tissue development, while the scaffold providing structural integrity until the cells/tissue fabricate their own natural matrix which can take over the mechanical load. Biocompatibility is also of paramount importance to avoid any unwanted reactions. To facilitate the cell seeding, attachment and diffusion of cells and nutrients, optimum porosity and an adequate pore size of the scaffold components are necessary. In addition to provision of structural support, ECM is also responsible for mediating cell-to-cell crosstalk via integration of cell signalling. This modulates the cellular behaviour, phenotype and function. Being able to injectable into an in vivo tissue is an added advantage, specifically when it comes to clinical use.

Scaffolds in Dental Pulp Tissue Engineering

Among the various types of scaffold materials that have been examined in pulp regeneration studies both in vitro and in vivo, natural biomaterials such as reconstituted collagen [14] and synthetic biomaterials such as poly(lactic) acid (PLA) [28] and poly(glycotic) acid (PGA) [29] are the most commonly investigated so far. However, neither of these materials tested are proven to have all the properties, and the structure of an ideal material, which should closely resemble the natural extracellular matrix (ECM) of the pulp [30]. Recently, hydrogels have been introduced to the field of pulp regeneration due to their favourable biological and mechanical properties.

Hydrogels

Hydrogels are known as water insoluble, crosslinked, three-dimensional networks of polymer chains plus water that fills the voids between polymer chains [31]. Hydrogels consist mostly of water with a much greater mass fraction than that of the polymer. Crosslinking enables insolubility in water and therefore, provides necessary mechanical strength and physical integrity. Hydrogels are mainly classified based on the method of cross-linking and the origin of the gel precursor molecules. Naturally occurring materials such as collagen, fibrin and matrigel are commonly used for their ability to provide crucial signals that are present within the natural cellular niche. In contrast, synthetic gels are not capable of providing specific cues for cellular functions, but they are readily optimized to allow a permissive niche for essential cell functions to occur. Polyethylene glycol (PEG), polyhydroxyethyl methacrylate (PHEMA), polyvinyl alcohol (PVA), and polyacrylic acid (PAA) are few macromolecules that are used in hydrogel synthesis.

DPSCs and Hydrogels in Dental Pulp Regeneration

Dental pulp is a loose, highly vascularized, soft connective tissue occupying the pulp chamber and root canals of the tooth. Histology of the dental pulp reveals that although it is composed of relatively less number of specific cell types, structural organization of these elements is quite distinct. Furthermore, due to the unique structure of the tooth with a pulp space encased within hard dentin, it gets a single blood supply only from a small apical opening located at the apex of the root canals. Additionally, varying size and contour of the root canals play a critical role and should get into serious consideration in the context of pulp regeneration.

Development and application of injectable hydrogel systems for pulp tissue engineering have been considered a promising approach in recent years, as they can be easily incorporated with growth factors and cells by simple mixing and can be adapted to the varying contour of the pulp chamber and root canals, following injection [30]. As the cross-linking of hydrogel precursors occur in situ after being injected into the target site, it can be injected to any irregular-shaped defect eliminating the necessity for custom-made scaffold designs.

Table 1 gives a summary of different types of hydrogels that have been utilized to encapsulate DPSCs for the purpose of investigating their potential in differentiating towards odontogenic lineage in vitro and engineering dental pulp tissue in vivo. Peptide-amphiphile, self-assembling peptide nanofibers/PuraMatrix, PEGylated fibrin/fibrinogen, collagen, and glycol-chitin are the predominant types of hydrogels that are used to encapsulate DPSCs in pulp regeneration studies. In these approaches, all the aspects of classical tissue engineering triad i.e. stem/progenitor cells, scaffold and growth factors, have been manipulated and tuned to augment the conditions for pulp regeneration.

In cell based regenerative approach, in order for pulp regeneration to occur, the stem/progenitor cells must proliferate and produce new matrix. In addition, there must be stimulation of the odontoblasts to proliferate and produce new dentin. Over the past few years there have been a significant advancement in dentin-pulp regeneration technology by stem cell based approach. Several studies, using small (mouse/rat) animal models, have shown that DPSC encapsulated hydrogel constructs can result in pulp/dentin tissue regeneration partially or completely in the root canals with enlarged apical openings of 0.7–3.0 mm.

Irrespective of the type of target tissue/organ, following implantation, an engineered three-dimensional tissue construct requires to develop rapid vasculature in order to meet the oxygen demand. However, immediately after implantation in vivo,

Table 1 Hydrogel types predominantly used as scaffolds in dental pulp regeneration

Type of hydrogel	Promising features	Experimental approach	Result	References
Peptide-Amphiphile (PA)	• Self-assemble into 3D networks • Easy to handle • Can be introduced into small defects	• Cell adhesion sequence RGD and an enzyme-cleavable site had been incorporated to tailor the cell–matrix interactions • DPSCs had been cultured in PA hydrogels for 4 weeks using osteogenic supplements	• DPSCs differentiated to osteoblast-like phenotype • Expressed osteoblast marker genes, and deposited mineral	Galler et al. [38]
PEGylated fibrin	• Injectable • PEGylation decelerates rapid degradation of fibrin	• Constructs of DPSC encapsulated PEGylated fibrin were transplanted in immunocompromised mice for 5 weeks and examined for new tissue formation	• Histologic analysis revealed vascularized soft connective tissue similar to dental pulp with degradation of fibrin and production of a collagenous matrix and mineral deposition	Galler et al. [39]
Self-assembling peptide nanofibers	• Customised for stem cell delivery • Growth factors can be incorporated • Ease of synthesis • Injectable • Biocompatible • Biodegradable	• DPSCs had been encapsulated together with fibroblast growth factor basic, transforming growth factor β1, and vascular endothelial growth factor via heparin binding • The hydrogel within dentin cylinders had been transplanted into immunocompromised mice	• A vascularized soft connective tissue similar to dental pulp was formed	Galler et al. [40] Galler et al. [30]

(continued)

Table 1 (continued)

Type of hydrogel	Promising features	Experimental approach	Result	References
Alginate	• Nontoxic • Biocompatible • Permeable to small molecular-weight proteins	• Exogenous TGFβ1 loaded alginate hydrogel constructs on human dental pulp repair in vitro	• Dentin matrix secretion • Odontoblast-like cell differentiation • Secretion of tubular dentin matrix	Dobie et al. [35]
		• DPSCs and HUVECs encapsulated within RGD-bearing alginate framework in 1:1 ratio was supplemented with VEGF and FGF • The contour of the construct was made to replicate the shape of gutta-percha points	• Both DPSCs and HUVECs showed high cell viability within alginate • The combination of vascular endothelial growth factor and fibroblast growth factor synergize to significantly up-regulate cell proliferation	Bhoj et al. [36]
Glycol Chitin-based Thermoresponsive (GC-TR) Hydrogel	• Biocompatible • Injectable • Conform more easily to the varying contours • Easy to manipulate as it maintains a sol phase unless the temperature reaches the body temperature	• Effects of GC-TRH on DPSC viability, expression of odontogenic/osteogenic differentiation markers were analysed	• Dentin sialophosphoprotein and dentin matrix protein-1 were expressed by cells cultured in GC-TRS at a higher level than that in cells cultured in collagen	Park et al. [41]
Collagen type I	• Biocompatible	• Radiolabelled rat pulp cells were added to polymerizing type I collagen hydrogel and was implanted in the emptied pulp chamber space in the upper first rat molar • Labelled cells were then tracked for a period of 3 weeks	• Histological analysis showed mitotically active fibroblastic cells as well as neo-angiogenesis and nervous fibres in pulp equivalents seeded with entire cells	Souron et al. [42]

PEG-fibrinogen (PF)	• Injectable • Mechanical properties can be tuned by the degree of cross-linking • The precursor solution can be injected into the dental pulp chamber before rapid gelation by photo-polymerization	• DPSCs were cultured within PF hydrogels with varying degree of cross-linking • Cell morphology, viability, odontogenic gene expression and mineralization were examined	• DPSCs cultured within the highest cross-linked hydrogel remained mostly rounded in aggregates and demonstrated the greatest enhancement in odontogenic gene expression and mineralization	Lu et al. [43]
Peptide Hydrogel (PuraMatrix)	• Biocompatible • Injectable	• DPSCs suspended in PuraMatrix was injected into the human tooth slices to verify DPSC differentiation, as measured by expression of DSPP and DMP-1	• After 21 days in tooth slices containing Puramatrix, DPSC cells expressed DMP-1 and DSPP and exhibited cytoplasmic elongations	Cavalcanti et al. [44]
		• DPSCs and HUVECs co-encapsulated in PuraMatrix was injected into the roots of human premolars and transplanted subcutaneously in SCID mice	• Formation of pulp-like tissue with higher vasculature and mineralizing capacity in prevascularized group compared with that of DPSC-alone group	Dissanayaka et al. [37]

tissue constructs depend solely on the oxygen supply diffuse from the nearest capillary, which could be only up to 200 μm away [32, 33]. If the distance form a capillary exceeds 200 μm, the majority of the cells undergo apoptosis [34]. Therefore, the survival of implanted tissue constructs of greater size requires the formation of a capillary network of its own which can deliver necessary nutrients for the cells. Although host blood vessels start to invade the implanted tissue construct, partly in response to the angiogenic factors secreted by the cells undergoing hypoxia, this process happens very slowly, growing only few tenths of micrometers per day [32]. Use of angiogenic growth factor incorporated hydrogel scaffolds [30, 35, 36] and co-culture of DPSCs with endothelial cells within a hydrogel scaffold [37] are two main approaches that have been investigated to promote rapid vascularization following implantation of a tissue/cellular construct.

In a similar study, we used the peptide hydrogel PuraMatrix as a scaffold system to investigate the role of DPSCs in prompting angiogenesis and the potential for regenerating vascularized pulp in vivo [37]. Human umbilical vein endothelial cells (HUVECs), DPSCs, or co-cultures of both cell types were encapsulated in three-dimensional PuraMatrix. The peptide nanofiber microenvironment supported cell survival, cell migration, and capillary network formation in the absence of exogenous growth factors (Fig. 1). Further, we demonstrated that DPSCs enhanced early vascular network formation by facilitating the migration of HUVECs and by increasing vascular endothelial growth factor (VEGF) expression. Both the DPSC-monoculture and co-culture transplanted groups exhibited vascularized pulp-like tissue with patches of osteodentin after transplantation in immunocompromised mice (Fig. 2). Interestingly, the co-cultured groups showed pulp-like tissues with more extracellular matrix, vascularization, and mineralization than that of the DPSC-monocultures in vivo. Findings of this study highlighted the crucial role of DPSCs in initial angiogenesis. Furthermore, this study convincingly showed the PuraMatrix hydrogel as a promising scaffold with a microenvironment to support cell–cell interactions and cell migration, which contribute to successful dental pulp regeneration.

Summary

DPSCs are considered as a promising population of stem/progenitor cells in regenerative medicine for their ready availability, ease of harvesting and stemness properties. Furthermore, DPSCs is the predominant type of stem cells that holds the potential to be used in dental pulp regeneration. Encapsulation of DPSCs within an injectable hydrogel scaffold provides a favourable approach for pulp regeneration as it can be injected to any irregular-shaped defect eliminating the necessity for custom-made scaffold designs. However, regeneration of a pulp-dentin complex that resembles the natural structural organization of cellular elements is yet to be achieved. With regards to this aspect, it is important to develop multiphase hydrogel systems that integrate different cell signalling and differentiation pathways.

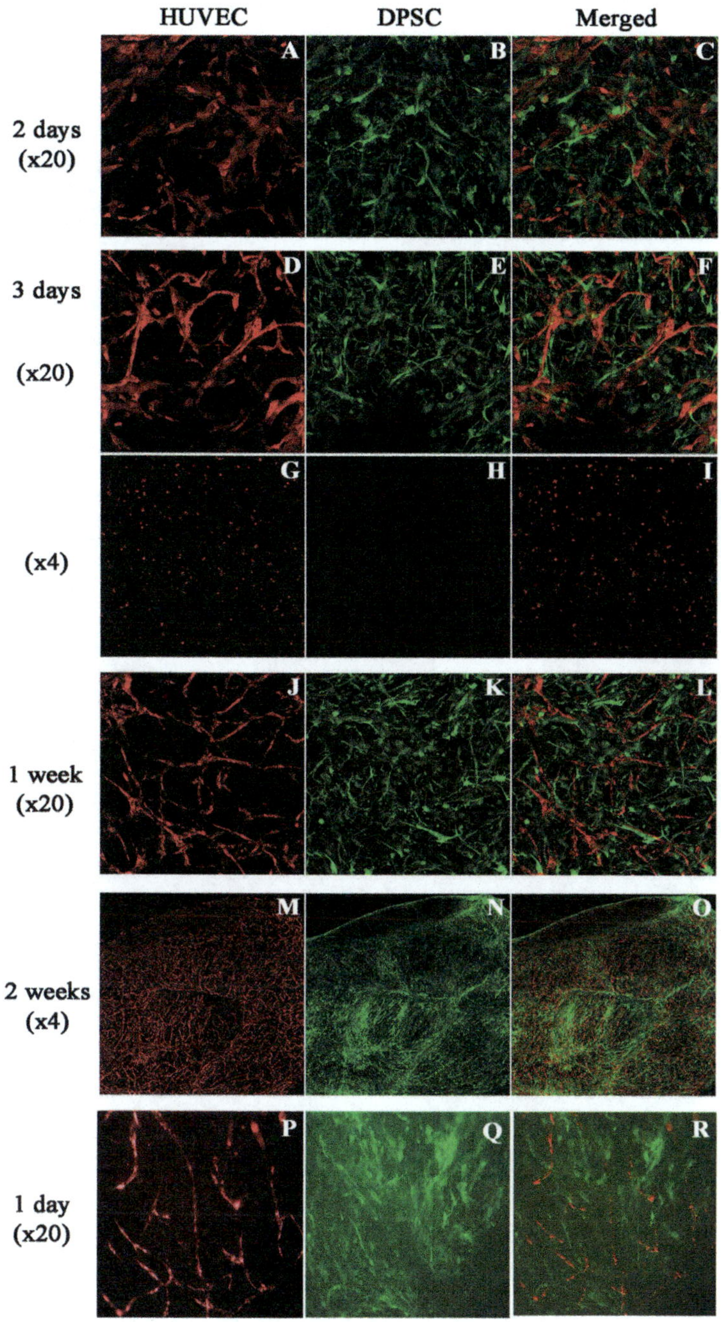

Fig. 1 Vessel-like structure formation by HUVECs that were cocultured as PuraMatrix constructs (Green: green fluorescent protein-expressing cells, Red: red fluorescent protein-expressing cells). (**a–c**) Coordinated cell migration was observed in the cocultures 48 h after seeding. (**d–f**) Vessel-like structure formation by HUVECs was observed in the cocultures 3 days after seeding, with DPSCs surrounding the nascent HUVEC networks. (**g–i**) HUVEC monocultures did not exhibit cell migration or vessel structure formation. (**j–o**) Vessel-like networks continued to be remodelled and to stabilize for 2 weeks. (**p–r**) Vascular structure formation was observed after 24 h when cocultured PuraMatrix was injected into root canals of root segments [37]

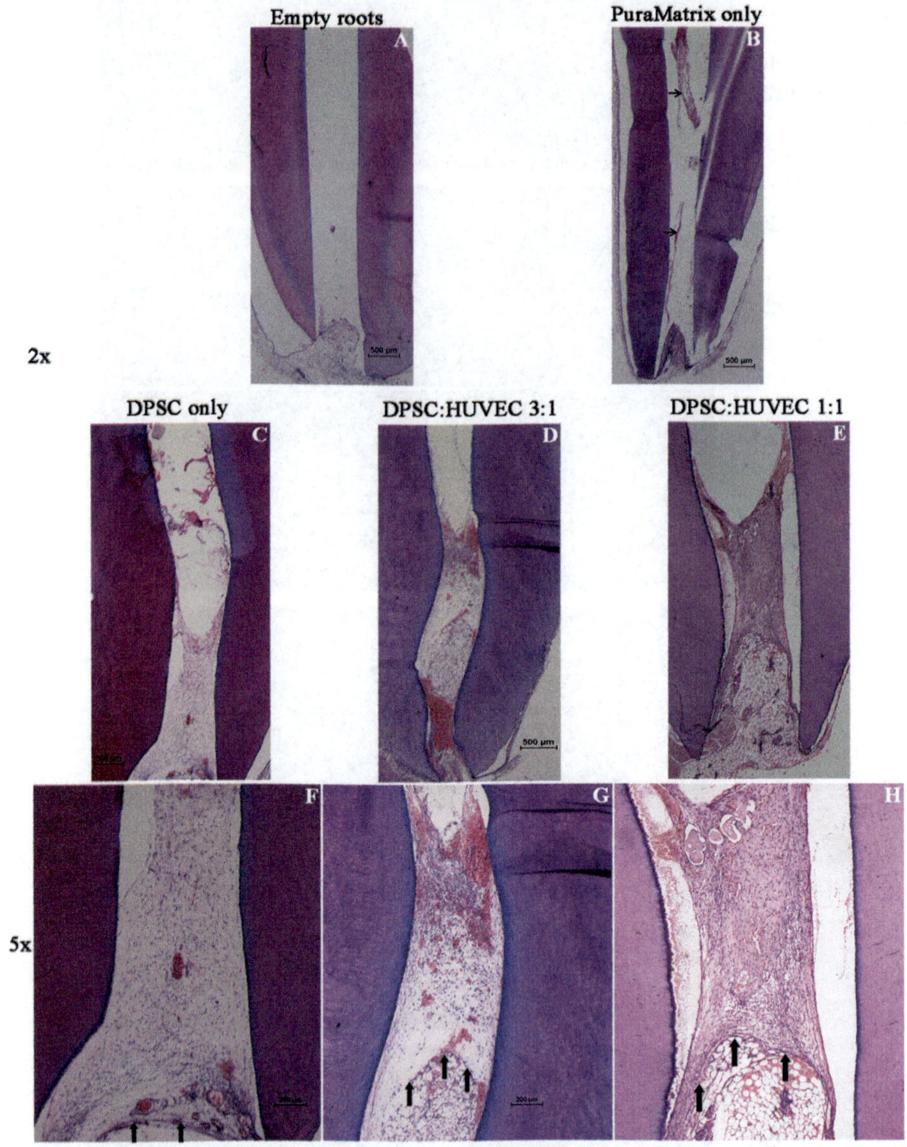

Fig. 2 Pulp regeneration in cell/PuraMatrix constructs in vivo 4 weeks after transplantation. (**a**) Empty roots, (**b**) PuraMatrix-alone in root fragments (*black arrows* indicate the remnants of PuraMatrix) (**c**, **f**) DPSC-alone in PuraMatrix, (**d**, **g**) DPSC: HUVEC 3:1 in PuraMatrix, (**e**, **h**) DPSC: HUVEC 1:1 in PuraMatrix. *Vertical black arrows* indicate the border between the transplanted tissue and host tissue [37]

References

1. Kitamura C, Kimura K, Nakayama T, Terashita M (1999) Temporal and spatial expression of c-jun and jun-B proto-oncogenes in pulp cells involved with reparative dentinogenesis after cavity preparation of rat molars. J Dent Res 78(2):673–680
2. Ruch JV (1998) Odontoblast commitment and differentiation. Biochem Cell Biol 76(6):923–938
3. Gronthos S, Mankani M, Brahim J, Robey PG, Shi S (2000) Postnatal human dental pulp stem cells (DPSCs) in vitro and in vivo. Proc Natl Acad Sci U S A 97(25):13625–13630
4. Shi S, Robey PG, Gronthos S (2001) Comparison of human dental pulp and bone marrow stromal stem cells by cDNA microarray analysis. Bone 29(6):532–539
5. Huang GT, Sonoyama W, Chen J, Park SH (2006) In vitro characterization of human dental pulp cells: various isolation methods and culturing environments. Cell Tissue Res 324(2): 225–236
6. Gronthos S, Brahim J, Li W, Fisher LW, Cherman N, Boyde A et al (2002) Stem cell properties of human dental pulp stem cells. J Dent Res 81(8):531–535
7. Laino G, Carinci F, Graziano A, d'Aquino R, Lanza V, De Rosa A et al (2006) In vitro bone production using stem cells derived from human dental pulp. J Craniofac Surg 17(3): 511–515
8. Shi S, Gronthos S (2003) Perivascular niche of postnatal mesenchymal stem cells in human bone marrow and dental pulp. J Bone Miner Res 18(4):696–704
9. Yang X, van der Kraan PM, van den Dolder J, Walboomers XF, Bian Z, Fan M et al (2007) STRO-1 selected rat dental pulp stem cells transfected with adenoviral-mediated human bone morphogenetic protein 2 gene show enhanced odontogenic differentiation. Tissue Eng 13(11):2803–2812
10. Yu J, Wang Y, Deng Z, Tang L, Li Y, Shi J et al (2007) Odontogenic capability: bone marrow stromal stem cells versus dental pulp stem cells. Biol Cell 99(8):465–474
11. Laino G, d'Aquino R, Graziano A, Lanza V, Carinci F, Naro F et al (2005) A new population of human adult dental pulp stem cells: a useful source of living autologous fibrous bone tissue (LAB). J Bone Miner Res 20(8):1394–1402
12. Graziano A, d'Aquino R, Laino G, Proto A, Giuliano MT, Pirozzi G et al (2008) Human CD34+ stem cells produce bone nodules in vivo. Cell Prolif 41(1):1–11
13. Waddington RJ, Youde SJ, Lee CP, Sloan AJ (2009) Isolation of distinct progenitor stem cell populations from dental pulp. Cells Tissues Organs 189(1–4):268–274
14. Iohara K, Imabayashi K, Ishizaka R, Watanabe A, Nabekura J, Ito M et al (2011) Complete pulp regeneration after pulpectomy by transplantation of CD105+ stem cells with stromal cell-derived factor-1. Tissue Eng Part A 17(15–16):1911–1920
15. Nakashima M, Iohara K, Sugiyama M (2009) Human dental pulp stem cells with highly angiogenic and neurogenic potential for possible use in pulp regeneration. Cytokine Growth Factor Rev 20(5–6):435–440
16. Kawanabe N, Murata S, Fukushima H, Ishihara Y, Yanagita T, Yanagita E et al (2012) Stage-specific embryonic antigen-4 identifies human dental pulp stem cells. Exp Cell Res 318(5): 453–463
17. Jiang L, Peng WW, Li LF, Yang Y, Zhu YQ (2012) Isolation and identification of CXCR4-positive cells from human dental pulp cells. J Endod 38(6):791–795
18. Iohara K, Zheng L, Ito M, Tomokiyo A, Matsushita K, Nakashima M (2006) Side population cells isolated from porcine dental pulp tissue with self-renewal and multipotency for dentinogenesis, chondrogenesis, adipogenesis, and neurogenesis. Stem Cells 24(11):2493–2503
19. Ishizaka R, Iohara K, Murakami M, Fukuta O, Nakashima M (2012) Regeneration of dental pulp following pulpectomy by fractionated stem/progenitor cells from bone marrow and adipose tissue. Biomaterials 33(7):2109–2118
20. Ishizaka R, Hayashi Y, Iohara K, Sugiyama M, Murakami M, Yamamoto T et al (2013) Stimulation of angiogenesis, neurogenesis and regeneration by side population cells from dental pulp. Biomaterials 34(8):1888–1897

21. Sugiyama M, Iohara K, Wakita H, Hattori H, Ueda M, Matsushita K et al (2011) Dental pulp-derived CD31(-)/CD146(-) side population stem/progenitor cells enhance recovery of focal cerebral ischemia in rats. Tissue Eng Part A 17(9–10):1303–1311
22. Chen YK, Huang AH, Chan AW, Shieh TY, Lin LM (2011) Human dental pulp stem cells derived from different cryopreservation methods of human dental pulp tissues of diseased teeth. J Oral Pathol Med 40(10):793–800
23. d'Aquino R, Graziano A, Sampaolesi M, Laino G, Pirozzi G, De Rosa A et al (2007) Human postnatal dental pulp cells co-differentiate into osteoblasts and endotheliocytes: a pivotal synergy leading to adult bone tissue formation. Cell Death Differ 14(6):1162–1171
24. Kerkis I, Kerkis A, Dozortsev D, Stukart-Parsons GC, Gomes Massironi SM, Pereira LV et al (2006) Isolation and characterization of a population of immature dental pulp stem cells expressing OCT-4 and other embryonic stem cell markers. Cells Tissues Organs 184(3–4): 105–116
25. Kerkis I, Ambrosio CE, Kerkis A, Martins DS, Zucconi E, Fonseca SA et al (2008) Early transplantation of human immature dental pulp stem cells from baby teeth to golden retriever muscular dystrophy (GRMD) dogs: local or systemic? J Transl Med 6:35
26. Nakatsuka R, Nozaki T, Uemura Y, Matsuoka Y, Sasaki Y, Shinohara M et al (2010) 5-Aza-2′-deoxycytidine treatment induces skeletal myogenic differentiation of mouse dental pulp stem cells. Arch Oral Biol 55(5):350–357
27. Arthur A, Rychkov G, Shi S, Koblar SA, Gronthos S (2008) Adult human dental pulp stem cells differentiate toward functionally active neurons under appropriate environmental cues. Stem Cells 26(7):1787–1795
28. Huang GT, Yamaza T, Shea LD, Djouad F, Kuhn NZ, Tuan RS et al (2010) Stem/progenitor cell-mediated de novo regeneration of dental pulp with newly deposited continuous layer of dentin in an in vivo model. Tissue Eng Part A 16(2):605–615
29. Mooney DJ, Powell C, Piana J, Rutherford B (1996) Engineering dental pulp-like tissue in vitro. Biotechnol Prog 12(6):865–868
30. Galler KM, Hartgerink JD, Cavender AC, Schmalz G, D'Souza RN (2012) A customized self-assembling peptide hydrogel for dental pulp tissue engineering. Tissue Eng Part A 18(1–2): 176–184
31. Salinas CN, Anseth KS (2009) Mesenchymal stem cells for craniofacial tissue regeneration: designing hydrogel delivery vehicles. J Dent Res 88(8):681–692
32. Rouwkema J, Rivron NC, van Blitterswijk CA (2008) Vascularization in tissue engineering. Trends Biotechnol 26(8):434–441
33. Carmeliet P, Jain RK (2000) Angiogenesis in cancer and other diseases. Nature 407(6801): 249–257
34. Muschler GF, Nakamoto C, Griffith LG (2004) Engineering principles of clinical cell-based tissue engineering. J Bone Joint Surg Am 86-A(7):1541–1558
35. Dobie K, Smith G, Sloan AJ, Smith AJ (2002) Effects of alginate hydrogels and TGF-beta 1 on human dental pulp repair in vitro. Connect Tissue Res 43(2–3):387–390
36. Bhoj M, Zhang C, Green DW (2015) A first step in de novo synthesis of a living pulp tissue replacement using dental pulp mscs and tissue growth factors, encapsulated within a bioinspired alginate hydrogel. J Endod 41(7):1100–1107
37. Dissanayaka WL, Hargreaves KM, Jin L, Samaranayake LP, Zhang C (2015) The interplay of dental pulp stem cells and endothelial cells in an injectable peptide hydrogel on angiogenesis and pulp regeneration in vivo. Tissue Eng Part A 21(3–4):550–563
38. Galler KM, Cavender A, Yuwono V, Dong H, Shi S, Schmalz G et al (2008) Self-assembling peptide amphiphile nanofibers as a scaffold for dental stem cells. Tissue Eng Part A 14(12): 2051–2058
39. Galler KM, Cavender AC, Koeklue U, Suggs LJ, Schmalz G, D'Souza RN (2011) Bioengineering of dental stem cells in a PEGylated fibrin gel. Regen Med 6(2):191–200
40. Galler KM, D'Souza RN, Federlin M, Cavender AC, Hartgerink JD, Hecker S et al (2011) Dentin conditioning codetermines cell fate in regenerative endodontics. J Endod 37(11): 1536–1541

41. Park SJ, Li Z, Hwang IN, Huh KM, Min KS (2013) Glycol chitin-based thermoresponsive hydrogel scaffold supplemented with enamel matrix derivative promotes odontogenic differentiation of human dental pulp cells. J Endod 39(8):1001–1007
42. Souron JB, Petiet A, Decup F, Tran XV, Lesieur J, Poliard A et al (2014) Pulp cell tracking by radionuclide imaging for dental tissue engineering. Tissue Eng Part C Methods 20(3): 188–197
43. Lu Q, Pandya M, Rufaihah AJ, Rosa V, Tong HJ, Seliktar D et al (2015) Modulation of dental pulp stem cell odontogenesis in a tunable PEG-Fibrinogen hydrogel system. Stem Cells Int 2015:525367
44. Cavalcanti BN, Zeitlin BD, Nor JE (2013) A hydrogel scaffold that maintains viability and supports differentiation of dental pulp stem cells. Dent Mater 29(1):97–102

Dental Stem Cells for Pulp Regeneration

Nileshkumar Dubey, Kyung-san Min, and Vinicius Rosa

Introduction

Dental pulp is a highly vascularized connective tissue encapsulated in mineralized structure formed by enamel, dentin and cementum. It is responsible to keep the homeostasis of the tooth organ and possesses highly responsive sensory nervous system that detects unhealthy stimuli [1]. The apical foramen is the opening at the apex of the root that provides the dental pulp with blood supply for oxygen, nutrients and cells for defence. This limited accessibility and inflexible surroundings makes it difficult for the immune system to eradicate infections in the tissue owing to the lack of a collateral blood supply [2].

Clinically, odontoblasts may survive and continue to produce dentin to protect the pulp in cases of injuries by mild caries, moderate attrition or erosion [3]. On the other hand, dental pulp exposure due to fracture, caries in deep dentin or where the progression of the lesion is aggressive, the survival of primary odontoblasts is frequently compromised demanding root canal treatment. This therapy often involves total pulp extirpation and disinfection followed by compaction of a material within the root canal [4]. Though this is a well-established approach, it is based on the substitution of host tissues for synthetic, and in most of the cases, inert materials [5]. In immature teeth, the routine root canal treatment can allow for infection control, but often does not facilitate the completion of root formation or protects against external root resorption [6]. In these teeth, the root canal treatment is even more

N. Dubey • V. Rosa (✉)
Discipline of Oral Sciences, Faculty of Dentistry, National University of Singapore, Singapore, Singapore
e-mail: denvr@nus.edu.sg

K.-s. Min
Department of Conservative Dentistry, School of Dentistry, Chonbuk National University, Jeonju, South Korea

B. Zavan, E. Bressan (eds.), *Dental Stem Cells: Regenerative Potential*, Stem Cell Biology and Regenerative Medicine, DOI 10.1007/978-3-319-33299-4_8

challenging due to large apical opening which can have a divergent configuration and does not provide the mechanical stop necessary to confine the filling material within the root canal [6].

Although the success rate of endodontic treatments is relatively high [7], this modality of treatment present some shortcomings such as the high sensitivity to obstruction of the root canal space, breakage of instruments in the canals and perforations [8, 9]. The root canal instrumentation removes significant amount of tooth structure that results in thin dentin walls and large pulp chamber making the remaining structure weaker. Moreover, the devitalization may increase root brittleness increasing the chances of postoperative fracture [10, 11]. In addition, pulpless teeth do not have the ability to sense the aggressions arising from caries. In fact, endodontically treated teeth loss is higher than non-treated ones due to secondary caries [12].

The shift towards the regenerative endodontics may have started with the advent of vital pulp therapy. Clinically, it is based on the indirect pulp treatment in cases of deep dentinal cavities and direct pulp capping where the pulp tissue is exposed [13]. Indirect pulp capping is the procedure wherein the deepest layer of affected dentin is covered with a thin layer of protective agent in order to prevent pulp exposure [13]. In the direct pulp capping, the exposed vital pulp is covered with a protective dressing (such as calcium hydroxide and mineral trioxide aggregate) directly applied over the exposure site to prevent further injury and to stimulate dentin-pulp complex regeneration [14, 15].

One of the methods to treat the pulp of immature teeth is the apexification. This relies on the formation of an apical barrier to close the open apex [4]. For this, calcium hydroxide ($Ca(OH)_2$) must be inserted in the root canal and replaced periodically to stimulate the formation of a calcified structure. Although this strategy has been used clinically for long time, the outcomes are largely unpredictable and fracture rates of nearly 25 % have been reported after long-term $Ca(OH)_2$ treatment due to changes in mechanical properties of dentin [14]. Alternatively, apexification strategies include use of mineral trioxide aggregate (MTA) that is a material with osteogenic, cementogenic and odontogenic potentials that can be used for perforation repairs, pulp capping and pulpotomy [4]. Unfortunately, apexification by either $Ca(OH)_2$ or MTA may prevent further development of the root [14, 15]. This is specially complex in immature tooth where the apexification does not stimulate the reestablishment of a functional pulp tissue or the continued root development [16].

In existing endodontic treatment, teeth lose substantial amount of sound structure and their ability to detect the secondary infections. There has been no superior synthetic material to replace natural pulp and dentin in terms of chemical composition, physical, mechanical and biological properties. Thus, current research in endodontics has extended its focus on the development of biological therapies that are more biologically effective and mechanically reliable than traditional pulp therapies. Hence, the regeneration of pulp-dentin complex via tissue engineering has the potential to restore tooth functions compromised by the loss of pulp vitality.

Dental Pulp Tissue Engineering

Dental Stem Cells

Stem cells are undifferentiated cells the can differentiate into specialized cells. These cells are responsible for normal tissue renewal, healing and regeneration after injuries [5, 17]. Multipotent stem cells are able to cross lineage boundaries and differentiate into multiple, but limited number of cell types. For instance, a multipotent blood stem cell is a hematopoietic cell that can differentiate into several types of blood cell types (such as neutrophils and lymphocytes) but cannot differentiate into brain cells, bone cells or other non-blood cell types [17].

The oral environment harbours several types of stem cells [5, 17, 18]. The stem cells from oral sources can be considered a population of mesenchymal stem cells (MSCs) as they present positive expression of STRO-1, Nanog, Oct4, CD73, CD90, CD105, CD106, CD146 and others [19–25]. These cells are capable to differentiate into many types of specialized cells such as adipocytes, neurons, osteoblasts and myocytes. For dental pulp regeneration, the potentially suitable cell are those directly derived from pulp tissue or from the precursor of dental pulp [17, 18, 20, 26, 27].

The apical papilla refers to the soft tissue at the apices of developing permanent teeth that is source of undifferentiated cells in the process of root development [24]. This population of stem cells from the apical papilla (SCAP) can be only obtained in a limited window of time as after the formation of the crown the dental papilla becomes the dental pulp [24, 28]. The apical portion of the dental papilla is loosely attached to the apex of the developing root and it is separated from the differentiated pulp tissue by a cell rich zone [29]. As the apical papilla is poorly vascularized, the perivascular niche does not appear to be a major source of SCAP [24, 28].

It is possible that SCAP are derived from neural crest cells or at least associated with them [24]. They present positive expression of CD73, CD90, CD105, CD146 and other MSCs markers [24, 28, 30]. SCAP are capable to express positive staining for neural markers such as neurofilament M, neuronal nuclear antigen, nestin and βIII tubulin after being cultured with neurogenic medium for 4 weeks [24]. These cells can also accumulate calcium when cultured under osteogenic medium. Moreover, subpopulations of SCAP with positive expression of STRO-1–positive co-expressed several dentinogenic markers such as bone sialoprotein (BSP), osteocalcin (OCN) and matrix extracellular phosphoglycoprotein (MEPE) after stimulation with odontogenic conditioning medium [24].

Dental pulp stem cells (DPSC) were firstly derived by enzymatic digestion of dental pulp from adult human impacted third molars in 2000 [31]. These highly proliferative cells have their niche localized in the perivascular and perineural sheath areas and are negative for the odontoblast-specific marker dentin sialophosphoprotein, suggesting an undifferentiated phenotype [23, 31, 32]. DPSC have gene expression profiles similar to bone marrow stem cells (BMSC) for more than 4000 genes [33]. In fact, both BMSC and DPSC present similar expression for several markers such as fibroblast growth factor 2, alkaline phosphatase, type I collagen, osteocalcin, α-smooth muscle actin, vascular cell adhesion molecule 1 and others. Nonetheless, some markers such as bone matrix protein, bone sialoprotein is positive at low levels in BMSC and negative in DPSC [31]. DPSC also express neuronal and glial cell markers, which may be related to their neural crest-cell origin [34].

DPSC can differentiate into multiple cell lineages (e.g. adipocytes, chrondocytes, β cells of islet of pancreas) making these cells a versatile and interesting model for tissue engineering and regeneration [31, 35–38]. DPSC also have the potential to induce angiogenesis/vasculogenesis as a subfraction of a side population was identified to be positive for CD31, CD146, CD34 and vascular endothelial growth factor-2 (VEGF-2). The transplantation of these cells into mouse ischemia resulted in high density of capillary formation and increase in the blood flow into the graft [39]. DPSC express various neural markers making them an interesting model for neural differentiation [38]. The differentiation of DPSC into glial cells and neurons was also confirmed by the upregulation of neural gene markers (GFAP and βIII-tubulin) in the presence of FGF and epidermal growth factor [40]. DPSC presented neuron-like morphological changes and increased gene (Nestin and Tub3) and protein (MAP2) expression after being cultured for 5 days in serum- and growth factor-free medium [41]. After in vivo transplantation into the mesencephalon of chicken embryo, DPSC not only acquired a neuronal morphology but also express neuronal markers [42].

Although both DPSC and SCAP are located in anatomically adjacent sites, they are originated in distinct developmental stages and present differences regarding their growth potential and biochemical characteristics [24, 29, 31]. Morphologically, SCAP are smaller in size, fibroblast-like or stellate in shape, with numerous cytoplasmic processes [43]. They also present higher number of STRO-1–positive cells when compared with DPSC. Compared to DPSC, SCAP present higher expression of the anti-apoptotic protein surviving and human telomerase reverse transcriptase (hTERT) activity, a catalytic subunit of the enzyme telomerase that maintains the telomere length [30]. Considering the proliferation rate, SCAP cultures can present more than double the number of cells compared to DPSCs after an observation period of 4 days [43]. In spite of dentinogenic markers expressed by SCAP after ex vivo expansion, DPSC present higher expression levels of dentin sialoprotein (DSP), matrix extracellular phosphoglycoprotein (MEPE), fibroblast growth factor receptor 3 (FGFR3), transforming growth factor beta receptor II (TGF beta RII) and others [24]. Although both DPSC and SCAP present remarkable characteristics making them attractive models for dental pulp regeneration, the first present a wider window of opportunity to be isolated from fully developed teeth while SCAP can only be retrieved from teeth under development [5, 24, 31, 43].

Stem cells from human exfoliated deciduous teeth (SHED) were firstly isolated in 2003 from normal exfoliated human deciduous incisors and present positive expression of MSCs markers (STRO-1 and CD146), embryonic stem cells markers (Oct4, Nanog), stage-specific embryonic antigens (SSEA-3, SSEA-4) and tumour recognition antigens (TRA-1-60 and TRA-1-81) [22, 44]. The ability that these cells have to cross lineage boundaries expands the potential use of SHED for therapies involving a large number of tissues. SHED can differentiate into adipogenic, chondrogenic, neurons and endothelial cells [6, 22, 27]. SHED can also differentiate into osteoblasts and are positive for TGF-β, FGF and VEGF receptors [27, 45].

SHED present a remarkable advantage over DPSC as they can be retrieved from naturally exfoliated teeth, which are one of the only disposable post-natal human tissues [6, 22, 26, 46]. Although both DPSC and SHED cells are originated from the dental pulp, deciduous teeth are different from permanent teeth with regards to their developmental processes, tissue structure and function. Therefore, it is not a surprise to find out that SHED are distinct from DPSC with respect to their higher proliferation rate, increased cell-population doublings, sphere-like cell-cluster formation and osteoinductive capacity in vivo. SHED apparently represent a population of multipotent stem cells that are perhaps more immature than DPSC [18, 19, 22, 26, 27, 31, 33, 36]. For example, during osteogenic differentiation, SHED present higher levels of alkaline phosphatase activity and osteocalcin production than DPSC [36]. However, the ability to regenerate a dentin/pulp-like complex found in DPSC is also observed in SHED [19, 22, 27, 31].

Dental Stem Cells-Based Pulp Tissue Engineering

The use of synthetic restorative materials as substitutes for dental structures is a practice nearly as old as dentistry itself. To date, most of the procedures performed in dentistry are limited to the replacement of damaged tissues for biocompatible materials that may not present chemical, biological or physico-mechanical characteristics similar to the host tissues. These discrepancies, together with the hostile environment of the oral cavity, can result in frequent need for re-treatment [5]. Particularly in endodontics, despite of the development and introduction of new techniques, instruments and medicaments for the clinical management of the dental pulp, the fundamental principles of clinical practices are not drastically different than those of the time when first root canal instruments and gutta-percha were introduced in the 1800s [26]. Most of the endodontic treatments rely on the disinfection of the root canal followed by a hermetic three-dimensional obturation. Though this is a well-established approach it is based on the substitution of a living tissue for synthetic and, in most of the cases, inert materials.

In regards to biological-based therapies for dental pulp there cell-free and cell-based approaches. The first relies on the chemo attraction of host cells into the root canal [47, 48]. The second is based on the transplantation of responsive stem cells using a suitable scaffold into the root canal to promote tissue formation and regeneration [5, 49–51].

Cell-free approaches can make use of functionalized scaffolds or release of dentin-derived chemotactic factors to induce the migration of host cells from surrounding areas to regenerate the tissue [52]. One example of clinically viable cell-free therapy is the use of the blood clot as a scaffold to attract and maintain periapical (or circulating progenitor) cells inside the root canal [52–54]. The root canal revascularization via blood clotting depends on the establishment of bleeding into the canal system via over instrumentation of the apex [54]. This is an attractive alternative to be performed directly in clinical appointments and eliminate the needs of retrieving, culturing and transplanting cells as required in cell-based approaches [5, 26, 52, 54]. Interestingly, by promoting the bleeding inside the root canal the level of stem cells markers in the blood clot can be up to 600 times higher as compared with levels found in the systemic blood [55]. In fact, revascularization was reported to effectively induce complete root development of immature permanent teeth with apical periodontitis [53]. Other approaches involve the use of functionalized scaffolds and bioactive materials for chemo attraction of periapical cells [56, 57].

Though cell-free approaches are a clinical reality [53], host cells need to migrate long distances inside the root canal space through the apical foramen that narrows with time [52]. Moreover, the types and concentration of cells and composition of the fibrin clot are unpredictable [54]. Nonetheless, the revascularization approach can lead to dentinal wall thickening, apical closure and increased root length [14, 15].

Cell-based strategies aim to develop clinically approachable techniques that induce the regeneration of a functional dental pulp that is capable to deposit mineralized and organized matrix contributing for the development of the root structure. Dental pulp tissue engineering requires the fine orchestration of three elements: cells, scaffolds and morphogenesis/signalling pathways [5, 26, 52]. In this new paradigm, it is rather exciting that stem cells from oral sources play the focal role moving forward the field of regenerative endodontics.

Apart of all stem cells that can be obtained in the oral environment, three of them, namely SCAP, DPSC and SHED have great potential to be used for dental pulp tissue engineering. The first can undergo dentinogenic differentiation in vitro under proper stimulation. Moreover, when SCAP were mixed with hydroxyapatite/tricalcium phosphate (HA/TCP) scaffolds simulating a root canal and implanted in vivo, they presented the ability to regenerate vascularized dental pulp–like tissues in the hollow space and to produce new dentin-like tissue on the walls [24, 29, 30].

DPSC were shown to have great potential to induce pulp regeneration since they were firstly isolated and extensively characterized in the landmark paper from Gronthos et al. [31]. The DPSC isolated were expanded in vitro, mixed with HA/TCP ceramic powder and transplanted into the dorsum of immunocompromised mice. Remarkably, after 6 weeks from implantation, the surface of the ceramic particles was coated by an odontoblast-like layer and dentin-like structure (Fig. 1). The mineralized tissue formed was composed of a highly ordered collagenous matrix containing mainly collagen type I which is the major fibrous component of the dental pulp besides collagen type III [58]. Notably, there was also the formation of a pulp-like tissue permeated with blood vessels and the transplants were positive for human-specific markers for bone sialoprotein, osteocalcin and dentin sialophosphoprotein (DSPP) [31].

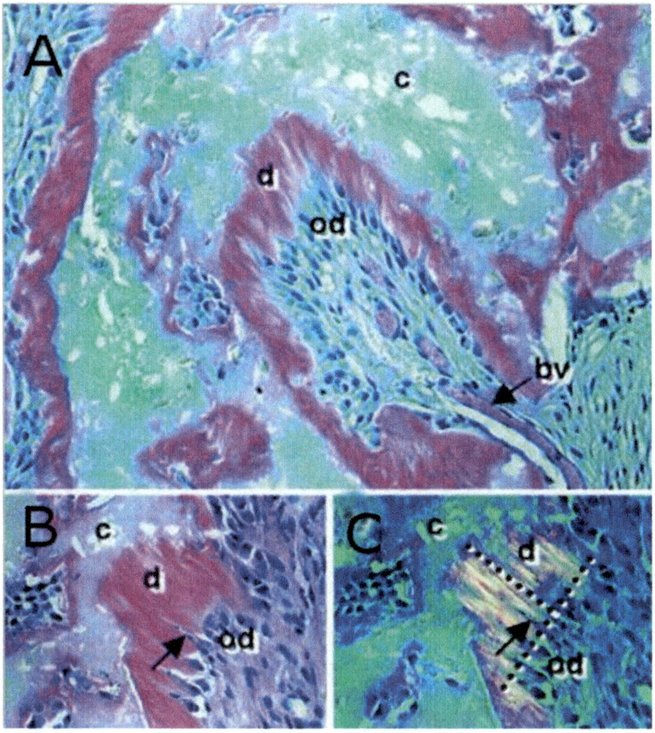

Fig. 1 Developmental potential of DPSC transplanted with HA/TCP carrier in vivo. (**a**) After 6 weeks, DPSC generated a dental pulp-like tissue within the HA/TCP carrier (c) with the presence of dentin-like matrix (d), odontoblast-like cells (od) and blood vessels (bv). (**b**) A magnified view of the dentin matrix (d) highlights the odontoblast-like layer (od) and odontoblast processes (*arrow*) (**c**). Polarized light demonstrates perpendicular alignment (*dashed lines*) of the collagen fibers to the forming surface. Adapted from Gronthos et al. [31] (Copyright (2000) National Academy of Sciences, U.S.A.)

The residual pulp of exfoliated deciduous teeth contains stem cells (SHED) that can regenerate a dental pulp-like tissue that can deposit mineralized tissue [6, 22, 27, 45]. These cells showed their potential to be used for dental pulp tissue regeneration in 2003 in a seminal paper published by Miura et al. There, SHED mixed with HA/TCP were capable to differentiate into odontoblasts after 8 weeks from implantation in immunocompromised mice. Nonetheless, the cells failed to regenerate a complete dentin–pulp-like complex in vivo as previously observed for DPSCs (Fig. 2) [22, 31]. Nonetheless, this ability was reported later [6, 27, 59].

Although SCAP is one of the candidates to be used in dental pulp tissue engineering, the majority of the research for such purpose focus on the use of stem cells obtained from the dental pulp itself (DPSC and SHED). Although SCAP are capable of forming odontoblast-like cells to produce dentin in vivo, they are likely to be the source of primary odontoblasts for the root dentin formation. This limits their

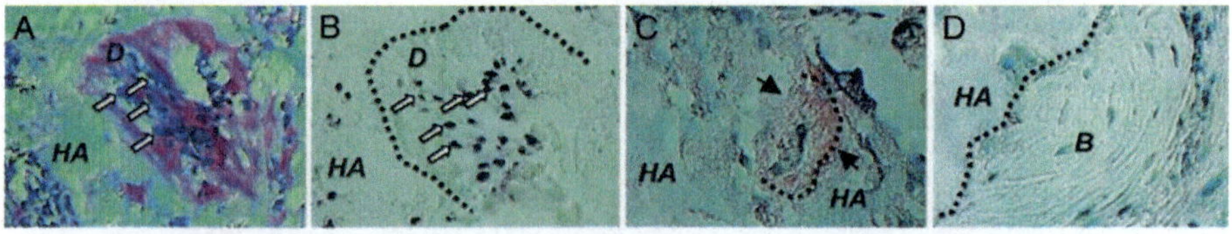

Fig. 2 (**a**) After 8 weeks of transplantation, SHED were able to differentiate into odontoblasts (*open arrows*) that were responsible for the dentin-like structure (D) formation on the surfaces of the HA/TCP scaffold (HA). (**b**) In situ hybridization showed that the tissue formed had human origin (*open arrows*). (**c**) The black dashed line represents interface between newly formed dentin. (**d**) bone (B) generated by host cells in the same SHED transplant shows no reactivity to the DSPP antibody. Adapted from Miura et al. [22] (Copyright (2003) National Academy of Sciences, U.S.A.)

isolation to the period when the root is not yet fully formed [24]. Thus, third molars are the most attractive source of SCAP as their root completion is only achieved by ages from 18 to 25. DPSC can be obtained from the dental pulp of permanent tooth. Here, pre-molars that are extracted due to orthodontic reasons are also an interesting source of DPSC. SHED can be retrieved from the only naturally disposable postnatal human tissue. This creates a window of approximately 11 years to obtain these stem cells mainly from deciduous central incisors and canines.

The success of cell-based regenerative endodontic strategies depends on techniques that will allow clinicians to create a functional pulp tissue within cleaned and shaped root canal systems [5, 17, 20, 52]. One of the first signs that this could be achieved was reported in 1996. The team led by David Mooney engineered pulp-like tissues in vitro utilizing fibroblast obtained from human adult dental pulp seeded in a synthetic extracellular matrix made of polyglycolic acid fiber-based meshes. Histological analysis performed after 60 days from implantation showed that the cells were capable to proliferate and form a new tissue with general appearance and cell density similar to native dental pulp [60]. In 2004, autogenous transplantation of pulp cells stimulated with bone morphogenetic protein 2 (BMP-2) into amputated pulp of dogs induced the formation of reparative dentin formation [61]. These and other studies shed a light that cell-based regenerative endodontics could be on the horizon of clinicians in the future. Since then, a series of proof-of principles research have been published showing the feasibility of regenerating functional human dental pulp using tissue engineering principles and dental stem cells.

To prove the potential of dental stem cells to induce pulp regeneration in vivo, several reports use the Tooth Slice/Scaffold of Dental Pulp Tissue Engineering model. This strategy allows for the generation of a dental-pulp-like tissue via the transplantation of human dental pulp stem cells seeded in a biodegradable scaffold cast within the pulp chamber of human tooth slices [62]. This model was used to show the ability of SHED to generate dental pulp-like tissue in vivo by the Nör's lab in 1996. There, SHED were seeded into poly-L-lactic acid (PLLA) scaffolds cast in 1 mm thick tooth slice and transplanted into the subcutaneous space of immunocompromised mice. After 4 weeks, a well-vascularized pulp-like tissue was formed in the pulp chamber space. Notably, there was a layer of odontoblast-like cells expressing dentin sialoprotein (DSP) lining the dentin surface. Although these cells were morphologically similar to odontoblasts, including the eccentric polarized position of the nucleus at the basal part of the cell body, no new dentin deposition was observed [46].

The capability of SHED to differentiate into fully functional odontoblasts capable of generating tubular dentin in vivo was observed later (Fig. 3). For this, SHED were seeded a poly-L-lactic acid (PLLA) scaffold cast in a tooth slice and implanted subcutaneously into the dorsum of mice, which received one injection of tetracycline hydrochloride every 5 days. After 32 days from implantation, the dentin disks presented well-defined fluorescent lines originated from the chelation of calcium ions in the newly deposited dentin. The dental pulp generated displayed positive protein expression for dentin sialophosphoprotein (DSPP) and dentin matrix protein 1 (DMP-1). Interestingly, the rate of dentin formation by SHED was far superior (14.1 μm/day) as compared to the one observed for the human dental pulp used as control (3.3 μm/day) [27]. Growth rates of 4–15 μm/day have been reported for primary dentin depending on the stage of development and age of the tooth [63].

DPSC were also used to regenerate pulp-like tissue using a similar approach. By seeding DPSC in a collagen-based scaffold along with a growth-factor (DMP-1) it was possible to engineer pulp-like tissue after 6 weeks from subcutaneous transplantation in mice. However, in this study, no odontoblast-like cells were found on the surface of pre-existing dentin surface and the quality of the pulp-like tissue obtained appeared to be less promising as the ones obtained using PLLA scaffold [64]. These drawbacks can be related to the contraction that collagen scaf-

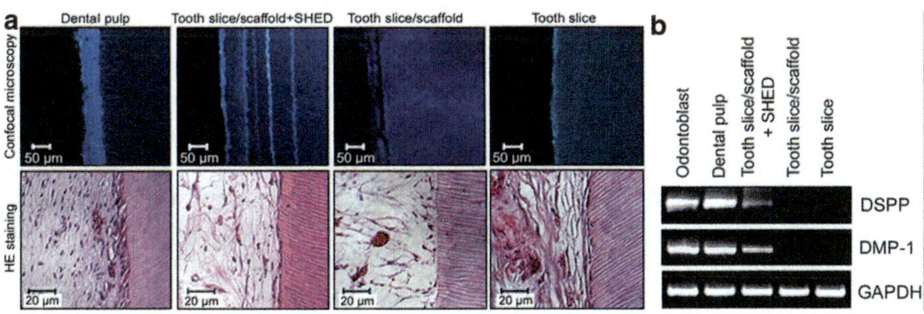

Fig. 3 SHED differentiate into functional odontoblasts that generate new tubular dentin in vivo. (**a**) The pictures show samples retrieved from mice after 32 days after implantation. (**b**) RT-PCR analysis showed the high expression of DSPP and DMP-1 on tissues engineered on tooth slice/scaffolds seeded with SHED. Reprinted from Journal of Dental Research, 89/8, VT Sakai, Z Zhang, Z Dong, KG Neiva, MA Machado, S Shi, CF Santos and JE Nör, SHED differentiate into functional odontoblasts and endothelium, 791–796, Copyright (2010), with permission from SAGE Publications

fold experiences leading to a size reduction proportional to the density of cells seeded in it [65].

Although these studies provide an exciting panorama for regenerative endodontics, the use of tooth substrates with few millimetres in thickness does not fully represent the challenges of regenerating a pulp tissue with in complete roots [6]. The larger dimension of the tissue that has to be engineered in addition to the fact that vascularization can be available only through the apex, pose important obstacles to be overcome. Hence, some studies aimed to prove that pulp regeneration can be achieved in three-dimensional environments.

An important step towards stem cells-based regenerative endodontics was obtained by a team of researchers who have seeded swine DPSC into the root segments with 5–6 mm in length and implanted into the jawbone of the adult minipigs. After 6 weeks from implantation, the proliferating cells were filling the space once left by the scaffolds made of collagen or poly(lactic-co-glycolic) acid (PLGA) undergoing degradation. At that time, new extracellular matrix had been occasionally deposited in a polar predentin-like pattern on the canal dentinal walls. After 10 weeks, a continuous layer of cells with columnar or spindle-shaped morphology along with the presence of newly formed organic matrix was observed [66]. An interesting fact about this study was that the swine DPSC used were cryopreserved for 1 year prior to implantation showing that the DPSC do not lose their potential to regenerate dentin/dental pulp-like tissue even after being cryopreserved.

Dental pulp tissue formation using DPSC was also achieved within tooth slices as long as 7 mm in length. Cells mixed with poly-D, L-lactide and glycolide (PLG) scaffolds were inserted into the canal with internal space ranging from 1 to 2.5 mm and one end sealed with MTA. Four months after the implantation into the subcutaneous of mice, the emptied canal was filled with regenerated pulp-like tissue. The entire pulp-like tissue was vascularized with a uniform cell density resembling the natural pulp. Though the odontoblast-like cells were not well organized and did not present characteristics of natural odontoblasts, a layer of mineralized tissue was deposited onto the dentinal walls and the MTA cement used to seal the canal [67]. However, as MTA is known to release calcium ions that induce the formation of calcified barriers [68] it was not possible to determine if DPSC alone could induce such extent of calcified tissue deposition. Nonetheless, that paper offered great perspectives on the capability of using dental stem cells to induce de novo synthesis of vascularized human pulp-like tissue capable to deposit dentin-like tissue onto the surface of root canals [67].

SHED are also capable to regenerate dental pulp within full-length root canals in vivo (Fig. 4). Cells were injected in empty roots using either a self-assembling hydrogel or recombinant human collagen type I. The roots were implanted in the subcutaneous of mice and after 28 days a human pulp-like tissue was observed throughout the extent of the root canals. More excitingly, the engineered pulp was capable of generating new dentin at a rate of approximately 10 μm/day. The high activity of the functional odontoblasts was supported by a high concentration of blood vessels close to the dentinal walls (Fig. 4b) [6]. One fact that deserves to be highlighted in this study is that cells were delivered inside the root canals using injectable scaffolds. As injection is one the most practiced procedures by dentists, the learning curve can be shortened for clinicians to apply this technology in patients.

Future Challenges

Stem cells-based therapies offer new perspectives targeting dental pulp regeneration and further development of root structure [17]. Besides that, the two branches of the tissue engineering triad (the development of suitable scaffolds to deliver cells and fine tuning of signalling pathways) still offer several challanges to be solved in order to improve the outcomes.

Scaffolds are frameworks that provide the support for cells to proliferate, differentiate and generate the desired tissue [5, 6]. Ideally, a scaffold must allow cell attachment and migration, permit the localized and sustained delivery of growth factors and enable the influx of oxygen to maintain the high metabolic demands of cells and tissues in formation [5].

The scaffolds physical and mechanical characteristics must be compatible with the surrounding tissues to support specific demands in vivo [69–71]. It is widely known that stiffness of the substrate play an important role in differentiation of stem cells thus. In fact, the physical and mechanical properties of the scaffolds can have direct impact in cell differentiation through mechano transduction [5, 69, 72]. Interestingly, the increase of scaffold' stiffness favours the differentiation of MSCs into neurons, myoblasts, and osteoblasts in that order [72]. DPSC seeded on soft matrices present upregulation of collagen I but downregulation of markers such as DSPP, DMP-1 [73]. In addition, physical features of the scaffolds (*e.g.* quantity and extension of pores) change the specific surface modifying its permeability and mechanical properties, influencing cell differentiation and tissue formation [5, 25, 74]. Noticeably, higher number and extension of pores can contribute significantly for cellularity but compromise scaffold strength [74, 75]. Pore interconnectivity also plays a crucial role to sustain tissue growth [75, 76]. Also, the rate of scaffold degradation is important to achieve success in tissue engineering therapies. The scaffold should ideally reabsorb once it has served its purpose of providing a template for tissue regeneration. The degradation ought happen at a rate compatible with the new tissue formation [5, 27, 71, 73, 76]. The by-products generated cannot be toxic and must be easily cleared or resorbed to minimize the risk of inflammatory response [77].

The third pillar to be tuned for successful tissue engineering is cell signalling [5, 52]. This is a complex system of communication that governs cell activities individually and it can be dramatically changed upon cellular interactions or external stimulation. This could be observed when proteins present in dentin disks [78], dentin extract in EDTA [79] or a tooth-germ conditioned extract [80] were found

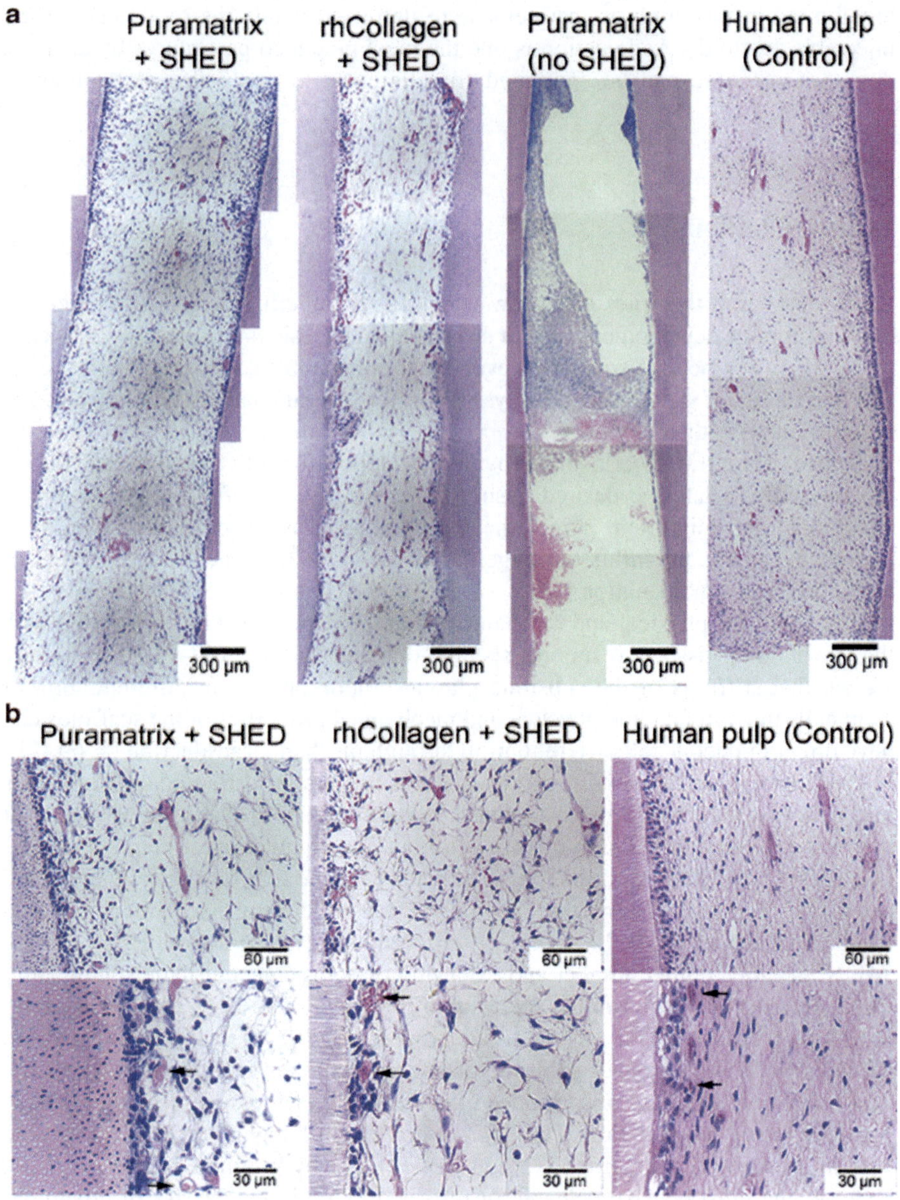

Fig. 4 Dental pulp tissue engineering with SHED injected into human root canals and transplanted into immunodeficient mice. (**a**) Low-magnification and (**b**) high-magnification images of tissues formed when SHED mixed with scaffolds (Puramatrix™, rhCollagen type I groups) were injected into full-length root canals of human premolars. A vascularized connective tissue occupied the full extension of the root canal. Cell densification and many blood vessels were observed along dentin walls. Scaffolds (Puramatrix™) injected into the root canals without cells were used as controls for SHED. Freshly extracted human premolars were used as tissue controls. *Black arrows* point to blood vessels close to the odontoblastic layer. Reprinted from Journal of Dental Research, 92/11, V Rosa, Z Zhang, RHM Grande and JE Nör, Dental Pulp Tissue Engineering in Full-length Human Root Canals, 970–975, Copyright (2013), with permission from SAGE Publications

to supplement the scaffolds as a mechanism of cellular induction. For example, there is evidence that the TGF-1 is released from the dentin after injuries [81] and dentin itself can induce the odontogenic differentiation by releasing embedded growth factors like transforming growth factor-β1 (TGF-β1) [82]. Moreover, BMP's are involved in the odontoblastic differentiation processes [59, 83] and both BMP-2 and 7 present inductive effects in reparative dentinogenesis [84, 85]. Nonetheless, SHED undergoing odontoblastic differentiation responds strongly to BMP-2 as compared to BMP-7 [59]. This complex system where molecules can trigger different responses creates a very challenging environment for researchers developing cell-based therapies.

Other challenge that needs to be addressed is the cost of therapies when available at large. Comparatively to routine root canal treatments, regenerative endodontics requires additional procedures and qualified manpower in order to isolate, expand, preserve and prepare cells for use. Undoubtedly, these add extra cost to the process that eventually will be directed to patients preventing popularization of the treatments. There are lots of research efforts to optimize process and minimize the number of procedures impacting the economic burden. History has shown that most of the revolutionary technologies became vertiginously popular as they also became more affordable. Hopefully, this also can be true for tissue engineering in endodontics (Fig. 5).

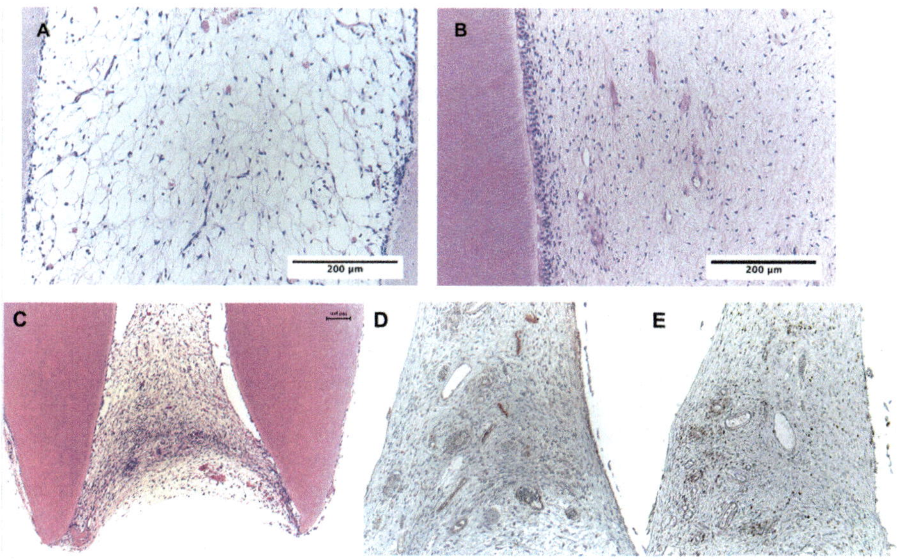

Fig. 5 Dental pulp tissue engineered inside root canal using SHED cells (**a**) and natural dental pulp from a young premolar (**b**). The engineered tissue occupies the apical portion (**c**). Immunohistochemistry with PCNA (proliferating cell nuclear activity) and Factor VIII show a proliferative tissue with established blood network (**d**, **e**). Reprinted from Dental Materials, 28/04, B Cavalcanti, A Della Bona and JE Nör, Tissue engineering: From research to dental clinics, 341–348, Copyright (2012), with permission from Elsevier

References

1. Nanci A (2007) Ten cate's oral histology-pageburst on vitalsource: development, structure, and function. Elsevier Health Sciences, St. Louis
2. Huang GTJ (2011) Dental pulp and dentin tissue engineering and regeneration: advancement and challenge. Frontiers in Bioscience (Elite Ed) 3:788–800
3. Zhang W, Yelick PC (2010) Vital pulp therapy—current progress of dental pulp regeneration and revascularization. Int J Dent 2010, 856087
4. Ingle JI (2008) Ingle's endodontics 6. PMPH-USA, Shelton, CT
5. Rosa V, Della Bona A, Cavalcanti BN, Nor JE (2012) Tissue engineering: from research to dental clinics. Dent Mater 28(4):341–348
6. Rosa V, Zhang Z, Grande RH, Nor JE (2013) Dental pulp tissue engineering in full-length human root canals. J Dent Res 92(11):970–975
7. Kojima K, Inamoto K, Nagamatsu K, Hara A, Nakata K, Morita I et al (2004) Success rate of endodontic treatment of teeth with vital and nonvital pulps. A meta-analysis. Oral Surg Oral Med Oral Pathol Oral Radiol Endod 97(1):95–99
8. McGuigan MB, Louca C, Duncan HF (2013) The impact of fractured endodontic instruments on treatment outcome. Br Dent J 214(6):285–289
9. Alhadainy HA (1994) Root perforations. A review of literature. Oral Surg Oral Med Oral Pathol 78(3):368–374
10. Dammaschke T, Steven D, Kaup M, Ott KHR (2003) Long-term survival of root-canal-treated teeth: a retrospective study over 10 years. J Endod 29(10):638–643
11. Reeh ES, Messer HH, Douglas WH (1989) Reduction in tooth stiffness as a result of endodontic and restorative procedures. J Endod 15(11):512–516
12. Caplan DJ, Cai J, Yin G, White BA (2005) Root canal filled versus non-root canal filled teeth: a retrospective comparison of survival times. J Public Health Dent 65(2):90–96
13. Fuks AB (2008) Vital pulp therapy with new materials for primary teeth: new directions and treatment perspectives. J Endod 34(7 Suppl):S18–S24
14. Jeeruphan T, Jantarat J, Yanpiset K, Suwannapan L, Khewsawai P, Hargreaves KM (2012) Mahidol study 1: comparison of radiographic and survival outcomes of immature teeth treated with either regenerative endodontic or apexification methods: a retrospective study. J Endod 38(10):1330–1336
15. Bose R, Nummikoski P, Hargreaves K (2009) A retrospective evaluation of radiographic outcomes in immature teeth with necrotic root canal systems treated with regenerative endodontic procedures. J Endod 35(10):1343–1349
16. Cvek M (1992) Prognosis of luxated non-vital maxillary incisors treated with calcium hydroxide and filled with gutta-percha. A retrospective clinical study. Dent Traumatol 8(2):45–55
17. Rosa V (2013) What and where are the stem cells for dentistry? Singapore Dent J 34(1): 13–18
18. Bojic S, Volarevic V, Ljujic B, Stojkovic M (2014) Dental stem cells: characteristics and potential. Histol Histopathol 29(6):699–706
19. Gronthos S, Brahim J, Li W, Fisher LW, Cherman N, Boyde A et al (2002) Stem cell properties of human dental pulp stem cells. J Dent Res 81(8):531–535
20. Huang GT (2011) Dental pulp and dentin tissue engineering and regeneration: advancement and challenge. Front Biosci (Elite Ed) 3:788–800
21. Kemoun P, Laurencin-Dalicieux S, Rue J, Farges JC, Gennero I, Conte-Auriol F et al (2007) Human dental follicle cells acquire cementoblast features under stimulation by BMP-2/-7 and enamel matrix derivatives (EMD) in vitro. Cell Tissue Res 329(2):283–294
22. Miura M, Gronthos S, Zhao M, Lu B, Fisher LW, Robey PG et al (2003) SHED: stem cells from human exfoliated deciduous teeth. Proc Natl Acad Sci U S A 100(10):5807–5812
23. Shi S, Gronthos S (2003) Perivascular niche of postnatal mesenchymal stem cells in human bone marrow and dental pulp. J Bone Miner Res 18(4):696–704

24. Sonoyama W, Liu Y, Yamaza T, Tuan RS, Wang S, Shi S et al (2008) Characterization of the apical papilla and its residing stem cells from human immature permanent teeth: a pilot study. J Endod 34(2):166–171
25. Xie H, Cao T, Viana Gomes JC, Castro Neto AH, Rosa V (2015) Two and three-dimensional graphene substrates to magnify osteogenic differentiation of periodontal ligament stem cells. Carbon 93:266–275
26. Rosa V, Botero TM, Nor JE (2011) Regenerative endodontics in light of the stem cell paradigm. Int Dent J 61(Suppl 1):23–28
27. Sakai VT, Zhang Z, Dong Z, Neiva KG, Machado MA, Shi S et al (2010) SHED differentiate into functional odontoblasts and endothelium. J Dent Res 89(8):791–796
28. Ruparel NB, de Almeida JF, Henry MA, Diogenes A (2013) Characterization of a stem cell of apical papilla cell line: effect of passage on cellular phenotype. J Endod 39(3):357–363
29. Huang GT, Sonoyama W, Liu Y, Liu H, Wang S, Shi S (2008) The hidden treasure in apical papilla: the potential role in pulp/dentin regeneration and bioroot engineering. J Endod 34(6):645–651
30. Sonoyama W, Liu Y, Fang D, Yamaza T, Seo BM, Zhang C et al (2006) Mesenchymal stem cell-mediated functional tooth regeneration in swine. PLoS One 1, e79
31. Gronthos S, Mankani M, Brahim J, Robey PG, Shi S (2000) Postnatal human dental pulp stem cells (DPSCs) in vitro and in vivo. Proc Natl Acad Sci U S A 97(25):13625–13630
32. Tecles O, Laurent P, Zygouritsas S, Burger AS, Camps J, Dejou J et al (2005) Activation of human dental pulp progenitor/stem cells in response to odontoblast injury. Arch Oral Biol 50(2):103–108
33. Shi S, Robey PG, Gronthos S (2001) Comparison of human dental pulp and bone marrow stromal stem cells by cDNA microarray analysis. Bone 29(6):532–539
34. Chai Y, Jiang X, Ito Y, Bringas P Jr, Han J, Rowitch DH et al (2000) Fate of the mammalian cranial neural crest during tooth and mandibular morphogenesis. Development 127(8): 1671–1679
35. Kiraly M, Porcsalmy B, Pataki A, Kadar K, Jelitai M, Molnar B et al (2009) Simultaneous PKC and cAMP activation induces differentiation of human dental pulp stem cells into functionally active neurons. Neurochem Int 55(5):323–332
36. Koyama N, Okubo Y, Nakao K, Bessho K (2009) Evaluation of pluripotency in human dental pulp cells. J Oral Maxillofac Surg 67(3):501–506
37. Govindasamy V, Ronald VS, Abdullah AN, Nathan KR, Ab Aziz ZA, Abdullah M et al (2011) Differentiation of dental pulp stem cells into islet-like aggregates. J Dent Res 90(5):646–652
38. Kim BC, Bae H, Kwon IK, Lee EJ, Park JH, Khademhosseini A et al (2012) Osteoblastic/cementoblastic and neural differentiation of dental stem cells and their applications to tissue engineering and regenerative medicine. Tissue Eng Part B Rev 18(3):235–244
39. Iohara K, Zheng L, Wake H, Ito M, Nabekura J, Wakita H et al (2008) A novel stem cell source for vasculogenesis in ischemia: subfraction of side population cells from dental pulp. Stem Cells 26(9):2408–2418
40. Takeyasu M, Nozaki T, Daito M (2006) Differentiation of dental pulp stem cells into a neural lineage. Pediatr Dent J 16:154
41. Zainal Ariffin SH, Kermani S, Zainol Abidin IZ, Megat Abdul Wahab R, Yamamoto Z, Senafi S et al (2013) Differentiation of dental pulp stem cells into neuron-like cells in serum-free medium. Stem Cells Int 2013, 250740
42. Arthur A, Rychkov G, Shi S, Koblar SA, Gronthos S (2008) Adult human dental pulp stem cells differentiate toward functionally active neurons under appropriate environmental cues. Stem Cells 26(7):1787–1795
43. Bakopoulou A, Leyhausen G, Volk J, Tsiftsoglou A, Garefis P, Koidis P et al (2011) Comparative analysis of in vitro osteo/odontogenic differentiation potential of human dental pulp stem cells (DPSCs) and stem cells from the apical papilla (SCAP). Arch Oral Biol 56(7):709–721

44. Kerkis I, Kerkis A, Dozortsev D, Stukart-Parsons GC, Gomes Massironi SM, Pereira LV et al (2006) Isolation and characterization of a population of immature dental pulp stem cells expressing OCT-4 and other embryonic stem cell markers. Cells Tissues Organs 184(3–4): 105–116
45. Seo BM, Sonoyama W, Yamaza T, Coppe C, Kikuiri T, Akiyama K et al (2008) SHED repair critical-size calvarial defects in mice. Oral Dis 14(5):428–434
46. Cordeiro MM, Dong Z, Kaneko T, Zhang Z, Miyazawa M, Shi S et al (2008) Dental pulp tissue engineering with stem cells from exfoliated deciduous teeth. J Endod 34(8):962–969
47. Smith JG, Smith AJ, Shelton RM, Cooper PR (2012) Recruitment of dental pulp cells by dentine and pulp extracellular matrix components. Exp Cell Res 318(18):2397–2406
48. Smith AJ, Scheven BA, Takahashi Y, Ferracane JL, Shelton RM, Cooper PR (2012) Dentine as a bioactive extracellular matrix. Arch Oral Biol 57(2):109–121
49. Nakashima M, Iohara K (2011) Regeneration of dental pulp by stem cells. Adv Dent Res 23(3):313–319
50. Yen AH, Sharpe PT (2008) Stem cells and tooth tissue engineering. Cell Tissue Res 331(1): 359–372
51. Albuquerque MT, Valera MC, Nakashima M, Nor JE, Bottino MC (2014) Tissue-engineering-based strategies for regenerative endodontics. J Dent Res 93(12):1222–1231
52. Botero TM, Nör JE (2015) Tissue engineering strategies for endodontic regeneration. In: Vishwakarma A, Sharpe P, Songtao S, Ramalingam M (eds) Stem cell biology and tissue engineering in dental sciences, 1st edn. Academic, London, pp 419–430
53. Ding RY, Cheung GS, Chen J, Yin XZ, Wang QQ, Zhang CF (2009) Pulp revascularization of immature teeth with apical periodontitis: a clinical study. J Endod 35(5):745–749
54. Murray PE, Garcia-Godoy F, Hargreaves KM (2007) Regenerative endodontics: a review of current status and a call for action. J Endod 33(4):377–390
55. Lovelace TW, Henry MA, Hargreaves KM, Diogenes A (2011) Evaluation of the delivery of mesenchymal stem cells into the root canal space of necrotic immature teeth after clinical regenerative endodontic procedure. J Endod 37(2):133–138
56. Galler KM, Aulisa L, Regan KR, D'Souza RN, Hartgerink JD (2010) Self-assembling multi-domain peptide hydrogels: designed susceptibility to enzymatic cleavage allows enhanced cell migration and spreading. J Am Chem Soc 132(9):3217–3223
57. Liu X, Wang X, Wang X, Ren H, He J, Qiao L et al (2013) Functionalized self-assembling peptide nanofiber hydrogels mimic stem cell niche to control human adipose stem cell behavior in vitro. Acta Biomater 9(6):6798–6805
58. Linde A (1985) The extracellular matrix of the dental pulp and dentin. J Dent Res 64: 523–529
59. Casagrande L, Demarco FF, Zhang Z, Araujo FB, Shi S, Nor JE (2010) Dentin-derived BMP-2 and odontoblast differentiation. J Dent Res 89(6):603–608
60. Mooney DJ, Powell C, Piana J, Rutherford B (1996) Engineering dental pulp-like tissue in vitro. Biotechnol Prog 12(6):865–868
61. Iohara K, Nakashima M, Ito M, Ishikawa M, Nakasima A, Akamine A (2004) Dentin regeneration by dental pulp stem cell therapy with recombinant human bone morphogenetic protein 2. J Dent Res 83(8):590–595
62. Sakai VT, Cordeiro MM, Dong Z, Zhang Z, Zeitlin BD, Nor JE (2011) Tooth slice/scaffold model of dental pulp tissue engineering. Adv Dent Res 23(3):325–332
63. Kawasaki K, Tanaka S, Ishikawa T (1977) On the incremental lines in human dentine as revealed by tetracycline labeling. J Anat 123(Pt 2):427–436
64. Prescott RS, Alsanea R, Fayad MI, Johnson BR, Wenckus CS, Hao J et al (2008) In vivo generation of dental pulp-like tissue by using dental pulp stem cells, a collagen scaffold, and dentin matrix protein 1 after subcutaneous transplantation in mice. J Endod 34(4):421–426
65. Huang GT, Sonoyama W, Chen J, Park SH (2006) In vitro characterization of human dental pulp cells: various isolation methods and culturing environments. Cell Tissue Res 324(2): 225–236

66. Kodonas K, Gogos C, Papadimitriou S, Kouzi-Koliakou K, Tziafas D (2012) Experimental formation of dentin-like structure in the root canal implant model using cryopreserved swine dental pulp progenitor cells. J Endod 38(7):913–919
67. Huang GT, Yamaza T, Shea LD, Djouad F, Kuhn NZ, Tuan RS et al (2010) Stem/progenitor cell-mediated de novo regeneration of dental pulp with newly deposited continuous layer of dentin in an in vivo model. Tissue Eng A 16(2):605–615
68. Natu VP, Dubey N, Loke GC, Tan TS, Ng WH, Yong CW et al (2015) Bioactivity, physical and chemical properties of MTA mixed with propylene glycol. J Appl Oral Sci 23(4):405–411
69. Kemppainen JM, Hollister SJ (2010) Tailoring the mechanical properties of 3D-designed poly(glycerol sebacate) scaffolds for cartilage applications. J Biomed Mater Res A 94(1): 9–18
70. Saito E, Kang H, Taboas JM, Diggs A, Flanagan CL, Hollister SJ (2010) Experimental and computational characterization of designed and fabricated 50:50 PLGA porous scaffolds for human trabecular bone applications. J Mater Sci Mater Med 21(8):2371–2383
71. Qazi TH, Mooney DJ, Pumberger M, Geissler S, Duda GN (2015) Biomaterials based strategies for skeletal muscle tissue engineering: existing technologies and future trends. Biomaterials 53:502–521
72. Engler AJ, Sen S, Sweeney HL, Discher DE (2006) Matrix elasticity directs stem cell lineage specification. Cell 126(4):677–689
73. Lu Q, Pandya M, Rufaihah AJ, Rosa V, Tong HJ, Seliktar D et al (2015) Modulation of dental pulp stem cell odontogenesis in a tunable peg-fibrinogen hydrogel system. Stem Cells Int 2015:525367
74. Salerno A, Zeppetelli S, Di Maio E, Iannace S, Netti PA (2011) Processing/structure/property relationship of multi-scaled PCL and PCL-HA composite scaffolds prepared via gas foaming and NaCl reverse templating. Biotechnol Bioeng 108(4):963–976
75. Loh QL, Choong C (2013) Three-dimensional scaffolds for tissue engineering applications: role of porosity and pore size. Tissue Eng B Rev 19(6):485–502
76. Salerno A, Iannace S, Netti PA (2008) Open-pore biodegradable foams prepared via gas foaming and microparticulate templating. Macromol Biosci 8(7):655–664
77. Nof M, Shea LD (2002) Drug-releasing scaffolds fabricated from drug-loaded microspheres. J Biomed Mater Res 59(2):349–356
78. Goncalves SB, Dong Z, Bramante CM, Holland GR, Smith AJ, Nor JE (2007) Tooth slice-based models for the study of human dental pulp angiogenesis. J Endod 33(7):811–814
79. Liu J, Jin T, Ritchie HH, Smith AJ, Clarkson BH (2005) In vitro differentiation and mineralization of human dental pulp cells induced by dentin extract. In Vitro Cell Dev Biol Anim 41(7):232–238
80. Yu J, Deng Z, Shi J, Zhai H, Nie X, Zhuang H et al (2006) Differentiation of dental pulp stem cells into regular-shaped dentin-pulp complex induced by tooth germ cell conditioned medium. Tissue Eng 12(11):3097–3105
81. Magloire H, Romeas A, Melin M, Couble ML, Bleicher F, Farges JC (2001) Molecular regulation of odontoblast activity under dentin injury. Adv Dent Res 15:46–50
82. Tziafas D, Kodonas K (2010) Differentiation potential of dental papilla, dental pulp, and apical papilla progenitor cells. J Endod 36(5):781–789
83. Nakashima M, Reddi AH (2003) The application of bone morphogenetic proteins to dental tissue engineering. Nat Biotechnol 21(9):1025–1032
84. Nakashima M (1994) Induction of dentine in amputated pulp of dogs by recombinant human bone morphogenetic proteins-2 and -4 with collagen matrix. Arch Oral Biol 39(12): 1085–1089
85. Goldberg M, Six N, Decup F, Buch D, Soheili Majd E, Lasfargues JJ et al (2001) Application of bioactive molecules in pulp-capping situations. Adv Dent Res 15:91–95

Stem Cells for Periodontal Regeneration

Giorgio Pagni

Background

The Importance of Maintaining an Adequate Oral Health Status

A poor oral health status has been associated with systemic diseases, such as diabetes and risk factors for oral diseases, such as tobacco use, are shared with other systemic diseases. Poor oral health also impacts diet and nutrition and affects social activities, such as school and work [1]. Significant disadvantages are consequently associated with oral diseases and tooth loss.

As an indirect indicator of oral health conditions, we know that since the 1970s, fewer adults have experienced tooth loss and the prevalence of complete edentulism among adults has been consistently declining; however, important variations remain among subgroups of the population. Age is strongly related to tooth loss, gender is not related after adjustment for age, and race-ethnicity is consistently after adjustment for age and gender [1]. While the prevalence of edentulism continues to decline, the rate of aging of the western civilization and the longer life expectancy will keep increasing the need for dental treatment in order to prevent tooth loss.

Periodontal Diseases

Periodontal diseases include a variety of pathologies affecting the supporting structures of teeth and eventually leading to tooth loss. Periodontal disease affects almost half of the population in western countries [2] and severe periodontitis is the sixth

G. Pagni (✉)
Private Practice, Via Lamarmora 29, 50121 Florence, Italy

University of Milan, Milan, Italy
e-mail: giorgio.pagni@gmail.com

B. Zavan, E. Bressan (eds.), *Dental Stem Cells: Regenerative Potential*, Stem Cell Biology and Regenerative Medicine, DOI 10.1007/978-3-319-33299-4_9

most prevalent disease of mankind [3]. More than 70 % of the population over 65 years old is affected by periodontal disease. The prevalence is higher in males than females and highest in Mexican–Americans than in African–Americans and in African–Americans in relation to Caucasians. In smokers the prevalence of periodontal disease is 64.2 %. Research has shown that periodontal disease is associated with other chronic inflammatory diseases as diabetes and cardiovascular disease [2]. Periodontitis is also one of the main causes of tooth loss followed by tooth decay [4–6].

Chronic Periodontitis, Aggressive Periodontitis, Periodontitis as a Manifestation of Systemic Diseases, Necrotizing Periodontal Diseases, Abscesses of the Periodontium, Periodontitis Associated With Endodontic lesions and Other Developmental and Acquired Deformities are the main groups of pathologies possibly affecting the attachment apparatus [7]. All these conditions may have different etiology and contributing factors, but they all share loss of clinical attachment as the main characteristic bringing them together. Once the etiological and contributing factors are controlled the clinician often finds himself in the difficult position of correcting the deformities caused by the disease in order to facilitate oral hygiene maintenance and reducing the chance for disease recurrence and tooth loss [8].

Challenges in Periodontal Tissue Regeneration

The tooth attachment apparatus is a complex organ including cementum, periodontal ligament (PDL) and boundle bone. Embryologically speaking, all these structures have ectodermal origin making them different in relation to bone tissues normally originating from the mesoderm, later on we'll discuss on the possible importance of this different origin. The cementum is a thin layer of a mineralized tissue covering the tooth root dentin and in which the PDL fibers are inserted. The PDL is the tooth shock absorber system. It is made of collagen fibers called Sharpey's fibers but it also includes vessels and nerves bringing nourishment and Proprioceptivity to the surrounding structures. The boundle bone is the layer of bone, which is closer to the root surface. Sharpey's fibers are inserted in it, which make it different in relation to the alveolar bone. Many teeth positioned outside the alveolar bone housing are covered by boundle bone only and not any alveolar bone [9].

When attachment loss occurs, the PDL and boundle bone are resorbed by tissue enzymes and the cementum is left of the tooth surface and covered by bacterial plaque and calculus [10, 11]. Removal of plaque and calculus results in the closure of the periodontal pocket via formation of a long junctional epithelium [12]. Other traditional surgical approaches showed different repair models.

In 1976 Melcher [13] hypothesized that the repair model was dictated by the proliferation speed of different cell populations. In other words once the surgical trauma occurred and bacterial plaque was removed, a competition between soft tissue cells and hard tissue cells would take place. Being epithelial cells the fastest they would colonize the periodontal defects first.

Nyman and coworkers decided therefore to exclude soft tissue cells from this competition by attempting the first human case of guided periodontal tissue regeneration using a Millipore filter between the flap and the previously diseased root surface [14]. They showed, for the first time in humans, the regeneration of all of the three attachment apparatus structures: cementum, a functionally oriented PDL and boundle bone. After their pioneering work several authors have researched on guided tissue regeneration procedures.

Following studies shifted the attention from the competition theory to the importance of blood clot stabilization: that is that if the adhesion between the blood clot and the root surface is preserved epithelial cells would not infiltrate between the clot and the root surface resulting in periodontal regeneration [15, 16].

Periodontal Regeneration: State of the Art

Modern approaches to periodontal regeneration demonstrated better clinical results when adopting surgical techniques minimizing surgical invasiveness by reducing flap extension and mobilization and preserving the soft tissues as much as possible. Papilla preservation techniques were first introduced in 1985 [17] and then refined in 1995 [18], 1999 [19] and 2001 [20]. Later on surgical techniques using a single flap approach were introduced [21–24] and other flaps boosting the biological performance of blood clot stabilization have been developed [25].

Biomaterials in Periodontal Regeneration

Following Nyman's first guided tissue regeneration case with a Millipore filter several biomaterials have been evaluated with the intent of enhancing the clinical performance of periodontal regeneration surgeries.

Barrier membranes have been developed attempting to maintain soft tissue separated from the underlying structures during wound healing. Non-resorbable e-PTFE membranes were widely adopted either with or without a titanium frame reinforcement. Soon after resorbable membranes were introduced to overcome the need for a second surgical intervention to remove the e-PTFE membrane. Synthetic materials as polylactic acid, polyglycolic acid and combinations or xenogenic collagen membranes with different resorption rates have been used and evaluated widely in recent decades. Other than the barrier effect, membranes maintain the capability to support and stabilize the underlying tissues thus improving the clinical outcome of the regeneration procedure.

Autogenous bone and bone substitutes as synthetic materials, xenogenic bone substitutes and allogenic bone substitutes such as freeze-dried bone allograft (FDBA) and demineralized freeze-dried bone allograft (DFDBA) have also been thoroughly evaluated. Autogenous bone has been considered the gold standard as it maintains all the characteristics required for a grafting material: a scaffolding effect, an osteoconductive effect, presence of inductive proteins as the Bone Morphogenic Proteins capable of inducing the differentiation of neighboring undifferentiated cells toward the osteoblast pathway (osteoinduction) and the presence of cells which could directly initiate bone formation (osteogenic capability). Other bone substitutes have demonstrated good osteoconductive capabilities but only DFDBA has shown limited osteoinductive capabilities.

The clinical efficacy of membranes and biomaterials including autogenous bone has been widely documented when using extended flaps. With the advancements in flap manipulation and surgical techniques the use of biomaterials have been questioned in contenitive defects as their impact on clinical performance may be limited [26, 27]. Biomaterials are still considered important when treating larger non-contenitive defects. Moreover, tissue engineering approaches are currently evaluated in order to facilitate, enhance and speed up periodontal regeneration and possibly make it more predictable even when treating more challenging defects.

Tissue Engineering in Periodontal Tissue Regeneration

Tissue engineering is a broad term that describes very different applications. In a comprehensive manner MacArthur and Oreffo defined it as the understanding of the principles of tissue growth, and its application to produce functional replacement tissue for clinical use [28]. In its common view though tissue engineering is closely associated with the repair or regeneration of portions or whole tissues. In the medical field its weight is directly related to the importance that the ability of regenerating an organ or a tissue has to the survival or quality of life of an individual. In dentistry tissue engineering techniques find practical applications in the regeneration of lost bone, periodontal ligament, dental pulp or keratinized mucosa.

Regenerative strategies can be categorized in three major classes: Conductive approaches utilize biomaterials in a passive manner to facilitate the growth or regenerative capacity of existing tissue (guided tissue regeneration in periodontal dentistry is an example of this approach). Another tissue engineering strategy is Induction. The original concept lying behind this mechanism was explored with the discovery of bone morphogenetic proteins (BMPs), molecules that are able to promote the differentiation of uncommitted mesenchymal cells through the osteoblastic pathway inducing the formation of bone even in an ectopic (nonmineralizing) tissue. Cell transplantation is the third class of tissue engineering strategies. The collaboration of surgeons and bioengineers is critical for the delivery of cells that will interface with the host cells for the regeneration of the new natural tissue [29].

As previously mentioned most bone substitute materials used nowadays are adopted for their conductive capability in the regeneration of periodontal tissues; other inductive approaches as protein therapy and gene therapy and of cell transplantation have been evaluated and are currently studied in the attempt to generate better grafting solution for periodontal regeneration.

Protein Therapy: Growth Factor Delivery

The incorporation of bioactive molecules as Growth Factors (GFs) into scaffolding materials has been used to stimulate precursor cells and enhance wound healing and bone formation. The concept lying behind the use of GFs is related on the knowledge about cell-to-cell and platelets-to-cells biochemical interactions during wound healing.

A variety of factors have been studied that we can mainly group into two categories: *growth factors* act primarily through increasing cell proliferation (mitogenic) and recruitment of cells (chemotactic) while *morphogens* act through the alteration of cellular phenotype. BMPs are an example of morphogens; they have one of the most outstanding inductive properties in tissue regeneration: the ability of induce the differentiation of stem cells into bone formatting cells in a process known as *osteoinduction* [30]. We will now focus on proteins that have shown the most promising result in periodontal and implant related tissue engineering.

Platelet-Rich Plasma

The first attempt of using growth factors in tissue regeneration was by the use of platelet-rich plasma commonly known with its acronym: PRP.

Platelets are one of the body's best sources of growth factors. When activated they release α-granules containing, between others, Platelet Derived Growth Factor (PDGF), Transforming Growth Factor-β (TGF-β) Epidermal Growth factor (EGF) and Endothelial Cell Growth Factor (ECGF). These factors initiate several pathways during wound healing in all tissues. PRP was originally used for treatment of diabetic ulcers, decubitus ulcers, treatment of sternal wounds after cardiac surgery, non-healing fractures and it was then introduced its use in the dental field in 1998 [31]. PRGF is a technique similar to the production of PRP using although only one centrifugation and depleting the plasma from all red blood cells [32]. Other techniques have been described with significant variations in centrifugations protocols and final product indications.

A review by Boyapati and Wang critically evaluated the available literature on animal and human application of PRP. The authors concluded that due to the paucity of critical scientific data regarding the effects of PRP in sinus augmentation procedures and the poor study design of the available trials the use of PRP cannot be supported in sinus augmentation procedures [33]. A systematic review from Plachokova et al. came to similar conclusions about the materials and methods of the analyzed articles. When considering periodontal tissue regeneration although the results seem to be a little more encouraging [34].

With centrifugation techniques increases in platelet concentration between 2 and 10 times can be achieved. Protein therapy approaches have been introduced which are able to provide growth factors concentrations of thousands times higher than that carried by the blood.

Platelet Derived Growth Factor (PDGF)

Platelet-Derived Growth Factor (rh PDGF-BB) is one of the numerous growth factors, or proteins that regulate cell growth and division. In particular, it plays a significant role in blood vessel formation (angiogenesis), the growth of blood vessels from already existing blood vessel tissue. Mitogenic and chemotactic effects have been shown on calvarial osteogenic cells, periosteal cells from long bones, trabecular bone cells and bone marrow stromal cells. PDGF also increases expression of angiogenic molecules such as VEGF, hepatocyte growth factor/scatter factor, and of inflammatory factors as IL-6. PDGFs also increase osteogenic cells' responsiveness to BMPs [30].

The use of PDGF-BB in combination with other growth factors as the insulin-like Growth Factor-1 (IGF-1) has shown increased bone density and alveolar bone formation when compared to controls in periodontal defects and increased bone to implant contact (BIC) when used around implants immediately placed in extraction sockets [35–38]. When used in combination with allograft it has also shown encouraging results on the treatment of Class-II furcations and infrabony interproximal pockets [39, 40]. Recombinant human platelet-derived growth factor (rhPDGF-BB) mixed with a synthetic beta-tricalcium phosphate (beta-TCP) has been favorably used in the treatment of periodontal osseous defects [41, 42].

Although good results have been showed in eliciting bone regeneration some authors have hypothesized that suboptimal tooth-supporting tissue regeneration is related to the short half-life of PDGF in vivo. This could be a result of proteolytic degradation, rapid diffusion, or the solubility of the delivery matrix within periodontal lesions [43]. Increasing the substantivity of the molecule in vivo may result in improved periodontal regeneration.

Bone Morphogenetic Proteins (BMPs)

Bone morphogenetic proteins (BMPs), members of the transforming growth factor-β superfamily, have been shown to be responsible for post-fetal bone induction, including normal bone remodeling healing and repair. BMPs have been shown to be able to induce heterotopic bone formation [44]. Although all BMPs are involved in bone and cartilage formation by stimulating cellular events and mesenchymal progenitor cells, only a subset of BMPs (i.e. BMP-2, -4, -7 and -9) has osteoinductive properties [45]. New bone growth following use of BMPs results from the differentiation of pluripotent mesenchymal cells along osteoblastic pathways [46]. RhBMP-2 in combination to an absorbable collagen sponge have been tested in maxillary sinus elevation and for local alveolar ridge preservation or augmentation [36, 37, 46–48]. Lee et al. have tested ceramic barriers impregnated with BMP-2 greatly to improve craniofacial bone regeneration in vivo [49]. Bone repair has been documented in periodontal defects when using rhBMP-2/ACS although ankylosis was a frequent observation and only occasionally was a regeneration with functionally oriented PDL fibers observed [50]. Bone morphogenetic protein-7 (BMP-7), is a multifunctional member of the BMP family with multiple effects on bone formation and regeneration. BMP-7 stimulates bone regeneration around teeth [51], around endosseous dental implants [52], and in maxillary sinus floor augmentation procedures [53, 54]. Wikesjo reported in an animal study that while rhBMP-2/ACS is able to induce more robust bone regeneration rhBMP-12/ACS can induce more proper PDL regeneration with periodontal fibers inserting perpendicularly to the root surface into the newly regenerated cementum and supporting bone [55].

One of the limitations related to the use of BMPs in oral applications is the super-physiological dose of BMPs implanted in the wound site in certain trials resulting in a dose-dumping of potent morphogens [47, 48, 54]. There is a concern that this delivery method could lead to systemic toxicity in example by diffusion of potent morphogens through the bloodstream. The effect and safety of BMPs in vivo may benefit from a more controlled delivery method able to provide optimal concentration of these molecules through an adequate period of time.

Gene Therapy

In vivo growth factors have a transient effect since the half-life of these molecules is in the order of minutes to hours. Therefore, the induced production of growth factor via gene transfer is could potentially increase the effect of GFs by releasing them in a sustained manner. The latter is called Gene therapy and consists of the insertion of genes into individual's cells either directly or indirectly with a matrix to promote a specific biological effect. Gene therapy typically aims to supplement a defective mutant allele with a functional one in a therapeutic approach but can also be used to induce a more favorable host response. Targeting cells for gene therapy requires the use of vectors or direct delivery methods to transfect them [56]. Tissue engineering of alveolar bone using gene therapeutic approaches may offer potential for optimizing the delivery of growth-promoting molecules such as bone morphogenetic proteins (BMPs) at implant osteotomy sites [43, 57].

The most commonly employed vectors for gene transfer are viruses as retrovirus, lentivirus, adenovirus or adeno-associated virus [58]. A possible concern of the use of viral vectors is the possibility of an accidental generation of replication competent virus during vector production and packaging, the mobilization of the engineered vector by endogenous retrovirus present in the human genome and the random insertion or mutagenesis of the vector leading to cancer or germ cell alteration resulting in vertical inheritance of the acquired genes [59].

Non-viral vectors include plasmid DNA and synthetic or natural polymers. These vectors are safer than the viral vectors but they have low and unpredictable gene-transfer efficiency and allow only a limited size of DNA insert to penetrate in the targeted cells.

Both viral and non-viral vectors can be used in in vivo or ex vivo. In vivo approaches require the delivery of the vector directly into the host where the gene-transfer will happen. This method is less safe, more unpredictable but less invasive than ex vivo approaches. In ex vivo applications the gene-transfer happens in vitro of previously harvested host cells or in cell lines. The efficiency of the gene-transfer can be controlled and the infected cells are then re-implanted in the host. Ex vivo gene therapy has been widely adopted in combination to stem cell therapy.

In vitro studies have shown promising results when using recombinant adenoviral vectors encoding for the PDGF-A gene on cloned cementoblasts [60]; Chen and Giannobile evaluated the effects of PDGF on cell signaling via gene therapy when applied to tissue engineering of skin and periodontal wounds. The results of this study showed a sustained activation of the transfected genome by ligand binding resulting in a prolonged production of PDGF-AA and a sustained stimulation of several signaling pathways [61]. Syngenic dermal fibroblasts (SDFs) transduced ex vivo with adenoviruses were transplanted into large mandibular defect in a rat model with significantly improved results compared to controls [62]. Sustained and targeted delivery of BMP-7 to large alveolar bone defects associated with dental implant fixtures in a rat model also demonstrated that in vivo gene transfer promotes bone regeneration [53].

In a rat model periodontal defects where treated with either a collagen matrix alone or with a matrix containing adenoviruses encoding luciferase, PDGF-A (PDGF-1308) or (4) PDGF-B. At 14 days the control groups showed limited bone bridging while a fourfold increase in bridging bone and sixfold increase in tooth-lining cemental repair was observed the Ad-PDGF-B-treated sites that presented greater evidence of vascularization [63].

Chang et al. demonstrated that the use of adenovirus encoding the PDGF-B gene (AdPDGF-B) delivered in a collagen matrix to periodontal lesions in rats exhibited acceptable safety profiles; AdPDGF-B was well contained within the localized osseous defect area without viremia or distant organ involvement [64]. The same group evaluated the efficacy of the gene therapy approach in two concentrations of Ad-PDGF-B, a control Ad encoding luciferase or the protein therapy with rhPDGF-BB. Bone repair was found to be accelerated by gene and protein therapy compared with control, with the high dose of Ad-PDGF-B more effective than the low dose. Interestingly the protein therapy approach was found to be more effective than the gene therapy approach. Typically, high concentrations of growth factors are required to promote tissue regeneration [65]. A possible explanation for this unexpected result is that the concentrations of growth factor released during healing in the Ad-PDGF-B treated sites remained below the effective levels of the rhPDGF-BB treatment.

Cell Therapy

To summarize a few concepts described above we can identify the needs for tissue regeneration in the following categories: cells, growth factors, scaffold, and new vessels [56]. In conventional regeneration therapies cells need to reach the area where the new tissue has to be created from the periphery of the grafted site.

Scaffolds are used both as a support for their migration and as a frame for cells to start building up the tissue matrix. The migration and the differentiation of the cells is regulated and stimulated by *growth factors* that modulate the cellular activity and provide stimuli to cells to differentiate and produce matrix toward the developing tissue. GFs also stimulate the production of angiogenic signals inducing the formation of a new *vascular network*, which is essential for working cells as a source of nutrients and oxygen and as a pathway for the removal of catabolites.

Cell therapy approaches provide an additional source of cells in the area of interest with the intent to be used as grafted cells (which will integrate into the patients body) or, when not intended for integration, as a source of growth factors.

Both cultivated somatic cells and stem cells have been studied and tested in different branches of contemporary medicine and periodontology. Dogan et al. seeded cells derived from R to promote the regeneration in artificial furcation defects of a dog model and found this method to be superior than controls [66]. Cultivated fibroblasts have also been safely used for the treatment of interdental papillary insufficiency following a papilla priming procedure by McGuire and coworkers with satisfactory results [67]. Cell therapy has also been tested in mucogingival surgery [68–73]. Cloned cementoblasts have also demonstrated potential in periodontal therapy [74, 75]. Skin fibroblasts transduced by the BMP-7 gene promoted the tissue engineering of periodontal bone defects including new bone, functional PDL and tooth root cementum [62].

Stem Cells

Stem cells are cells that maintain an increased potential for renewal and differentiation in relation to cultivated cells. The benefit in the use of stem cells is that once transplanted in the grafted site they may differentiate in strictly osteogenic cells as well as in "supportive osteogenic cells". Supportive osteogenic cells are defined, as cells that do not directly create bone but that facilitate bone depositions creating structures that are needed to allow this process (i.e. vascular network) [76, 77]. Mesenchymal stem cells also play a role in preventing inflammation in the grafted site [78] allowing faster regeneration of new tissues.

Mesenchymal stem cells (MSCs) were first discovered in bone marrow and then in other adult tissues such as liver, adipose tissue, and muscle, but the marrow stroma remains the most widely used sources of stem cells [79]. They are characterized by elevated renewal potency and the ability of growing bone, cartilage, midollar adipocytes, miocytes, fibrous tissues from a single colony forming unit-fibroblast (CFU-F) when transplanted in vivo [80]. In the differentiation pathway, limitative or inductive differentiation stimuli may lead to cells progressively characterized by lower renewal capacity and by an augmented degree of differentiation [79]. This path can be reversed so that an adult adipocyte may de-differentiate back to levels with higher generative capabilities and then differentiate through the osteogenic pathway [81].

Given the inconvenience of obtaining stem cells from adult tissues including the use of invasive harvesting techniques, researchers have succeeded in harvesting mesenchymal stem cells from dental-derived tissues. Dental Stem Cells (DSCs) are adult stem cells derived from dental-related tissues such as the periodontal ligament of extracted teeth [82], gingiva [83], dental follicles [84], dental pulp [85], apical papilla [86], and human exfoliated deciduous teeth [87, 88].

Dental stem cells populations still maintain similar properties as bone marrow derived stem cells [85, 89–96]. Although they are reported to derive from the neural crest during tooth development [90], in mature individuals they demonstrate stem cells properties similar to MSCs rather than neural crest cells [97]. Jo and coworkers isolated stem cells from dental pulp, periodontal ligament, periapical follicle, and the surrounding mandibular bone marrow and found that they were able to differentiate into osteoblasts, adipocytes, and other kinds of cells with varying efficiency [98]. Because of their MSCs-like characteristics DSCs have been used in the regeneration of distant tissues and have been suggested as excellent candidates for cell-based therapy to treat liver diseases [99].

It has to be considered though that despite having similar properties each single stem cell type is different to the other. One important feature of dental pulp stem cells (DPSCs), in example, is their odontoblastic differentiation potential making them ideal to be seeded onto dentin and for dentin regeneration strategies. In endodontic therapy interesting results come from the autogenous transplantation of the BMP2-treated pellet culture of pulp progenitor/stem cells onto the amputated pulp. The treatment stimulated reparative dentin formation in a dog model [100]. Stem Cells from Human Exfoliated Deciduous Teeth (SHED) proliferate faster with greater population doublings than DPSCs and BMMSCs; unlike DPSCs, SHED are unable to regenerate a complete dentin-pulp-like complex in vivo. They showed to maintain the ability to differentiate into osteoblasts and induce bone formation, generate dentin and dental pulp [92, 101, 102]. Their ability to differentiate into bone-forming cells is not a property attributed to DPSCs following transplantation in vivo [90]. Similar to DPSCs and SHED, ex vivo expanded Stem Cells from Apical Papilla (SCAP) can undergo odontogenic differentiation in vitro. SCAP also demonstrate the capacity to undergo adipogenic differentiation following induction in vitro and show positive staining for several neural markers even without neurogenic stimulation [90, 103] and they are able to differentiate into functional dentinogenic cells. Periodontal Ligament Stem Cells (PDLSCs) seem to be the most promising dental derived stem cells for periodontal regeneration purposes as they can differentiate into either cementum-forming cells (cementoblasts) or bone-forming cells (osteoblasts) [104, 105]. They still demonstrate osteogenic, adipogenic, and chondrogenic characteristics under defined culture conditions [72, 73, 89, 106] but their capacity to form all of the PDL-related structures in vivo [82] makes of them the ideal candidates in periodontal regeneration [107]. In recent studies evaluating the osteogenic activity of different stem cell populations, the expression of RUNX2—a gene playing a crucial role in osteoblast differentiation [108]—peaked in BMSCs at day 14, while DPSCs and GSCs peaked at day 21 and PDLSCs peaked at day 28. The authors concluded that a possible explanation for the early RUNX2 expression in BMSCs may be that the embryological niche in the mesoderm area is more conductive for mineralization as compared to the ectodermal origin of PDLSC, GSC and DPSC [109, 110]. Moreover it has been shown that hPDLSC produce negative regulators of osteogenesis such as chordin and PDL-associated protein-1 or aspirin. In particular chondrin specifically binds to BMP-2 suppressing its osteoinductive capacity [111–115]. Despite this piece of knowledge may seem conflicting with the PDLSC task of regenerating a damaged periodontium, a physiological explanation may be hypothesized. Should a damaged PDL come in contact predominantly with bone MSCs in an osteoinductive environment ankylosis and root resorption have been documented [116]. The PDLSC ability to suppress osteoinduction may be needed to maintain normal tooth tissues homeostasis and prevent remodeling of the root.

From the information provided above it seems clear that despite their plasticity and flexibility not every stem cell is the same but one particular stem cell population may be more suitable to a particular task than another. Other cells showing capacity to form periodontal tissues are the Dental Follicle Precursor Cells (DFPCs): progenitor cells that form the periodontium, i.e., cementum, PDL, and alveolar bone which have been isolated from human dental follicles of impacted third molars. Kémoun et al. demonstrated expression of cementum attachment protein and cementum protein-23 (CP-23), two putative cementoblast markers, in EMD-stimulated whole dental follicle and in cultured DFPCs stimulated with EMD or BMP-2 and BMP-7 [117].

Even within a single cell source differences may be greater than one would think. PDLSCs have been isolated from the alveolar socket following extraction, from the root surface after extraction [95, 96], from deciduous [118–120] or permanent teeth [118, 119], from supernumerary teeth [120] and from periodontal granulation tissue [121, 122] each source showed different characteristics and regenerative potential [123], therefore strict identification criteria should be used when reporting and comparing results from each particular cell type.

Other important consideration are related to the culture medium, the oxygen concentration during cell culturing or the combined use of growth factors, which may enhance or reduce their regenerative potential and modify their characteristics [123].

The use of dental derived stem cells in periodontal therapy has important advantages. First of all that, because of their origin, they may be more adequate in the regeneration of periodontal tissues as opposed to other MSCs as dental stem cells appear to be more committed to odontogenic rather than osteogenic development [90]. Secondly, when autologous therapy is considered, the same dentist providing treatment to the patient could also be harvesting the cells beforehand. The problem in periodontal therapy is that not always the patient requiring periodontal regeneration will also need tooth extraction so not all cell types may be available. Periodontitis defects need a large number of cells (about 10^7 cells for one defect), which we may not be able to obtain from a single subject when using autologous stem cells from oral tissues [124]. Also at the time of treatment most periodontal patients are not young and the potential of their stem cells may be reduced [125]. Luckily because of immunoregulatory potential of DPSCs [126] SCAP and PDLSCs [90] allogenic dental stem cell therapy may be feasible as demonstrated in animal models [127, 128] therefore stem cells collected from deciduous teeth or teeth extracted for orthodontic reasons in younger individuals may be used in older patients. This immunoregulatory capacity is interesting as it has been suggested that it may play a role in reducing the inflammatory reaction to biomaterials used in tissue regeneration thus stimulating faster bone formation [129]. Other solutions to improve stem cell availability could be the cryopreservation of one's stem cells for future autologous use. Cryopreservation of extracted third molars was found to be an adequate and cost-effective method of preservation of MSC cells. After recovery, these cells maintained the characteristics of mesenchymal stem cells (ability to differentiate into osteogenic, adipogenic, and chondrogenic pathways) showing the possibility of banking MSC with minimal processing [130]. Finally, the immortalization of adult cells via gene therapy has been suggested [131] although raising questions on the costs and safety of these therapies.

On this path, adult adipocytes are known to be able to de-differentiated back to levels with higher generative capabilities and because of the large amounts of human lipoaspirates readily available, and the fact that their procurement induces only low morbidity, ASCs may be useful in future clinical cell-based therapy for periodontal disease. Adipose tissue is the richest source of MSCs, 100 times more than bone marrow [132]. The use of adipose-derived stem cells (ASCs) for periodontal tissue regeneration in vivo has been tested in animal models. These studies demonstrated the ability to promote tissue regeneration with this cell construct [133–136].

iPSCs may also represent an alternative source of stem cells. These are induced pluripotent stem (iPS) cells were first created reprogramming mice somatic cells through viral introduction of four transcription factors (Oct3/4, Sox2, Klf4, and c-Myc) from a group from Kyoto University [137]. iPS cells demonstrated to be completely similar to the most widely used Stem Cells and can be created reprogramming adult human dermal fibroblasts into a pluripotent state or practically any human somatic cell source [138]. IPSC lines were derived from human somatic cells reprogram somatic cell nuclei to an undifferentiated state [139, 140]. This technology may allow the implantation of Stem Cells without the drawbacks of the surgery required for their harvesting from the patient. Moreover, in alternative to reprogramming of patients' somatic cells, encouraging results have been conducted towards the establishment of an iPS cell bank consisting of various human leukocyte antigen (HLA) types. Dental Pulp Cells (DPCs) could represent an excellent source for iPS cell banks using retroviral vector for delivery of the exogenous genes [141].

Once proved as safe and effective, the possibility to generate an HLA-type banking of human pluripotent stem cell (hPSC) lines would greatly increase cost-effectiveness and practicality with the medical and oral health care cell therapy delivery systems, as well as safety and quality control measures of these therapies. Studies should determine whether IPSCs are better than DSCs in terms of safety and efficacy in periodontal regeneration.

An important issue recently emerging is the use of xenogenic components and growth factors in culture media, which may transmit animal pathogens [132].

Luckily researchers have been developing xeno-free culture system of human periodontal ligament stem cells, which after culture showed the morphological features of stem cells, expressed the markers associated with pluripotency, and a normal karyotype encouraging the use of this system in future human research experiments [142].

Applications in Periodontal Regeneration

In 2006 ex vivo expanded bone marrow mesenchymal stem cells (BMMSCs) in a PRP gel were shown to facilitate periodontal tissue regeneration, and suggested as a possible treatment solution for aesthetically sensitive sites, while reducing patient morbidity [114, 115]. Despite being a case report with no control the study draw significant attention on cell-based regenerative periodontal therapy.

Li et al. demonstrated that cryopreservation of bone marrow mesenchymal stem cells (BMSCs) had no negative effect on their growth and differentiation in vitro and that both freshly isolated and cryopreserved BMSC induced significantly better periodontal regeneration compared with collagen sponge alone [143].

Autologous PDLSCs were used to treat periodontal defects in a minipig model showing successful regeneration of the PDL and recovery of the heights of the alveolar bone [91].

Ding et al. evaluated both autologous and allogeneic periodontal ligament stem cells (PDLSCs) in the treatment of periodontal defects in a miniature pig periodontitis model. Both the autologous and the allogeneic PDLSCs demonstrated significant periodontal tissue regeneration with no marked difference between the autologous and allogeneic PDLSCs transplantation groups. The authors also observed lack of immunological rejections in the animals that received the allogeneic PDLSCs transplantation [127].

Cultured mandibular periosteum-derived cell sheets with PRP have also been used in the treatment of chronic periodontitis without [112] or with HA in patients suffering from advanced chronic periodontitis [144].

Recently, human osteoprotegerin (hOPG)-transfected periodontal ligament stem cells (PDLSCs) seeded on beta-tricalcium phosphate (β-TCP) were used to enhance periodontal regeneration in New Zealand rabbits. Unfortunately the adopted model was not adequate to evaluate periodontal regeneration but more for alveolar bone regeneration [145].

Future Directions

Several therapies have been evaluated for periodontal regeneration including guided tissue regeneration, and bone grafting. These techniques have been shown to improve clinical outcomes in relation to their control treatments although some authors have expressed concerns regarding the possible interference of mineralized grafts with bone repair in periodontal defects and some others have shown the long-lasting persistence of grafted particles in the regenerated area [146] arising questions on the clinical implications that these remaining particles might have. Recently an increased interest has been put on tissue engineering approaches, and in particular on the therapeutic potential of different stem cells populations for periodontal regenerative applications.

In the dental field different animal studies have shown encouraging results in bone regeneration. Marei et al. regenerated bone by implanting an engineered porous scaffold seeded with bone marrow mesenchymal stem cells (BMSCs) in an extraction socket in rabbits [129]. De Kok et al. in canine extraction sockets found that MSCs remained in the specified site contributing to bone formation as differentiated osteoblastic cells and that MSC-mediated bone repair was greater than bone repair in controls [147]. Some studies utilized fresh bone marrow aspirates [148] although, when using an autologous bone marrow grafting procedure, it should be kept in mind that while some of the cells populating bone marrow have strong osteogenic capacity, others have more of a regulatory effect and inhibit bone formation. Culturing protocols must therefore be adopted, which should be aimed at expanding those cells with osteogenic capacity while reducing inhibitor cells. Bone Repair Cells (BRCs) are an example of a culturing method of bone marrow aspirates successful in highly enriching Thy1 cells (CD90+) and CD105+ cells [58, 149]. BRC technology has been shown to be able not only to provide mesenchymal cells at an early stage of differentiation but also to produce significant concentrations of cytokines and were shown to maintain their ability to differentiate into endothelial cells and produce angiogenic factors providing supportive osteogenic cells that do not directly create bone but that facilitate bone depositions creating structures that are needed for the regeneration to take place (i.e. vascular network) [76, 77]. In the dental field, this cell construct has been already successfully tested in humans in a socket grafting procedure [150] and maxillary sinus grafting [151] while awaiting for even more advanced applications.

In periodontology mesenchymal stem cells (MSCs) isolated from a patient's iliac crest marrow aspirates in combination with platelet-rich plasma (PRP) were successful in the regeneration of periodontal tissue [114, 115].

Clinicians at times tend to underestimate the importance of a more rapid tissue formation. In periodontology the importance of boosting tissue regeneration may dictate the difference between succeeding or not in the preservation of teeth therefore novel tissue regeneration approaches must be evaluated in order for the periodontal field to advance.

On the other side, though case reports and clinical trials are promising, their results should be interpreted with caution [124]: while the potential of stem cells to induce tissue regeneration is more and more consolidated, the mechanisms involved in these processes are not clearly understood [152, 153]. Also determining which patient would greatly benefit from this advanced therapy has to be properly determined as several factors influence the disease and the clinician's ability to treat it [154]. The type of periodontal defects (number of walls), the type of periodontitis, the type of cell source and culturing method, the scaffold used, adjunctive growth factors, specific culture medium, or the enhancements of cultured cells via gene therapy, all these factors have to be properly evaluated before stem cell therapy can be applied in clinical practice.

References

1. Dye BA, Tan S, Smith V, Lewis BG, Barker LK, Thornton-Evans G, Eke PI, Beltran-Aguilar ED, Horowitz AM, Li CH (2007) Trends in oral health status: United States, 1988–1994 and 1999–2004. Vital Health Stat 11(248):1–92
2. Eke PI, Dye BA, Wei L, Thornton-Evans GO, Genco RJ, CDC Periodontal Disease Surveillance Workgroup, Beck J, Douglass G, Page R (2012) Prevalence of periodontitis in adults in the United States: 2009 and 2010. J Dent Res 91(10):914–920
3. Kassebaum NJ, Bernabe E, Dahiya M, Bhandari B, Murray CJ, Marcenes W (2014) Global burden of severe periodontitis in 1990–2010: a systematic review and meta-regression. J Dent Res 93(11):1045–1053
4. Axelsson P, Nystrom B, Lindhe J (2004) The long-term effect of a plaque control program on tooth mortality, caries and periodontal disease in adults. Results after 30 years of maintenance. J Clin Periodontol 31(9):749–757
5. Phipps KR, Stevens VJ (1995) Relative contribution of caries and periodontal disease in adult tooth loss for an HMO dental population. J Public Health Dent 55(4):250–252
6. Reich E, Hiller KA (1993) Reasons for tooth extraction in the western states of Germany. Community Dent Oral Epidemiol 21(6):379–383
7. Armitage GC (1999) Development of a classification system for periodontal diseases and conditions. Ann Periodontol 4(1):1–6
8. Matuliene G, Pjetursson BE, Salvi GE, Schmidlin K, Bragger U, Zwahlen M, Lang NP (2008) Influence of residual pockets on progression of periodontitis and tooth loss: results after 11 years of maintenance. J Clin Periodontol 35(8):685–695
9. Nevins M, Camelo M, De Paoli S, Friedland B, Schenk RK, Parma-Benfenati S, Simion M, Tinti C, Wagenberg B (2006) A study of the fate of the buccal wall of extraction sockets of teeth with prominent roots. Int J Periodontics Restorative Dent 26(1):19–29
10. Waerhaug J (1978) Healing of the dento-epithelial junction following subgingival plaque control. I. As observed in human biopsy material. J Periodontol 49(1):1–8
11. Waerhaug J (1978) Healing of the dento-epithelial junction following subgingival plaque control. II. As observed on extracted teeth. J Periodontol 49(3):119–134
12. Caton JG, Zander HA (1979) The attachment between tooth and gingival tissues after periodic root planing and soft tissue curettage. J Periodontol 50(9):462–466
13. Melcher AH (1976) On the repair potential of periodontal tissues. J Periodontol 47(5): 256–260
14. Nyman S, Lindhe J, Karring T, Rylander H (1982) New attachment following surgical treatment of human periodontal disease. J Clin Periodontol 9(4):290–296
15. Wikesjo UM, Claffey N, Egelberg J (1991) Periodontal repair in dogs. Effect of heparin treatment of the root surface. J Clin Periodontol 18(1):60–64
16. Wikesjo UM, Nilveus R (1990) Periodontal repair in dogs: effect of wound stabilization on healing. J Periodontol 61(12):719–724
17. Takei HH, Han TJ, Carranza FA Jr, Kenney EB, Lekovic V (1985) Flap technique for periodontal bone implants. Papilla preservation technique. J Periodontol 56(4):204–210
18. Cortellini P, Prato GP, Tonetti MS (1995) The modified papilla preservation technique. A new surgical approach for interproximal regenerative procedures. J Periodontol 66(4):261–266
19. Cortellini P, Prato GP, Tonetti MS (1999) The simplified papilla preservation flap. A novel surgical approach for the management of soft tissues in regenerative procedures. Int J Periodontics Restorative Dent 19(6):589–599

20. Cortellini P, Tonetti MS (2001) Microsurgical approach to periodontal regeneration. Initial evaluation in a case cohort. J Periodontol 72(4):559–569
21. Cortellini P, Tonetti MS (2009) Improved wound stability with a modified minimally invasive surgical technique in the regenerative treatment of isolated interdental intrabony defects. J Clin Periodontol 36(2):157–163
22. Trombelli L, Farina R, Franceschetti G (2007) Uso di un solo lembo in chirurgia ricostruttiva. Dent Cadmos 8:19
23. Trombelli L, Farina R, Franceschetti G, Calura G (2009) Single-flap approach with buccal access in periodontal reconstructive procedures. J Periodontol 80(2):353–360
24. Trombelli L, Farina R, Franceschetti G, Minenna L (2008) Single Flap Approach. Un caso clinico paradigmatico. Dent Clin 2(3):1–7
25. Rasperini G, Acunzo R, Barnett A, Pagni G (2013) The soft tissue wall technique for the regenerative treatment of non-contained infrabony defects: a case series. Int J Periodontics Restorative Dent 33(3):e79–e87
26. Cortellini P, Tonetti MS (2011) Clinical and radiographic outcomes of the modified minimally invasive surgical technique with and without regenerative materials: a randomized-controlled trial in intra-bony defects. J Clin Periodontol 38(4):365–373
27. Mishra A, Avula H, Pathakota KR, Avula J (2013) Efficacy of modified minimally invasive surgical technique in the treatment of human intrabony defects with or without use of rhPDGF-BB gel: a randomized controlled trial. J Clin Periodontol 40(2):172–179
28. MacArthur BD, Oreffo RO (2005) Bridging the gap. Nature 433(7021):19
29. Kaigler D, Mooney D (2001) Tissue engineering's impact on dentistry. J Dent Educ 65(5):456–462
30. Lynch SE (2008) Tissue engineering: applications in oral and maxillofacial surgery and periodontics, 2nd edn. Quintessence Publishing Co. Ltd., Chicago
31. Marx RE, Carlson ER, Eichstaedt RM, Schimmele SR, Strauss JE, Georgeff KR (1998) Platelet-rich plasma: growth factor enhancement for bone grafts. Oral Surg Oral Med Oral Pathol Oral Radiol Endod 85(6):638–646
32. Anitua E (1999) Plasma rich in growth factors: preliminary results of use in the preparation of future sites for implants. Int J Oral Maxillofac Implants 14(4):529–535
33. Boyapati L, Wang HL (2006) The role of platelet-rich plasma in sinus augmentation: a critical review. Implant Dent 15(2):160–170
34. Plachokova AS, Nikolidakis D, Mulder J, Jansen JA, Creugers NH (2008) Effect of platelet-rich plasma on bone regeneration in dentistry: a systematic review. Clin Oral Implants Res 19(6):539–545
35. Becker W, Lynch SE, Lekholm U, Becker BE, Caffesse R, Donath K, Sanchez R (1992) A comparison of ePTFE membranes alone or in combination with platelet-derived growth factors and insulin-like growth factor-I or demineralized freeze-dried bone in promoting bone formation around immediate extraction socket implants. J Periodontol 63(11):929–940
36. Howell TH, Fiorellini J, Jones A, Alder M, Nummikoski P, Lazaro M, Lilly L, Cochran D (1997) A feasibility study evaluating rhBMP-2/absorbable collagen sponge device for local alveolar ridge preservation or augmentation. Int J Periodontics Restorative Dent 17(2):124–139
37. Howell TH, Fiorellini JP, Paquette DW, Offenbacher S, Giannobile WV, Lynch SE (1997) A phase I/II clinical trial to evaluate a combination of recombinant human platelet-derived growth factor-BB and recombinant human insulin-like growth factor-I in patients with periodontal disease. J Periodontol 68(12):1186–1193
38. Stefani CM, Machado MA, Sallum EA, Sallum AW, Toledo S, Nociti FH Jr (2000) Platelet-derived growth factor/insulin-like growth factor-1 combination and bone regeneration around implants placed into extraction sockets: a histometric study in dogs. Implant Dent 9(2):126–131
39. Camelo M, Nevins ML, Schenk RK, Lynch SE, Nevins M (2003) Periodontal regeneration in human Class II furcations using purified recombinant human platelet-derived growth factor-BB (rhPDGF-BB) with bone allograft. Int J Periodontics Restorative Dent 23(3):213–225

40. Nevins M, Camelo M, Nevins ML, Schenk RK, Lynch SE (2003) Periodontal regeneration in humans using recombinant human platelet-derived growth factor-BB (rhPDGF-BB) and allogenic bone. J Periodontol 74(9):1282–1292
41. McGuire MK, Kao RT, Nevins M, Lynch SE (2006) rhPDGF-BB promotes healing of periodontal defects: 24-month clinical and radiographic observations. Int J Periodontics Restorative Dent 26(3):223–231
42. Nevins M, Giannobile WV, McGuire MK, Kao RT, Mellonig JT, Hinrichs JE, McAllister BS, Murphy KS, McClain PK, Nevins ML, Paquette DW, Han TJ, Reddy MS, Lavin PT, Genco RJ, Lynch SE (2005) Platelet-derived growth factor stimulates bone fill and rate of attachment level gain: results of a large multicenter randomized controlled trial. J Periodontol 76(12):2205–2215
43. Giannobile WV (1996) Periodontal tissue engineering by growth factors. Bone 19(1 Suppl):23S–37S
44. Urist MR, Iwata H, Ceccotti PL, Dorfman RL, Boyd SD, McDowell RM, Chien C (1973) Bone morphogenesis in implants of insoluble bone gelatin. Proc Natl Acad Sci U S A 70(12):3511–3515
45. Cheng H, Jiang W, Phillips FM, Haydon RC, Peng Y, Zhou L, Luu HH, An N, Breyer B, Vanichakarn P, Szatkowski JP, Park JY, He TC (2003) Osteogenic activity of the fourteen types of human bone morphogenetic proteins (BMPs). J Bone Joint Surg Am 85-A(8): 1544–1552
46. Fiorellini JP, Howell TH, Cochran D, Malmquist J, Lilly LC, Spagnoli D, Toljanic J, Jones A, Nevins M (2005) Randomized study evaluating recombinant human bone morphogenetic protein-2 for extraction socket augmentation. J Periodontol 76(4):605–613
47. Boyne PJ, Marx RE, Nevins M, Triplett G, Lazaro E, Lilly LC, Alder M, Nummikoski P (1997) A feasibility study evaluating rhBMP-2/absorbable collagen sponge for maxillary sinus floor augmentation. Int J Periodontics Restorative Dent 17(1):11–25
48. Cochran DL, Jones AA, Lilly LC, Fiorellini JP, Howell H (2000) Evaluation of recombinant human bone morphogenetic protein-2 in oral applications including the use of endosseous implants: 3-year results of a pilot study in humans. J Periodontol 71(8):1241–1257
49. Lee YM, Nam SH, Seol YJ, Kim TI, Lee SJ, Ku Y, Rhyu IC, Chung CP, Han SB, Choi SM (2003) Enhanced bone augmentation by controlled release of recombinant human bone morphogenetic protein-2 from bioabsorbable membranes. J Periodontol 74(6):865–872
50. Selvig KA, Sorensen RG, Wozney JM, Wikesjo UM (2002) Bone repair following recombinant human bone morphogenetic protein-2 stimulated periodontal regeneration. J Periodontol 73(9):1020–1029
51. Giannobile WV, Ryan S, Shih MS, Su DL, Kaplan PL, Chan TC (1998) Recombinant human osteogenic protein-1 (OP-1) stimulates periodontal wound healing in class III furcation defects. J Periodontol 69(2):129–137
52. Rutherford RB, Sampath TK, Rueger DC, Taylor TD (1992) Use of bovine osteogenic protein to promote rapid osseointegration of endosseous dental implants. Int J Oral Maxillofac Implants 7(3):297–301
53. Dunn CA, Jin Q, Taba M Jr, Franceschi RT, Bruce Rutherford R, Giannobile WV (2005) BMP gene delivery for alveolar bone engineering at dental implant defects. Mol Ther 11(2):294–299
54. van den Bergh JP, ten Bruggenkate CM, Groeneveld HH, Burger EH, Tuinzing DB (2000) Recombinant human bone morphogenetic protein-7 in maxillary sinus floor elevation surgery in 3 patients compared to autogenous bone grafts. A clinical pilot study. J Clin Periodontol 27(9):627–636
55. Wikesjo UM, Sorensen RG, Kinoshita A, Jian Li X, Wozney JM (2004) Periodontal repair in dogs: effect of recombinant human bone morphogenetic protein-12 (rhBMP-12) on regeneration of alveolar bone and periodontal attachment. J Clin Periodontol 31(8):662–670
56. Taba M Jr, Jin Q, Sugai JV, Giannobile WV (2005) Current concepts in periodontal bioengineering. Orthod Craniofac Res 8(4):292–302

57. Reddi AH (1998) Role of morphogenetic proteins in skeletal tissue engineering and regeneration. Nat Biotechnol 16(3):247–252
58. Kaigler D, Cirelli JA, Giannobile WV (2006) Growth factor delivery for oral and periodontal tissue engineering. Expert Opin Drug Deliv 3(5):647–662
59. Connolly JB (2002) Lentiviruses in gene therapy clinical research. Gene Ther 9(24): 1730–1734
60. Giannobile WV, Lee CS, Tomala MP, Tejeda KM, Zhu Z (2001) Platelet-derived growth factor (PDGF) gene delivery for application in periodontal tissue engineering. J Periodontol 72(6):815–823
61. Chen QP, Giannobile WV (2002) Adenoviral gene transfer of PDGF downregulates gas gene product PDGFalphaR and prolongs ERK and Akt/PKB activation. Am J Physiol Cell Physiol 282(3):C538–C544
62. Jin QM, Anusaksathien O, Webb SA, Rutherford RB, Giannobile WV (2003) Gene therapy of bone morphogenetic protein for periodontal tissue engineering. J Periodontol 74(2): 202–213
63. Jin Q, Anusaksathien O, Webb SA, Printz MA, Giannobile WV (2004) Engineering of tooth-supporting structures by delivery of PDGF gene therapy vectors. Mol Ther 9(4):519–526
64. Chang PC, Cirelli JA, Jin Q, Seol YJ, Sugai JV, D'Silva NJ, Danciu TE, Chandler LA, Sosnowski BA, Giannobile WV (2009) Adenovirus encoding human platelet-derived growth factor-B delivered to alveolar bone defects exhibits safety and biodistribution profiles favorable for clinical use. Hum Gene Ther 20(5):486–496
65. Fang J, Zhu YY, Smiley E, Bonadio J, Rouleau JP, Goldstein SA, McCauley LK, Davidson BL, Roessler BJ (1996) Stimulation of new bone formation by direct transfer of osteogenic plasmid genes. Proc Natl Acad Sci U S A 93(12):5753–5758
66. Dogan A, Ozdemir A, Kubar A, Oygur T (2002) Assessment of periodontal healing by seeding of fibroblast-like cells derived from regenerated periodontal ligament in artificial furcation defects in a dog: a pilot study. Tissue Eng 8(2):273–282
67. McGuire MK, Scheyer ET (2007) A randomized, double-blind, placebo-controlled study to determine the safety and efficacy of cultured and expanded autologous fibroblast injections for the treatment of interdental papillary insufficiency associated with the papilla priming procedure. J Periodontol 78(1):4–17
68. Izumi K, Feinberg SE, Iida A, Yoshizawa M (2003) Intraoral grafting of an ex vivo produced oral mucosa equivalent: a preliminary report. Int J Oral Maxillofac Surg 32(2):188–197
69. McGuire MK, Scheyer ET, Nunn ME, Lavin PT (2008) A pilot study to evaluate a tissue-engineered bilayered cell therapy as an alternative to tissue from the palate. J Periodontol 79(10):1847–1856
70. Morelli T, Neiva R, Nevins ML, McGuire MK, Scheyer ET, Oh TJ, Braun TM, Nor JE, Bates D, Giannobile WV (2011) Angiogenic biomarkers and healing of living cellular constructs. J Dent Res 90(4):456–462
71. Song J, Izumi K, Lanigan T, Feinberg SE (2004) Development and characterization of a canine oral mucosa equivalent in a serum-free environment. J Biomed Mater Res A 71(1):143–153
72. Xu J, Wang W, Kapila Y, Lotz J, Kapila S (2009) Multiple differentiation capacity of STRO-1+/CD146+ PDL mesenchymal progenitor cells. Stem Cells Dev 18(3):487–496
73. Xu Q, Izumi K, Tobita T, Nakanishi Y, Feinberg SE (2009) Constitutive release of cytokines by human oral keratinocytes in an organotypic culture. J Oral Maxillofac Surg 67(6):1256–1264
74. Somerman MJ, Ouyang HJ, Berry JE, Saygin NE, Strayhorn CL, D'Errico JA, Hullinger T, Giannobile WV (1999) Evolution of periodontal regeneration: from the roots' point of view. J Periodontal Res 34(7):420–424
75. Zhao M, Jin Q, Berry JE, Nociti FH Jr, Giannobile WV, Somerman MJ (2004) Cementoblast delivery for periodontal tissue engineering. J Periodontol 75(1):154–161
76. Kaigler D, Pagni G, Park CH, Tarle SA, Bartel RL, Giannobile WV (2010) Angiogenic and osteogenic potential of bone repair cells for craniofacial regeneration. Tissue Eng A 16(9):2809–2820

77. Pagni G, Kaigler D, Rasperini G, Avila-Ortiz G, Barte R, Giannobile WV (2012) Bone repair cells for craniofacial regeneration. Adv Drug Deliv Rev 64(12):1310–1319
78. Grande DA, Southerland SS, Manji R, Pate DW, Schwartz RE, Lucas PA (1995) Repair of articular cartilage defects using mesenchymal stem cells. Tissue Eng 1(4):345–353
79. Owen P, Connaghan DG, Holland HK, Steis RG (1998) Bone marrow transplantation: cancer therapy comes of age. J Med Assoc Ga 87(2):145–146, 148
80. Prockop DJ (1997) Marrow stromal cells as stem cells for nonhematopoietic tissues. Science 276(5309):71–74
81. Park SR, Oreffo RO, Triffitt JT (1999) Interconversion potential of cloned human marrow adipocytes in vitro. Bone 24(6):549–554
82. Seo BM, Miura M, Gronthos S, Bartold PM, Batouli S, Brahim J, Young M, Robey PG, Wang CY, Shi S (2004) Investigation of multipotent postnatal stem cells from human periodontal ligament. Lancet 364(9429):149–155
83. Zhang Q, Shi S, Liu Y, Uyanne J, Shi Y, Shi S, Le AD (2009) Mesenchymal stem cells derived from human gingiva are capable of immunomodulatory functions and ameliorate inflammation-related tissue destruction in experimental colitis. J Immunol 183(12): 7787–7798
84. Morsczeck C, Gotz W, Schierholz J, Zeilhofer F, Kuhn U, Mohl C, Sippel C, Hoffmann KH (2005) Isolation of precursor cells (PCs) from human dental follicle of wisdom teeth. Matrix Biol 24(2):155–165
85. Gronthos S, Mankani M, Brahim J, Robey PG, Shi S (2000) Postnatal human dental pulp stem cells (DPSCs) in vitro and in vivo. Proc Natl Acad Sci U S A 97(25):13625–13630
86. Sonoyama W, Liu Y, Yamaza T, Tuan RS, Wang S, Shi S, Huang GT (2008) Characterization of the apical papilla and its residing stem cells from human immature permanent teeth: a pilot study. J Endod 34(2):166–171
87. Hynes K, Menicanin D, Gronthos S, Bartold PM (2012) Clinical utility of stem cells for periodontal regeneration. Periodontol 2000 59(1):203–227
88. Park JY, Jeon SH, Choung PH (2011) Efficacy of periodontal stem cell transplantation in the treatment of advanced periodontitis. Cell Transplant 20(2):271–285
89. Gay IC, Chen S, MacDougall M (2007) Isolation and characterization of multipotent human periodontal ligament stem cells. Orthod Craniofac Res 10(3):149–160
90. Huang GT, Gronthos S, Shi S (2009) Mesenchymal stem cells derived from dental tissues vs. those from other sources: their biology and role in regenerative medicine. J Dent Res 88(9): 792–806
91. Liu Y, Zheng Y, Ding G, Fang D, Zhang C, Bartold PM, Gronthos S, Shi S, Wang S (2008) Periodontal ligament stem cell-mediated treatment for periodontitis in miniature swine. Stem Cells 26(4):1065–1073
92. Miura M, Gronthos S, Zhao M, Lu B, Fisher LW, Robey PG, Shi S (2003) SHED: stem cells from human exfoliated deciduous teeth. Proc Natl Acad Sci U S A 100(10):5807–5812
93. Shi S, Gronthos S (2003) Perivascular niche of postnatal mesenchymal stem cells in human bone marrow and dental pulp. J Bone Miner Res 18(4):696–704
94. Treves-Manusevitz S, Hoz L, Rachima H, Montoya G, Tzur E, Vardimon A, Narayanan AS, Amar S, Arzate H, Pitaru S (2013) Stem cells of the lamina propria of human oral mucosa and gingiva develop into mineralized tissues in vivo. J Clin Periodontol 40(1):73–81
95. Wang F, Yu M, Yan X, Wen Y, Zeng Q, Yue W, Yang P, Pei X (2011) Gingiva-derived mesenchymal stem cell-mediated therapeutic approach for bone tissue regeneration. Stem Cells Dev 20(12):2093–2102
96. Wang L, Shen H, Zheng W, Tang L, Yang Z, Gao Y, Yang Q, Wang C, Duan Y, Jin Y (2011) Characterization of stem cells from alveolar periodontal ligament. Tissue Eng A 17(7–8): 1015–1026
97. Kaku M, Komatsu Y, Mochida Y, Yamauchi M, Mishina Y, Ko CC (2012) Identification and characterization of neural crest-derived cells in adult periodontal ligament of mice. Arch Oral Biol 57(12):1668–1675

98. Jo YY, Lee HJ, Kook SY, Choung HW, Park JY, Chung JH, Choung YH, Kim ES, Yang HC, Choung PH (2007) Isolation and characterization of postnatal stem cells from human dental tissues. Tissue Eng 13(4):767–773
99. Ikeda E, Yagi K, Kojima M, Yagyuu T, Ohshima A, Sobajima S, Tadokoro M, Katsube Y, Isoda K, Kondoh M, Kawase M, Go MJ, Adachi H, Yokota Y, Kirita T, Ohgushi H (2007) Multipotent cells from the human third molar: feasibility of cell-based therapy for liver disease. Differentiation 75(5):495–505
100. Iohara K, Zheng L, Ito M, Tomokiyo A, Matsushita K, Nakashima M (2006) Side population cells isolated from porcine dental pulp tissue with self-renewal and multipotency for dentinogenesis, chondrogenesis, adipogenesis, and neurogenesis. Stem Cells 24(11):2493–2503
101. Cordeiro MM, Dong Z, Kaneko T, Zhang Z, Miyazawa M, Shi S, Smith AJ, Nor JE (2008) Dental pulp tissue engineering with stem cells from exfoliated deciduous teeth. J Endod 34(8):962–969
102. Laino G, Graziano A, d'Aquino R, Pirozzi G, Lanza V, Valiante S, De Rosa A, Naro F, Vivarelli E, Papaccio G (2006) An approachable human adult stem cell source for hard-tissue engineering. J Cell Physiol 206(3):693–701
103. Sonoyama W, Liu Y, Fang D, Yamaza T, Seo BM, Zhang C, Liu H, Gronthos S, Wang CY, Wang S, Shi S (2006) Mesenchymal stem cell-mediated functional tooth regeneration in swine. PLoS One 1, e79
104. Isaka J, Ohazama A, Kobayashi M, Nagashima C, Takiguchi T, Kawasaki H, Tachikawa T, Hasegawa K (2001) Participation of periodontal ligament cells with regeneration of alveolar bone. J Periodontol 72(3):314–323
105. McCulloch CA, Bordin S (1991) Role of fibroblast subpopulations in periodontal physiology and pathology. J Periodontal Res 26(3 Pt 1):144–154
106. Lindroos B, Maenpaa K, Ylikomi T, Oja H, Suuronen R, Miettinen S (2008) Characterisation of human dental stem cells and buccal mucosa fibroblasts. Biochem Biophys Res Commun 368(2):329–335
107. Seo BM, Miura M, Sonoyama W, Coppe C, Stanyon R, Shi S (2005) Recovery of stem cells from cryopreserved periodontal ligament. J Dent Res 84(10):907–912
108. Lian JB, Javed A, Zaidi SK, Lengner C, Montecino M, van Wijnen AJ, Stein JL, Stein GS (2004) Regulatory controls for osteoblast growth and differentiation: role of Runx/Cbfa/AML factors. Crit Rev Eukaryot Gene Expr 14(1–2):1–41
109. Gay I, Cavender A, Peto D, Sun Z, Speer A, Cao H, Amendt BA (2014) Differentiation of human dental stem cells reveals a role for microRNA-218. J Periodontal Res 49(1):110–120
110. Sharpe PT (2001) Neural crest and tooth morphogenesis. Adv Dent Res 15:4–7
111. Barboza EP, Duarte ME, Geolas L, Sorensen RG, Riedel GE, Wikesjo UM (2000) Ridge augmentation following implantation of recombinant human bone morphogenetic protein-2 in the dog. J Periodontol 71(3):488–496
112. Mizuno N, Ozeki Y, Shiba H, Kajiya M, Nagahara T, Takeda K, Kawaguchi H, Abiko Y, Kurihara H (2008) Humoral factors released from human periodontal ligament cells influence calcification and proliferation in human bone marrow mesenchymal stem cells. J Periodontol 79(12):2361–2370
113. Tomoeda M, Yamada S, Shirai H, Ozawa Y, Yanagita M, Murakami S (2008) PLAP-1/asporin inhibits activation of BMP receptor via its leucine-rich repeat motif. Biochem Biophys Res Commun 371(2):191–196
114. Yamada S, Ozawa Y, Tomoeda M, Matoba R, Matsubara K, Murakami S (2006) Regulation of PLAP-1 expression in periodontal ligament cells. J Dent Res 85(5):447–451
115. Yamada Y, Ueda M, Hibi H, Baba S (2006) A novel approach to periodontal tissue regeneration with mesenchymal stem cells and platelet-rich plasma using tissue engineering technology: a clinical case report. Int J Periodontics Restorative Dent 26(4):363–369
116. Karring T, Nyman S, Lindhe J, Sirirat M (1984) Potentials for root resorption during periodontal wound healing. J Clin Periodontol 11(1):41–52
117. Kémoun P, Laurencin-Dalicieux S, Rue J, Farges JC, Gennero I, Conte-Auriol F, Briand-Mesange F, Gadelorge M, Arzate H, Narayanan AS, Brunel G, Salles JP (2007) Human

dental follicle cells acquire cementoblast features under stimulation by BMP-2/-7 and enamel matrix derivatives (EMD) in vitro. Cell Tissue Res 329(2):283–294
118. Ji K, Liu Y, Lu W, Yang F, Yu J, Wang X, Ma Q, Yang Z, Wen L, Xuan K (2013) Periodontal tissue engineering with stem cells from the periodontal ligament of human retained deciduous teeth. J Periodontal Res 48(1):105–116
119. Silverio KG, Rodrigues TL, Coletta RD, Benevides L, Da Silva JS, Casati MZ, Sallum EA, Nociti FH Jr (2010) Mesenchymal stem cell properties of periodontal ligament cells from deciduous and permanent teeth. J Periodontol 81(8):1207–1215
120. Song JS, Kim SO, Kim SH, Choi HJ, Son HK, Jung HS, Kim CS, Lee JH (2012) In vitro and in vivo characteristics of stem cells derived from the periodontal ligament of human deciduous and permanent teeth. Tissue Eng A 18(19–20):2040–2051
121. Hung TY, Lin HC, Chan YJ, Yuan K (2012) Isolating stromal stem cells from periodontal granulation tissues. Clin Oral Investig 16(4):1171–1180
122. Ronay V, Belibasakis GN, Attin T, Schmidlin PR, Bostanci N (2014) Expression of embryonic stem cell markers and osteogenic differentiation potential in cells derived from periodontal granulation tissue. Cell Biol Int 38(2):179–186
123. Zhu W, Liang M (2015) Periodontal ligament stem cells: current status, concerns, and future prospects. Stem Cells Int 2015:972313
124. Trofin EA, Monsarrat P, Kémoun P (2013) Cell therapy of periodontium: from animal to human? Front Physiol 4:325
125. Zheng W, Wang S, Ma D, Tang L, Duan Y, Jin Y (2009) Loss of proliferation and differentiation capacity of aged human periodontal ligament stem cells and rejuvenation by exposure to the young extrinsic environment. Tissue Eng A 15(9):2363–2371
126. Pierdomenico L, Bonsi L, Calvitti M, Rondelli D, Arpinati M, Chirumbolo G, Becchetti E, Marchionni C, Alviano F, Fossati V, Staffolani N, Franchina M, Grossi A, Bagnara GP (2005) Multipotent mesenchymal stem cells with immunosuppressive activity can be easily isolated from dental pulp. Transplantation 80(6):836–842
127. Ding G, Liu Y, Wang W, Wei F, Liu D, Fan Z, An Y, Zhang C, Wang S (2010) Allogeneic periodontal ligament stem cell therapy for periodontitis in swine. Stem Cells 28(10): 1829–1838
128. Mrozik KM, Wada N, Marino V, Richter W, Shi S, Wheeler DL, Gronthos S, Bartold PM (2013) Regeneration of periodontal tissues using allogeneic periodontal ligament stem cells in an ovine model. Regen Med 8(6):711–723
129. Marei MK, Nouh SR, Saad MM, Ismail NS (2005) Preservation and regeneration of alveolar bone by tissue-engineered implants. Tissue Eng 11(5–6):751–767
130. Perry BC, Zhou D, Wu X, Yang FC, Byers MA, Chu TM, Hockema JJ, Woods EJ, Goebel WS (2008) Collection, cryopreservation, and characterization of human dental pulp-derived mesenchymal stem cells for banking and clinical use. Tissue Eng C Methods 14(2):149–156
131. Fujii S, Maeda H, Wada N, Kano Y, Akamine A (2006) Establishing and characterizing human periodontal ligament fibroblasts immortalized by SV40T-antigen and hTERT gene transfer. Cell Tissue Res 324(1):117–125
132. Philippe B, Luc S, Valerie PB, Jerome R, Alessandra BR, Louis C (2010) Culture and use of mesenchymal stromal cells in phase I and II clinical trials. Stem Cells Int 2010:503593
133. Akita D, Morokuma M, Saito Y, Yamanaka K, Akiyama Y, Sato M, Mashimo T, Toriumi T, Arai Y, Kaneko T, Tsukimura N, Isokawa K, Ishigami T, Honda MJ (2014) Periodontal tissue regeneration by transplantation of rat adipose-derived stromal cells in combination with PLGA-based solid scaffolds. Biomed Res 35(2):91–103
134. Tobita M, Uysal AC, Ogawa R, Hyakusoku H, Mizuno H (2007) Periodontal tissue regeneration with adipose-derived stem cells. Tissue Eng 14(6):945–953
135. Tobita M, Uysal AC, Ogawa R, Hyakusoku H, Mizuno H (2008) Periodontal tissue regeneration with adipose-derived stem cells. Tissue Eng A 14(6):945–953
136. Tobita M, Uysal CA, Guo X, Hyakusoku H, Mizuno H (2013) Periodontal tissue regeneration by combined implantation of adipose tissue-derived stem cells and platelet-rich plasma in a canine model. Cytotherapy 15(12):1517–1526

137. Takahashi K, Yamanaka S (2006) Induction of pluripotent stem cells from mouse embryonic and adult fibroblast cultures by defined factors. Cell 126(4):663–676
138. Takahashi K, Tanabe K, Ohnuki M, Narita M, Ichisaka T, Tomoda K, Yamanaka S (2007) Induction of pluripotent stem cells from adult human fibroblasts by defined factors. Cell 131(5):861–872
139. Park IH, Zhao R, West JA, Yabuuchi A, Huo H, Ince TA, Lerou PH, Lensch MW, Daley GQ (2008) Reprogramming of human somatic cells to pluripotency with defined factors. Nature 451(7175):141–146
140. Yu J, Vodyanik MA, Smuga-Otto K, Antosiewicz-Bourget J, Frane JL, Tian S, Nie J, Jonsdottir GA, Ruotti V, Stewart R, Slukvin II, Thomson JA (2007) Induced pluripotent stem cell lines derived from human somatic cells. Science 318(5858):1917–1920
141. Tamaoki N, Takahashi K, Tanaka T, Ichisaka T, Aoki H, Takeda-Kawaguchi T, Iida K, Kunisada T, Shibata T, Yamanaka S, Tezuka K (2010) Dental pulp cells for induced pluripotent stem cell banking. J Dent Res 89(8):773–778
142. Trubiani O, Piattelli A, Gatta V, Marchisio M, Diomede F, D'Aurora M, Merciaro I, Pierdomenico L, Maraldi NM, Zini N (2015) Assessment of an efficient xeno-free culture system of human periodontal ligament stem cells. Tissue Eng C Methods 21(1):52–64
143. Li H, Yan F, Lei L, Li Y, Xiao Y (2009) Application of autologous cryopreserved bone marrow mesenchymal stem cells for periodontal regeneration in dogs. Cells Tissues Organs 190(2):94–101
144. Okuda K, Yamamiya K, Kawase T, Mizuno H, Ueda M, Yoshie H (2009) Treatment of human infrabony periodontal defects by grafting human cultured periosteum sheets combined with platelet-rich plasma and porous hydroxyapatite granules: case series. J Int Acad Periodontol 11(3):206–213
145. Su F, Liu SS, Ma JL, Wang DS, E LL, Liu HC (2015) Enhancement of periodontal tissue regeneration by transplantation of osteoprotegerin-engineered periodontal ligament stem cells. Stem Cell Res Ther 6:22
146. Mordenfeld A, Hallman M, Johansson CB, Albrektsson T (2010) Histological and histomorphometrical analyses of biopsies harvested 11 years after maxillary sinus floor augmentation with deproteinized bovine and autogenous bone. Clin Oral Implants Res 21(9):961–970
147. De Kok IJ, Drapeau SJ, Young R, Cooper LF (2005) Evaluation of mesenchymal stem cells following implantation in alveolar sockets: a canine safety study. Int J Oral Maxillofac Implants 20(4):511–518
148. Pelegrine AA, da Costa CE, Correa ME, Marques JF Jr (2010) Clinical and histomorphometric evaluation of extraction sockets treated with an autologous bone marrow graft. Clin Oral Implants Res 21(5):535–542
149. Dennis JE, Esterly K, Awadallah A, Parrish CR, Poynter GM, Goltry KL (2007) Clinical-scale expansion of a mixed population of bone-marrow-derived stem and progenitor cells for potential use in bone-tissue regeneration. Stem Cells 25(10):2575–2582
150. Kaigler D, Pagni G, Park CH, Braun TM, Holman LA, Yi E, Tarle SA, Bartel RL, Giannobile WV (2013) Stem cell therapy for craniofacial bone regeneration: a randomized, controlled feasibility trial. Cell Transplant 22(5):767–777
151. Kaigler D, Avila-Ortiz G, Travan S, Taut AD, Padial-Molina M, Rudek I, Wang F, Lanis A, Giannobile WV (2015) Bone engineering of maxillary sinus bone deficiencies using enriched CD90+ stem cell therapy: a randomized clinical trial. J Bone Miner Res 30(7):1206–1216
152. Chen JS, Wong VW, Gurtner GC (2012) Therapeutic potential of bone marrow-derived mesenchymal stem cells for cutaneous wound healing. Front Immunol 3:192
153. Hoogduijn MJ, Dor FJ (2013) Mesenchymal stem cells: are we ready for clinical application in transplantation and tissue regeneration? Front Immunol 4:144
154. Chen FM, Jin Y (2010) Periodontal tissue engineering and regeneration: current approaches and expanding opportunities. Tissue Eng B Rev 16(2):219–255

Dental Stem Cells for Tooth Regeneration

Thimios A. Mitsiadis, Giovanna Orsini, and Lucia Jimenez-Rojo

Tooth Formation

Odontogenesis involves a precisely orchestrated series of cellular, morphogenetic, and molecular events [1]. Generation of teeth relies upon sequential and reciprocal interactions between the oral epithelium and mesenchymal cells that are derived from the cranial neural crest (CNC) [2]. CNC-derived cells migrate from the dorsal part of the neural tube and subsequently generate craniofacial structures of unique morphology and function such as teeth (Fig. 1a) [2–4]. Genetic markers, cell labelling, and lineage tracer techniques have shown that CNC-derived cells are responsible for tooth mesenchyme formation [5–7]. Classical tissue recombination experiments between oral epithelium and CNC-derived mesenchyme have identified the oral epithelium as providing the instructive information for tooth initiation [8, 9]. Tooth formation starts at precise locations of the oral epithelium that thickens and gradually invaginates the underlying mesenchyme (Fig. 1b). The invaginating oral epithelium gives rise to the dental epithelium that progressively acquires characteristic shapes such as the bud, cap, and bell configurations. After tooth initiation the odontogenic potential shifts to the mesenchyme that can instruct any kind of epithelium to form tooth-specific structures [9, 10]. These and other tissue recombination experiments have provided evidence of an essential role of the mesenchyme in establishing tooth shape.

At the bell stage, the mesenchyme gives rise to the dental pulp. Pulp cells adjacent to the dental epithelium start to differentiate into odontoblasts, while epithelial

T.A. Mitsiadis (✉) • L. Jimenez-Rojo
Orofacial Development and Regeneration, Institute of Oral Biology, Centre for Dental Medicine, Medical Faculty, University of Zurich, Plattenstrasse 11, 8032 Zurich, Switzerland
e-mail: thimios.mitsiadis@zzm.uzh.ch

G. Orsini
Department of Clinical Sciences and Stomatology, Polytechnic University of Marche, Ancona, Italy

B. Zavan, E. Bressan (eds.), *Dental Stem Cells: Regenerative Potential*, Stem Cell Biology and Regenerative Medicine, DOI 10.1007/978-3-319-33299-4_10

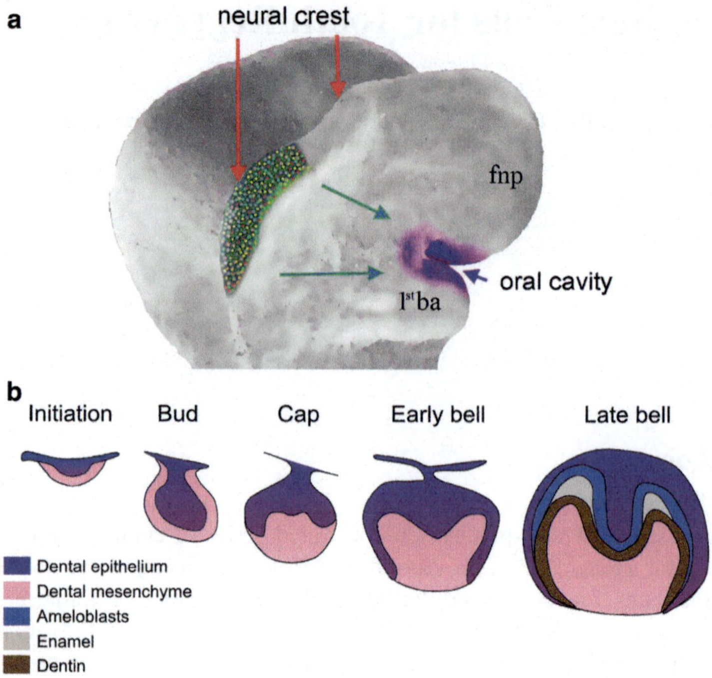

Fig. 1 Schematic representation of (**a**) cranial neural crest cell populations (*red arrows*) represented by the various colors and their migratory paths towards the oral cavity territory (*blue–green arrows*), and (**b**) the various stages of embryonic tooth development. *fnp* frontonasal process, *1st ba* first brachial arch

cells facing the dental pulp differentiate into ameloblasts [11, 12]. Odontoblasts are polarized columnar cells with processes that participate in the secretion of dentin matrix and minerals. Dentin matrix is composed mainly of collagen (90 %) and various non-collagenous proteins such as dentin sialophosphoprotein (DSPP) and dentin matrix protein 1 (DMP1) that represent the 10 % of the matrix. Apatite minerals are deposited on the matrix forming the mature calcified dentin. Following initial dentin deposition, the ameloblasts polarize and start to secrete the enamel matrix along the dentin-enamel junction. Enamel matrix is essentially composed of tissue specific hydrophobic proteins such as amelogenin, ameloblastin, enamelin, amelotin, tuftelin and ODAM (odontogenic ameloblast associated proteins) [12, 13].

Molecular Control of Odontogenesis

Over the last years a big effort has been made to understand the molecular mechanisms that dictate tooth development, pathology, and repair [1, 14]. To this contributed the constantly increasing knowledge on genetics, stem cells, and tissue engineering.

Signaling molecules control all steps of odontogenesis by coordinating cell proliferation, differentiation, apoptosis, extracellular matrix synthesis, and mineral deposition. The same molecules are repetitively used during the different stages of odontogenesis and are regulated according to a precise timing mechanism [1, 15–17]. Signals produced at a wrong time lead to inappropriate cell proliferation, differentiation, and apoptosis, thus affecting the overall tooth development and morphology [1].

The territory where teeth will form is defined early, before any obvious sign of tooth development. The earliest marker of the odontogenic territory within the oral epithelium is the transcription factor Pitx2 [18, 19]. Dental fields are also established by morphogenic signals that determine the display and fate of the CNC cells, thus leading to the generation of distinct tooth shapes [20, 21]. For example, ectodysplasin (Eda) molecules are involved in the determination of dental fields size that controls teeth number [22]. In fact, increased Eda signaling leads to supernumerary teeth formation, while Eda deletion often leads to tooth number reduction [1, 22].

Bone morphogenic proteins (BMPs), Wnts, sonic hedgehog (shh), and fibroblast growth factors (FGFs) regulate the epithelial-mesenchymal interactions during tooth initiation [1, 17]. BMP4 and FGF8 molecules constitute essential early oral epithelial signals that activate tooth specific genes (e.g. *Msx1*, *Msx2*, *Barx1*, *Dlx1*, *Dlx2*, *Pax9*) in the underlying mesenchyme [1, 18, 19, 23], thus controlling tooth patterning (i.e., BMP4 directs incisors shape while FGF8 molars shape) [2, 21].

Signals Specifying Dental Cell Fates

The specification of the various dental cell-types involves genes with restricted expression patterns during odontogenesis. It has been proposed that the determination of cell fates in teeth occurs via inhibitory interactions between adjacent cells through the Notch signaling pathway [17, 24]. Indeed, previous studies have shown that molecules of the Notch signaling pathway are involved in tooth development and dental cell specification [25–29]. In order to influence developmental decisions, molecules of the Notch signaling pathway must obviously interact with other signaling pathways. Notch-dependent cell fate acquisition between non-equivalent dental precursor cells is influenced by extrinsic signals such as BMP and FGF molecules. BMP and FGF molecules have opposite effects on the expression of Notch receptors and ligands in dental tissues [25, 26, 29], indicating that cell fate choices during odontogenesis are under the concomitant control of the Notch and BMP/FGF signaling pathways. Furthermore, it has been shown that Notch-mediated lateral inhibition has a pivotal role in the establishment of the tooth morphology, as shown in *Jagged2* mutant mice where the overall development and structure of their teeth is severely affected [28, 29]. Several genetic findings have shown that Tbx1 plays a significant role for the early determination of epithelial cells to adopt the ameloblast fate. Indeed, hypoplastic incisors that lack enamel are observed in mice lacking the *Tbx1* gene [30, 31]. Mesenchyme-derived FGF molecules activate the expression of *Tbx1* in dental epithelium [32, 33].

Dental Pathologies

Traumatic injuries and external harmful agents such as bacteria and acids jeopardize tooth integrity. Periodontal disease and caries are mainly responsible for pathologies affecting the dental and alveolar bone tissues. Dental pathologies, combined with age, constitute important factors that can trigger tooth loss [34].

In most cases, the reparative mechanisms following dental and periodontal injuries involve a series of highly conserved processes that share genetic programs that occur throughout embryogenesis [14]. For example, after severe injury or inflammation of the dental pulp (i.e., pulpitis), the dying odontoblasts are replaced by progenitor cells, which differentiate into a new generation of odontoblasts and produce the reparative dentin [14, 35]. Signaling molecules released at the injury sites may attract progenitor cells, thus initiating the healing processes.

The increased knowledge on the reparative events within dental tissues has contributed to the proposal of alternative methods for the treatment of dental pathologies. However, traditional treatments continue to be applied in dental clinics since most of these novel approaches are at the experimental level and not well established yet.

Traditional Dental Treatments and Main Limitations

Restorative dentistry has undergone significant advances in recent years [36, 37]. Contemporary techniques to replace partially lost hard tissues consist of direct or indirect tooth-colored restorations (e.g., resin-based composites, porcelain inlays and onlays). The resin-based composites are bonded to the dental tissues through elaborated adhesive materials. While adhesion to enamel is stable over time, adhesion to dentin is weaker and unstable, mainly because of the higher levels of organic matrix of dentin [38]. Therefore, the goal of the adhesive procedures is the widespread infiltration and encapsulation of the demineralized collagen fibrils of dentin with the monomeric resin [37, 39]. Dentin biomodification strategies have as goal to enhance the properties of dentin matrix, thus inducing remarkable resin resistance against degradation that leads to short durability of restorations [39–41].

Endodontic therapy is a procedure implying the removal of contaminated or necrotic dental tissues within the pulp, thus eliminating future tooth contamination. Clinically, vital pulp therapy can be divided into two main groups: indirect and direct pulp capping. Indirect pulp capping is achieved by applying a protective agent on the thin layer of dentin remaining over a nearly exposed pulp, in order to allow the underlying dental pulp to repair [42]. Direct pulp capping necessitates the use of a protective agent directly on the exposed pulp, thus allowing pulp regeneration [43]. In case of large pulp exposure or infection, the coronal pulp has to be removed before direct pulp capping at the tooth root level (i.e., pulpotomy) [44]. This method preserves the vitality of pulp located at the root canal, thus allowing the accomplishment of the root growth [45]. Apexogenesis is another procedure of vital pulp treatment that promotes physiological development and formation of the root using mineral trioxide aggregates [46]. When the whole pulp is damaged or infected is replaced with inorganic materials such as guttapercha after thorough root canal treatment. However, root canal treatment results in the loss of dental pulp vitality. Dental pulp is an important element of the tooth since it provides nutrition, sensation, and defense against the various pathogens. Therefore, devitalized teeth are subsequent to various complications causing tooth strength reduction, increased fragility, and predisposition to postoperative fracture [47]. In terms of aesthetics, endodontic treatment often results in tooth crown discoloration due to either the filling material or the unsuccessful pulp chamber cleaning [48]. Thus, it becomes obvious that the maintaining of dental pulp vitality is of prime importance and this is highlighted by the emergence of new cell-based techniques focusing on partial or total pulp regeneration.

Missing teeth have traditionally been replaced with removable dentures, fixed bridges, and dental implants [49]. Currently, dental implants offer the most valuable option for tooth replacement, but a prerequisite for successful implant treatment is the quantity and quality of the remaining alveolar bone. Indeed, dental implant retention requires the close contact between the alveolar bone and the surface of the implant, a process called osseointegration. Dental implants are made of biocompatible titanium and they are inserted into the bone after surgical intervention. Survival rates of osseointegrated dental implants have been reported to be of over 95 % [50]. Apart alveolar bone quality and dimensions, the clinical success of implants also depends on primary implant stability, surgical methods, time of loading, infections, and implant surface characteristics [51]. When the dental implant is not well anchored to the bone, soft tissue encapsulation can occur, thus leading to unsuccessful treatment. Nanotechnology offer the opportunity to develop new implant surfaces that increased implant osseointegration [52], thus allowing the shortening of healing period without jeopardizing implant success [49]. However, the risk of infection of the tissues surrounding the implant (i.e., peri-implantitis) is still high and has to be taken seriously into account [53].

Recently, dental implants have been benefited from outcomes from stem cell biology and tissue engineering in order to maximize their survival rate in cases of poor bone quality. New regenerative technologies using scaffolds, stem cells, drug and growth factor delivery, and gene therapy are expected to enhance host tissue response and implant osseointegration [54]. Indeed, numerous pre-clinical studies have been performed already in various large animal models for guided bone regeneration around implants using growth factors and protein delivery [55, 56]. More recently, clinical studies have been realized using growth factor delivery for osseous and periodontal regeneration [57–59]. Pre-clinical applications for oral tissue regeneration using stem cell and gene therapies have been realized in rat models [60, 61]. Presently, safe and effective cell-based regenerative therapies have started to be applied in the dental clinics [54]. Clinical trials have documented that stem cells seeded in specific scaffolds are capable of generating sufficient amounts of bone in order to achieve primary implant stability [62–66]. However, all these new approaches should be further studied using controlled randomized clinical trials.

Alternative Dental Treatments

Novel cell-based therapeutic approaches have been proposed the last two decades [34, 67–69]. These treatments are promising, challenging, and complement traditional restorative or surgical techniques that are currently used in dentistry. The administration to patients of stem cells in conjunction with scaffolds and bioactive molecules might accelerate and increase the repair process of the impaired dental tissues. Although most of the studies focus on partial dental tissue regeneration, few attempts have been also made for the regeneration of entire teeth.

Here we report on selected dental stem cell populations that are currently used for experimental regenerative purposes in dentistry.

Stem Cells in Tooth Homeostasis and Regeneration

Stem cells play a critical role in tissue homeostasis and repair. Their fate is regulated by cell intrinsic determinants and signals from a specialized microenvironment [70–73]. Molecules of the Wnt, Notch, and BMP pathways have been shown to control stem cell fate specification [71, 74, 75]. The growing interest in the molecular regulation of dental stem cells arises from the potential to influence their fate and consequently their functions during tooth repair.

A considerable effort has been produced the last years to isolate epithelial and mesenchymal stem cell populations from deciduous and adult human teeth. Dental mesenchymal stem cells (DMSCs) were found in the dental pulp of permanent and exfoliated deciduous human teeth (Fig. 2) [11, 34, 69]. DMSCs were also identified in the apical part of dental papilla, the dental follicle, and the periodontal ligament. Because dental epithelial stem cells (DESCs) are very rare in adult human teeth, the current knowledge on DESCs has been obtained from studies in rodent incisors, which are continuously growing teeth [24, 76].

Stem Cell Populations Within the Dental Pulp

Activation of Notch signaling in various dental pulp cell populations during the dynamic processes triggered by dental injury may reflect the existence of specific stem cell pools [71, 77, 78]. Indeed, Notch expression is activated in endothelial cells, in cells located close to injury sites, as well as in cells located at the apex of the roots [71, 77]. Currently there are available protocols for the isolation and in vitro culture adult DMSCs [79–81]. These protocols are continuously being improved in order to further enrich the DMSC populations. Mesenchymal stem cell markers such as STRO-1, CD146 and CD44 are commonly used for the isolation of dental pulp stem cells (DPSCs) [11, 34, 79]. These cells are capable to differentiate

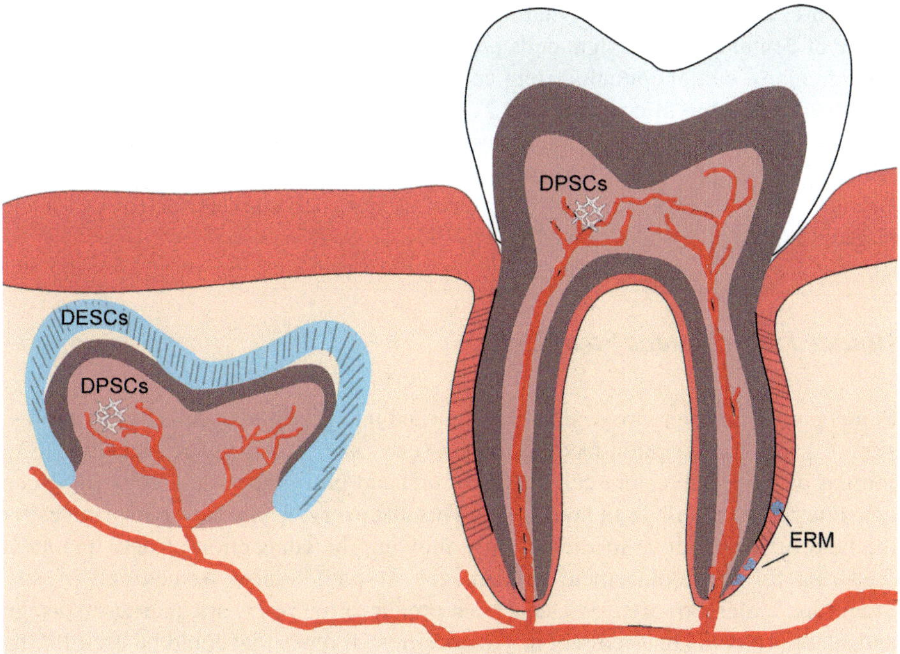

Fig. 2 Schematic representation of the locations of epithelial stem cells (i.e., hDESCs and ERM) and dental pulp stem cells that can be found in unerupted and erupted teeth of adult humans. *ERM* epithelial rests of Malassez, *hDESCs* human dental epithelial stem cells, *DPSCs* dental pulp stem cells

into odontogenic, osteogenic, chondrogenic, adipogenic, myogenic, and neurogenic cells in vitro and in vivo [11, 34]. Pulp-dentin tissues have been generated after ectopic transplantation of DPSCs mixed with hydroxyapatite/tricalcium phosphate [79]. Furthermore, clinical trials using autologous human DPSCs combined with collagen scaffolds for alveolar bone reconstruction have been successfully performed several years ago [82].

Dental Epithelial Stem Cells

In mice, cells responsible for the continual growth of the incisor are located in the cervical loop area (posterior part of the incisor), which is a putative stem cell niche [76]. Cells originated from the cervical loop region are clonogenic and express Notch receptors and stem cell markers such as Sox2 and $p75^{NTFR}$ [26, 28, 76, 83]. This expression pattern indicates that Notch signaling plays a role in the maintenance of the dental epithelial stem cells. FGF3 and FGF10 are important mesenchymal signals involved in the maintenance and creation of the adult stem cell compartment within the cervical [84]. In the FGF10 null mice, the incisors are smaller due to the lack of the stem cell compartment in the cervical loop [84].

Therefore, FGF and Notch signaling pathways appear to interact for the maintenance of dental epithelial stem cells [76, 83].

In humans, dental epithelial stem cells (hDESCs) can be isolated from the third molars that develop after birth (Fig. 2) [11, 68]. Another source of hDESCs is the epithelial root sheath that disintegrates and forms the epithelial rests of Malassez (ERM) (Fig. 2) [11, 68]. ERM cells express epithelial stem cell markers such as Bmi-1, E-CAM, and p75, as well as embryonic stem cell markers such as Oct-4 and Nanog [24, 69].

Induced Pluripotent Stem Cells

Somatic cells can be reprogrammed and turned into pluripotent cells by the expression of a four transcription factors cocktail (i.e., Oct4, Sox2, c-Myc and Klf4) [85]. Similar to embryonic stem cells (ESCs), induced pluripotent stem cells (iPS) can generate cells from all germ layers [85]. This discovery opened new perspectives for the field of regenerative medicine, thus allowing the conception of new treatment scenarios for the replacement of damaged or pathological tissues and organs. Therefore, a plethora of efforts have been produced the last years in order to design and adapt protocols that redirect iPS to specific cell types that could be used for the regeneration of a precise organ. In fact, the first clinical trial using iPS technology has been recently initiated: a sheet of iPS-derived retinal cells has been transplanted in a patient for the successful treatment of the age-related macular degeneration (AMD) pathology [86].

For dental-linked purposes, iPS have been generated from different dental cell types. Furthermore, iPS derived from skin fibroblasts have been used to produce various dental mesenchymal and epithelial cell populations [87].

Experimental Approaches for Tooth Regeneration

Regeneration of Dentin-Pulp Complex

Dental pulp is composed by a highly innervated and vascularized connective tissue containing a variety of cell types such as fibroblasts, immune cells, odontoblasts, and stem cells [88]. DPSCs show a limited potential to regenerate dentin, which becomes insufficient after tooth damage such as to caries or trauma. Several attempts have already been made in order to use DMSCs for the regeneration of the dentin-pulp complex in a variety of animal models. The first ever experiment realized using DMSCs isolated from human teeth has demonstrated that these cells are able to give rise to odontoblast-like cells, which form a dentin-like structure when transplanted together with HA/TCP ceramic powder into the dorsal surface of immunocompromised mice (Fig. 3a) [79]. Infiltration of mouse-derived blood vessels was observed in the dentin-pulp complex generated under these conditions.

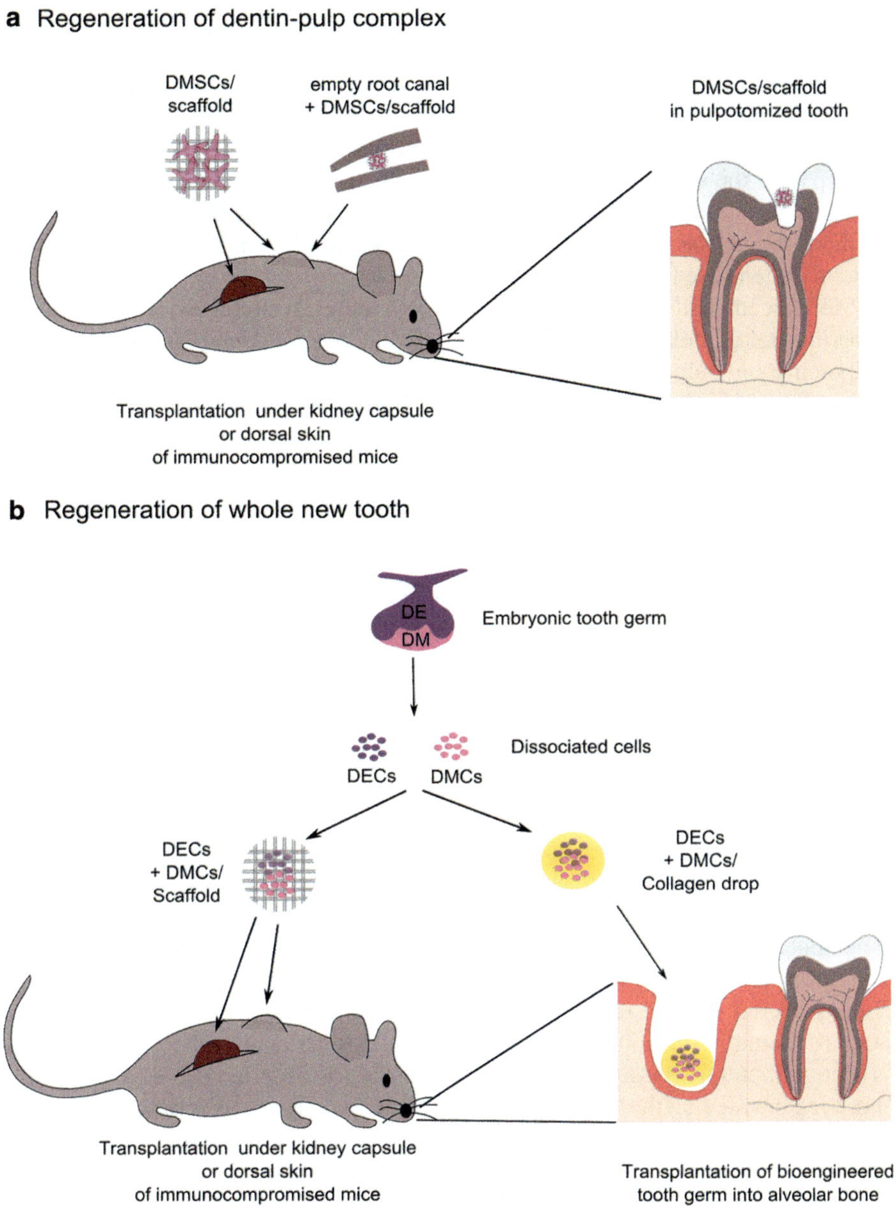

Fig. 3 Summary of different strategies for the regeneration of (**a**) dentin-pulp complex, and (**b**) an entire tooth. *DE* dental epithelium, *DM* dental mesenchyme, *DECs* dental epithelial cells, *DMCs* dental mesenchymal cells, *DMSCs* dental mesenchymal stem cells

More recent studies, using human DPSCs and SCAPs seeded on a poly-D,L-lactide/glycole scaffold and transplanted into the empty root canal space of mouse teeth, have demonstrated the ability of DMSCs to regenerate vascularized dentin-pulp complexes (Fig. 3a) [89–91].

Although being a useful in vivo model to assess the potency of DMSCs to form all or part of the various pulp tissues in vivo, the these experimental strategies were based on the transplantation of DMSCs in ectopic locations (Fig. 3a). Thus, those methods cannot be directly translated into a potential stem cell-based therapy in order to regenerate human dental tissues.

More recent studies have adopted more physiological strategies. For example, an experimental model consists of either the partial or total dental pulp removal (i.e., pulpotomy and pulpectomy, respectively), which is followed by fulfilling the empty space of the pulp chamber with DMSCs either alone or in combination with scaffold and bioactive molecules (Fig. 3a) [92–94]. For instance, autologous DPSCs transplanted together with granulocyte-colony stimulating factor (G-CSF) in pulpectomized teeth of dogs, gave rise to new dentin formation and vascularized and innervated pulp tissues [94]. Interestingly, partial pulp regeneration was also observed in pulpectomized teeth without DPSCs transplantation that were used as control, indicating that an endogenous population of DMSCs may be also participating in the regenerative process.

Regeneration of a Whole New Tooth

Severe dental pathologies can lead to tooth loss. In such cases, whole tooth regeneration could be the considered the right therapeutic approach. Functional teeth have been experimentally bioengineered by the re-association of disaggregated epithelial and mesenchymal cells from mouse embryonic tooth germs [67, 68, 95]. The bioengineered tooth germs can be subsequently transplanted into the alveolar bone and erupt into the oral cavity, thus replacing the missing teeth (Fig. 3b). Bioengineered erupted teeth possess enamel and dentin, are innervated, and can achieve normal occlusion indicating that are fully functional teeth [95]. However, it is quite difficult to control tooth size and shape during the in vitro development of these regenerated teeth. Indeed, the teeth formed using this experimental protocol were smaller when compared to the natural teeth. Thus, a big challenge is the generation of bioengineering functional teeth with the appropriate crown morphology. Therefore, the use of specifically designed scaffolds that will respect the crown anatomy of the various human teeth will greatly help to overcome that limitation [96–99]. Several attempts have been made towards this direction using dental epithelial and mesenchymal stem cells combined with tailor made scaffolds. Nevertheless, the result was the generation of teeth with irregular dental structures [96]. It is obvious that this method needs to be revisited and optimized before any application in dental clinics.

The previously mentioned approaches use embryonic dental stem cells in order to generate bioengineered teeth in the various small animal models. However, for the generation of teeth in humans the use of embryonic stem cells is not possible yet because of the various ethical and regulatory problems [34]. Therefore, the is an absolute need to find alternative sources of human stem cells that could serve for tooth regeneration purposes in the dental clinics. Ideally, those stem cells would be isolated from adult tissues. Although a sufficient number of DPSCs can be collected from the pulp of adult human teeth, harvesting DESCs from adult teeth for the generation of epithelial structures in the bioengineered teeth is almost impossible. In this sense, iPS technology could be of great help for the formation of dental epithelial tissues. Indeed, studies in mice have shown that mouse iPS can form ameloblast-like cells when co-cultured with dental epithelium [100]. In addition, a similar effect has been observed when mouse iPS were cultured in presence of ERM-conditioned medium [101] or recombined with dental embryonic mesenchyme [87]. However, it is still required to identify the precise molecules responsible for the redirection of iPS cells towards a dental epithelial fate. A better understanding of the regulation of iPS reprogramming towards dental cell types is essential in order to develop reproducible and safety strategies for new iPS-based therapies for regenerative purposes in dentistry.

Conclusion

The control of tooth regeneration necessitates a thorough understanding at the cellular and molecular level. The identification and characterization of dental epithelial and mesenchymal stem cell populations is a considerable accomplishment that can offer new perspectives in dentistry. Understanding when and how signaling molecules control the fate, migration, proliferation, and differentiation of dental stem cells will open new unexplored horizons for all dental disciplines. The acquired scientific knowledge together with novel approaches based on tissue engineering and nanotechnology are likely to instruct development of innovative dental therapies and materials.

Acknowledgements This work was supported by funds of the University of Zurich (TAM and LJR) and the Polytechnic University of Marche (GO). The authors contributed to the planning, writing, critical reading, and editing of the present book chapter. The authors confirm that there are no conflicts of interest associated with this work.

References

1. Mitsiadis TA, Luder HU (2011) Genetic basis for tooth malformations: from mice to men and back again. Clin Genet 80(4):319–329
2. Cobourne MT, Mitsiadis T (2006) Neural crest cells and patterning of the mammalian dentition. J Exp Zool B Mol Dev Evol 306(3):251–260

3. Anderson DJ (1993) Molecular control of cell fate in the neural crest: the sympathoadrenal lineage. Annu Rev Neurosci 16:129–158
4. Weissman IL, Anderson DJ, Gage F (2001) Stem and progenitor cells: origins, phenotypes, lineage commitments, and transdifferentiations. Annu Rev Cell Dev Biol 17:387–403
5. Imai H, Osumi-Yamashita N, Ninomiya Y, Eto K (1996) Contribution of early-emigrating midbrain crest cells to the dental mesenchyme of mandibular molar teeth in rat embryos. Dev Biol 176(2):151–165
6. Chai Y, Jiang X, Ito Y, Bringas P Jr, Han J, Rowitch DH, Soriano P, McMahon AP, Sucov HM (2000) Fate of the mammalian cranial neural crest during tooth and mandibular morphogenesis. Development 127(8):1671–1679
7. Diep L, Matalova E, Mitsiadis TA, Tucker AS (2009) Contribution of the tooth bud mesenchyme to alveolar bone. J Exp Zool B Mol Dev Evol 312B(5):510–517
8. Lumsden AG (1988) Spatial organization of the epithelium and the role of neural crest cells in the initiation of the mammalian tooth germ. Dev Suppl 103:155–169
9. Mina M, Kollar EJ (1987) The induction of odontogenesis in non-dental mesenchyme combined with early murine mandibular arch epithelium. Arch Oral Biol 32(2):123–127
10. Kollar EJ, Fisher C (1980) Tooth induction in chick epithelium: expression of quiescent genes for enamel synthesis. Science 207(4434):993–995
11. Bluteau G, Luder HU, De Bari C, Mitsiadis TA (2008) Stem cells for tooth engineering. Eur Cell Mater 16:1–9
12. Mitsiadis T, Papagerakis P (2013) Development and structure of teeth and periodontal tissues. In: Rosen CJ (ed) Primer on the metabolic bone diseases and disorders of mineral metabolism. Wiley-Blackwell, New Jersey (United States), pp 904–913
13. Sire JY, Davit-Beal T, Delgado S, Gu X (2007) The origin and evolution of enamel mineralization genes. Cells Tissues Organs 186(1):25–48
14. Mitsiadis TA, Rahiotis C (2004) Parallels between tooth development and repair: conserved molecular mechanisms following carious and dental injury. J Dent Res 83(12):896–902
15. Dorsky RI, Moon RT, Raible DW (2000) Environmental signals and cell fate specification in premigratory neural crest. Bioessays 22(8):708–716
16. Moss EG (2007) Heterochronic genes and the nature of developmental time. Curr Biol 17(11):R425–R434
17. Mitsiadis TA, Graf D (2009) Cell fate determination during tooth development and regeneration. Birth Defects Res C Embryo Today 87(3):199–211
18. Mitsiadis TA, Mucchielli ML, Raffo S, Proust JP, Koopman P, Goridis C (1998) Expression of the transcription factors Otlx2, Barx1 and Sox9 during mouse odontogenesis. Eur J Oral Sci 106(Suppl 1):112–116
19. Mucchielli ML, Mitsiadis TA, Raffo S, Brunet JF, Proust JP, Goridis C (1997) Mouse Otlx2/RIEG expression in the odontogenic epithelium precedes tooth initiation and requires mesenchyme-derived signals for its maintenance. Dev Biol 189(2):275–284
20. Green J (2002) Morphogen gradients, positional information, and *Xenopus*: interplay of theory and experiment. Dev Dyn 225(4):392–408
21. Mitsiadis TA, Smith MM (2006) How do genes make teeth to order through development? J Exp Zool B Mol Dev Evol 306(3):177–182
22. Mikkola ML (2008) TNF superfamily in skin appendage development. Cytokine Growth Factor Rev 19(3–4):219–230
23. Neubuser A, Peters H, Balling R, Martin GR (1997) Antagonistic interactions between FGF and BMP signaling pathways: a mechanism for positioning the sites of tooth formation. Cell 90(2):247–255
24. Jimenez-Rojo L, Granchi Z, Graf D, Mitsiadis TA (2012) Stem cell fate determination during development and regeneration of ectodermal organs. Front Physiol 3:107
25. Mitsiadis TA, Henrique D, Thesleff I, Lendahl U (1997) Mouse Serrate-1 (Jagged-1): expression in the developing tooth is regulated by epithelial-mesenchymal interactions and fibroblast growth factor-4. Development 124(8):1473–1483

26. Mitsiadis TA, Hirsinger E, Lendahl U, Goridis C (1998) Delta-notch signaling in odontogenesis: correlation with cytodifferentiation and evidence for feedback regulation. Dev Biol 204(2):420–431
27. Mitsiadis TA, Lardelli M, Lendahl U, Thesleff I (1995) Expression of Notch 1, 2 and 3 is regulated by epithelial-mesenchymal interactions and retinoic acid in the developing mouse tooth and associated with determination of ameloblast cell fate. J Cell Biol 130(2):407–418
28. Mitsiadis TA, Regaudiat L, Gridley T (2005) Role of the Notch signalling pathway in tooth morphogenesis. Arch Oral Biol 50(2):137–140
29. Mitsiadis TA, Graf D, Luder H, Gridley T, Bluteau G (2010) BMPs and FGFs target Notch signalling via jagged 2 to regulate tooth morphogenesis and cytodifferentiation. Development 137(18):3025–3035
30. Caton J, Luder HU, Zoupa M, Bradman M, Bluteau G, Tucker AS, Klein O, Mitsiadis TA (2009) Enamel-free teeth: Tbx1 deletion affects amelogenesis in rodent incisors. Dev Biol 328(2):493–505
31. Jerome LA, Papaioannou VE (2001) DiGeorge syndrome phenotype in mice mutant for the T-box gene, Tbx1. Nat Genet 27(3):286–291
32. Kettunen P, Laurikkala J, Itaranta P, Vainio S, Itoh N, Thesleff I (2000) Associations of FGF-3 and FGF-10 with signaling networks regulating tooth morphogenesis. Dev Dyn 219(3):322–332
33. Mitsiadis TA, Tucker AS, De Bari C, Cobourne MT, Rice DP (2008) A regulatory relationship between Tbx1 and FGF signaling during tooth morphogenesis and ameloblast lineage determination. Dev Biol 320(1):39–48
34. Caton J, Bostanci N, Remboutsika E, De Bari C, Mitsiadis TA (2011) Future dentistry: cell therapy meets tooth and periodontal repair and regeneration. J Cell Mol Med 15(5): 1054–1065
35. Mitsiadis TA, De Bari C, About I (2008) Apoptosis in developmental and repair-related human tooth remodeling: a view from the inside. Exp Cell Res 314(4):869–877
36. Ferracane JL (2011) Resin composite: state of the art. Dent Mater 27(1):29–38
37. Pashley DH, Tay FR, Breschi L, Tjaderhane L, Carvalho RM, Carrilho M, Tezvergil-Mutluay A (2011) State of the art etch-and-rinse adhesives. Dent Mater 27(1):1–16
38. Lehmann N, Debret R, Romeas A, Magloire H, Degrange M, Bleicher F, Sommer P, Seux D (2009) Self-etching increases matrix metalloproteinase expression in the dentin-pulp complex. J Dent Res 88(1):77–82
39. Breschi L, Mazzoni A, Ruggeri A, Cadenaro M, Di Lenarda R, De Stefano DE (2008) Dental adhesion review: aging and stability of the bonded interface. Dent Mater 24(1):90–101
40. Castellan CS, Bedran-Russo AK, Karol S, Pereira PN (2011) Long-term stability of dentin matrix following treatment with various natural collagen cross-linkers. J Mech Behav Biomed Mater 4(7):1343–1350
41. Bedran-Russo AK, Castellan CS, Shinohara MS, Hassan L, Antunes A (2011) Characterization of biomodified dentin matrices for potential preventive and reparative therapies. Acta Biomater 7(4):1735–1741
42. Fuks AB (2008) Vital pulp therapy with new materials for primary teeth: new directions and treatment perspectives. J Endod 34(7 Suppl):S18–S24
43. Dammaschke T (2008) The history of direct pulp capping. J Hist Dent 56(1):9–23
44. DeRosa TA (2006) A retrospective evaluation of pulpotomy as an alternative to extraction. Gen Dent 54(1):37–40
45. Bishop BG, Woollard GW (2002) Modern endodontic therapy for an incompletely developed tooth. Gen Dent 50(3):252–256, quiz 257–258
46. Nosrat A, Asgary S (2010) Apexogenesis treatment with a new endodontic cement: a case report. J Endod 36(5):912–914
47. Ricketts D (2001) Management of the deep carious lesion and the vital pulp dentine complex. Br Dent J 191(11):606–610
48. Goto Y, Ceyhan J, Chu SJ (2009) Restorations of endodontically treated teeth: new concepts, materials, and aesthetics. Pract Proced Aesthet Dent 21(2):81–89

49. Esposito M, Grusovin MG, Maghaireh H, Worthington HV (2013) Interventions for replacing missing teeth: different times for loading dental implants. Cochrane Database Syst Rev 3, CD003878
50. Temmerman A, Keestra JA, Coucke W, Teughels W, Quirynen M (2015) The outcome of oral implants placed in bone with limited bucco-oral dimensions: a 3-year follow-up study. J Clin Periodontol 42(3):311–318
51. Esposito M, Murray-Curtis L, Grusovin MG, Coulthard P, Worthington HV (2007) Interventions for replacing missing teeth: different types of dental implants. Cochrane Database Syst Rev 4, CD003815
52. Variola F, Brunski JB, Orsini G, Tambasco de Oliveira P, Wazen R, Nanci A (2011) Nanoscale surface modifications of medically relevant metals: state-of-the art and perspectives. Nanoscale 3(2):335–353
53. Singh P (2011) Understanding peri-implantitis: a strategic review. J Oral Implantol 37(5): 622–626
54. Pilipchuk SP, Plonka AB, Monje A, Taut AD, Lanis A, Kang B, Giannobile WV (2015) Tissue engineering for bone regeneration and osseointegration in the oral cavity. Dent Mater 31(4):317–338
55. Lynch SE, Buser D, Hernandez RA, Weber HP, Stich H, Fox CH, Williams RC (1991) Effects of the platelet-derived growth factor/insulin-like growth factor-I combination on bone regeneration around titanium dental implants. Results of a pilot study in beagle dogs. J Periodontol 62(11):710–716
56. Becker W, Lynch SE, Lekholm U, Becker BE, Caffesse R, Donath K, Sanchez R (1992) A comparison of ePTFE membranes alone or in combination with platelet-derived growth factors and insulin-like growth factor-I or demineralized freeze-dried bone in promoting bone formation around immediate extraction socket implants. J Periodontol 63(11):929–940
57. Windisch P, Stavropoulos A, Molnar B, Szendroi-Kiss D, Szilagyi E, Rosta P, Horvath A, Capsius B, Wikesjo UM, Sculean A (2012) A phase IIa randomized controlled pilot study evaluating the safety and clinical outcomes following the use of rhGDF-5/beta-TCP in regenerative periodontal therapy. Clin Oral Investig 16(4):1181–1189
58. Kitamura M, Akamatsu M, Machigashira M, Hara Y, Sakagami R, Hirofuji T, Hamachi T, Maeda K, Yokota M, Kido J, Nagata T, Kurihara H, Takashiba S, Sibutani T, Fukuda M, Noguchi T, Yamazaki K, Yoshie H, Ioroi K, Arai T, Nakagawa T, Ito K, Oda S, Izumi Y, Ogata Y, Yamada S, Shimauchi H, Kunimatsu K, Kawanami M, Fujii T, Furuichi Y, Furuuchi T, Sasano T, Imai E, Omae M, Yamada S, Watanuki M, Murakami S (2011) FGF-2 stimulates periodontal regeneration: results of a multi-center randomized clinical trial. J Dent Res 90(1):35–40
59. Bashutski JD, Eber RM, Kinney JS, Benavides E, Maitra S, Braun TM, Giannobile WV, McCauley LK (2010) Teriparatide and osseous regeneration in the oral cavity. N Engl J Med 363(25):2396–2405
60. Huang YC, Simmons C, Kaigler D, Rice KG, Mooney DJ (2005) Bone regeneration in a rat cranial defect with delivery of PEI-condensed plasmid DNA encoding for bone morphogenetic protein-4 (BMP-4). Gene Ther 12(5):418–426
61. Cirelli JA, Park CH, MacKool K, Taba M Jr, Lustig KH, Burstein H, Giannobile WV (2009) AAV2/1-TNFR:Fc gene delivery prevents periodontal disease progression. Gene Ther 16(3):426–436
62. Pelegrine AA, da Costa CE, Correa ME, Marques JF Jr (2010) Clinical and histomorphometric evaluation of extraction sockets treated with an autologous bone marrow graft. Clin Oral Implants Res 21(5):535–542
63. Gonshor A, McAllister BS, Wallace SS, Prasad H (2011) Histologic and histomorphometric evaluation of an allograft stem cell-based matrix sinus augmentation procedure. Int J Oral Maxillofac Implants 26(1):123–131
64. Rickert D, Sauerbier S, Nagursky H, Menne D, Vissink A, Raghoebar GM (2011) Maxillary sinus floor elevation with bovine bone mineral combined with either autogenous bone or autogenous stem cells: a prospective randomized clinical trial. Clin Oral Implants Res 22(3):251–258

65. Sauerbier S, Rickert D, Gutwald R, Nagursky H, Oshima T, Xavier SP, Christmann J, Kurz P, Menne D, Vissink A, Raghoebar G, Schmelzeisen R, Wagner W, Koch FP (2011) Bone marrow concentrate and bovine bone mineral for sinus floor augmentation: a controlled, randomized, single-blinded clinical and histological trial--per-protocol analysis. Tissue Eng A 17(17-18):2187–2197
66. Schmelzeisen R, Gutwald R, Oshima T, Nagursky H, Vogeler M, Sauerbier S (2011) Making bone II: maxillary sinus augmentation with mononuclear cells--case report with a new clinical method. Br J Oral Maxillofac Surg 49(6):480–482
67. Mitsiadis TA, Papagerakis P (2011) Regenerated teeth: the future of tooth replacement? Regen Med 6(2):135–139
68. Mitsiadis TA, Harada H (2015) Regenerated teeth: the future of tooth replacement. An update. Regen Med 10(1):5–8
69. Mitsiadis TA, Woloszyk A, Jimenez-Rojo L (2012) Nanodentistry: combining nanostructured materials and stem cells for dental tissue regeneration. Nanomedicine (Lond) 7(11):1743–1753
70. Moore KA, Lemischka IR (2006) Stem cells and their niches. Science 311(5769): 1880–1885
71. Mitsiadis TA, Feki A, Papaccio G, Caton J (2011) Dental pulp stem cells, niches, and notch signaling in tooth injury. Adv Dent Res 23(3):275–279
72. Pagella P, Jimenez-Rojo L, Mitsiadis TA (2014) Roles of innervation in developing and regenerating orofacial tissues. Cell Mol Life Sci 71(12):2241–2251
73. Pagella P, Neto E, Lamghari M, Mitsiadis TA (2015) Investigation of orofacial stem cell niches and their innervation through microfluidic devices. Eur Cell Mater 29:213–223
74. Betschinger J, Knoblich JA (2004) Dare to be different: asymmetric cell division in *Drosophila*, *C. elegans* and vertebrates. Curr Biol 14(16):R674–R685
75. Lechler T, Fuchs E (2005) Asymmetric cell divisions promote stratification and differentiation of mammalian skin. Nature 437(7056):275–280
76. Mitsiadis TA, Barrandon O, Rochat A, Barrandon Y, De Bari C (2007) Stem cell niches in mammals. Exp Cell Res 313(16):3377–3385
77. Mitsiadis TA, Fried K, Goridis C (1999) Reactivation of Delta-Notch signaling after injury: complementary expression patterns of ligand and receptor in dental pulp. Exp Cell Res 246(2):312–318
78. Mitsiadis TA, Romeas A, Lendahl U, Sharpe PT, Farges JC (2003) Notch2 protein distribution in human teeth under normal and pathological conditions. Exp Cell Res 282(2): 101–109
79. Gronthos S, Mankani M, Brahim J, Robey PG, Shi S (2000) Postnatal human dental pulp stem cells (DPSCs) in vitro and in vivo. Proc Natl Acad Sci U S A 97(25):13625–13630
80. Miura M, Gronthos S, Zhao M, Lu B, Fisher LW, Robey PG, Shi S (2003) SHED: stem cells from human exfoliated deciduous teeth. Proc Natl Acad Sci U S A 100(10):5807–5812
81. Sonoyama W, Liu Y, Yamaza T, Tuan RS, Wang S, Shi S, Huang GT (2008) Characterization of the apical papilla and its residing stem cells from human immature permanent teeth: a pilot study. J Endod 34(2):166–171
82. d'Aquino R, De Rosa A, Lanza V, Tirino V, Laino L, Graziano A, Desiderio V, Laino G, Papaccio G (2009) Human mandible bone defect repair by the grafting of dental pulp stem/progenitor cells and collagen sponge biocomplexes. Eur Cell Mater 18:75–83
83. Harada H, Kettunen P, Jung HS, Mustonen T, Wang YA, Thesleff I (1999) Localization of putative stem cells in dental epithelium and their association with Notch and FGF signaling. J Cell Biol 147(1):105–120
84. Harada H, Toyono T, Toyoshima K, Yamasaki M, Itoh N, Kato S, Sekine K, Ohuchi H (2002) FGF10 maintains stem cell compartment in developing mouse incisors. Development 129(6):1533–1541
85. Takahashi K, Yamanaka S (2006) Induction of pluripotent stem cells from mouse embryonic and adult fibroblast cultures by defined factors. Cell 126(4):663–676

86. Kamao H, Mandai M, Okamoto S, Sakai N, Suga A, Sugita S, Kiryu J, Takahashi M (2014) Characterization of human induced pluripotent stem cell-derived retinal pigment epithelium cell sheets aiming for clinical application. Stem Cell Rep 2(2):205–218
87. Otsu K, Kumakami-Sakano M, Fujiwara N, Kikuchi K, Keller L, Lesot H, Harada H (2014) Stem cell sources for tooth regeneration: current status and future prospects. Front Physiol 5:36
88. Nanci A (2008) Ten cate's oral histology: development, structure and function (Oral Anatomy), 7th edn. Mosby, St. Louis, p 411
89. Huang GT, Yamaza T, Shea LD, Djouad F, Kuhn NZ, Tuan RS, Shi S (2010) Stem/progenitor cell-mediated de novo regeneration of dental pulp with newly deposited continuous layer of dentin in an in vivo model. Tissue Eng A 16(2):605–615
90. Volponi AA, Pang Y, Sharpe PT (2010) Stem cell-based biological tooth repair and regeneration. Trends Cell Biol 20(12):715–722
91. Hayashi Y, Murakami M, Kawamura R, Ishizaka R, Fukuta O, Nakashima M (2015) CXCL14 and MCP1 are potent trophic factors associated with cell migration and angiogenesis leading to higher regenerative potential of dental pulp side population cells. Stem Cell Res Ther 6:111
92. Zheng Y, Wang XY, Wang YM, Liu XY, Zhang CM, Hou BX, Wang SL (2012) Dentin regeneration using deciduous pulp stem/progenitor cells. J Dent Res 91(7):676–682
93. Iohara K, Imabayashi K, Ishizaka R, Watanabe A, Nabekura J, Ito M, Matsushita K, Nakamura H, Nakashima M (2011) Complete pulp regeneration after pulpectomy by transplantation of CD105+ stem cells with stromal cell-derived factor-1. Tissue Eng A 17(15–16): 1911–1920
94. Iohara K, Murakami M, Takeuchi N, Osako Y, Ito M, Ishizaka R, Utunomiya S, Nakamura H, Matsushita K, Nakashima M (2013) A novel combinatorial therapy with pulp stem cells and granulocyte colony-stimulating factor for total pulp regeneration. Stem Cells Transl Med 2(7):521–533
95. Ikeda E, Tsuji T (2008) Growing bioengineered teeth from single cells: potential for dental regenerative medicine. Expert Opin Biol Ther 8(6):735–744
96. Oshima M, Tsuji T (2014) Functional tooth regenerative therapy: tooth tissue regeneration and whole-tooth replacement. Odontology 102(2):123–136
97. Young CS, Terada S, Vacanti JP, Honda M, Bartlett JD, Yelick PC (2002) Tissue engineering of complex tooth structures on biodegradable polymer scaffolds. J Dent Res 81(10): 695–700
98. Iwatsuki S, Honda MJ, Harada H, Ueda M (2006) Cell proliferation in teeth reconstructed from dispersed cells of embryonic tooth germs in a three-dimensional scaffold. Eur J Oral Sci 114(4):310–317
99. Sumita Y, Honda MJ, Ohara T, Tsuchiya S, Sagara H, Kagami H, Ueda M (2006) Performance of collagen sponge as a 3-D scaffold for tooth-tissue engineering. Biomaterials 27(17): 3238–3248
100. Arakaki M, Ishikawa M, Nakamura T, Iwamoto T, Yamada A, Fukumoto E, Saito M, Otsu K, Harada H, Yamada Y, Fukumoto S (2012) Role of epithelial-stem cell interactions during dental cell differentiation. J Biol Chem 287(13):10590–10601
101. Yoshida K, Sato J, Takai R, Uehara O, Kurashige Y, Nishimura M, Chiba I, Saitoh M, Abiko Y (2014) Differentiation of mouse iPS cells into ameloblast-like cells in cultures using medium conditioned by epithelial cell rests of Malassez and gelatin-coated dishes. Med Mol Morphol 48(3):138–145

Dental Stem Cells for Bone Regeneration

Evangelia Diamanti, Xenos Petridis, Amalia Kaparou, and Efthymia Kitraki

Introduction

Bone fractures related to trauma, neoplasia or metabolic disorders such as osteoporosis, are among the most common human injuries. Unfavorable conditions, such as extended bone fractures (accompanied with soft tissue damage and poor blood supply) or inappropriate fracture stabilization, may lead to delayed union, nonunions or permanent bone loss. These unfavorable outcomes are estimated at a rate of 5–10 %, but the number is expected to increase along with the increase of the aging population. Successful bone repair is thus of immense clinical importance. Notably, bone repair ends up with an osseous part identical to the intact bone. This is a unique feature of the skeleton, compared to soft tissues' repair that results in fibrous scar formation.

Bone regeneration is a highly coordinated physiological process with close similarity to that of skeletal formation during organogenesis. During fetal life, osteogenesis occurs through either endochondral or intramembranous ossification. Endochondral ossification gives rise to the majority of long bones, while intramembranous ossification produces the flat bones of skull, maxilla, mandible and clavicle. Similarly to fetal bone formation, both endochondral and intramembranous ossification contribute to bone restoration. Adult mesenchymal stem cells are driven by

E. Diamanti, DDS, MSc, PhD • A. Kaparou, DDS, MSc • E. Kitraki, PhD (✉)
Faculty of Dentistry, National and Kapodistrian University of Athens,
2 Thivon str., Athens 11527, Greece
e-mail: ekitraki@dent.uoa.gr

X. Petridis, DDS, MSc, PhD student
Faculty of Dentistry, National and Kapodistrian University of Athens,
2 Thivon str., Athens 11527, Greece

Center for Dentistry and Oral Hygiene, University Medical Center Groningen,
1 Antonius Deusinglaan str., Groningen 9713 AV, The Netherlands

B. Zavan, E. Bressan (eds.), *Dental Stem Cells: Regenerative Potential*, Stem Cell Biology and Regenerative Medicine, DOI 10.1007/978-3-319-33299-4_11

the same molecular signals that regulate fetal osteogenesis. However the amount and potential of adult stem cells is limited compared to the population of embryonic pluripotent and osteoprogenitor cells [1]. Moreover, adult bone repair further depends on inflammation signals and mechanical forces in load bearing skeletal parts. At the beginning of the healing procedure, inflammatory cells, recruited at the fracture cite by pro-inflammatory cytokines, stimulate angiogenesis to support the ossification process [2]. The exerted mechanical load is also critical for the healing of long bones in particular. Complete stabilization of the fracture site results in direct (intramebranous) bone formation, whereas motility is associated with the formation of a cartilage callus intermediate.

The current standard of care for complicated bone fractures is bone grafting. Bone progenitor cells stemming from the grafts can differentiate into osteogenic cells to promote bone regeneration. Due to limitations of the bone transplants, alternative therapeutic strategies have been looked for. To this direction, bone marrow-derived mesenchymal stem cells (BMSCs), first isolated by Friedenstein et al. [3, 4], have shown promising results when applied at the fracture site either alone or in combination with osteo-conductive matrices and growth signals. This cell-based therapy provides an alternative to bone grafting by preventing bone loss and reducing graft-related morbidity [5]. BMSCs-mediated bone repair has been tested in clinical trials to restore extended craniomaxillofacial defects or problematic long bone fractures with established results [6]. However, the use of BMSCs in regenerative medicine is accompanied by several drawbacks the most important being the remaining invasiveness and cost of their collection as well as their limited quantity. Consequently, the ongoing research for other postnatal (adult) stem cells has led to the discovery of a variety of mesenchymal stem cells (MSCs) from different tissues and organs that exhibit osteogenic potential. Among them, MSCs from tissues of dental origin, initially isolated from the adult dental pulp [7] represent a promising player in field of bone regeneration. Due to their simpler accessibility and the non-invasive and low-cost way of harvesting, the dental stem cells could possibly substitute BMSCs in the cell-based approaches for bone healing.

This chapter aims to provide a concise presentation of the different types and properties of stem cells that derive from dental tissues with emphasis on their osteogenic characteristics. Subsequently, evidence for the in vivo applications of dental stem cells in bone repair will be described separately for the craniofacial and long bone defects.

Stem Cells of Dental Tissue Origin

Dental tissues comprised of the tooth pulp, the periodontium, the gingiva and also developmental tissues such as the apical papilla and the dental follicle are identified sources of stem cells. These stem cells originate embryologically from the neural crest, a cell population that comes from the dorsal margins of the closing neural folds and migrates under the induction of several signals to various locations in the embryo. Several distinct populations of dental tissue-derived stem cells have been isolated and characterized. All of them share common properties with adult MSCs from other tissues. They exhibit self-renewal capacity, multilineage differentiation potential and immunomodulatory properties [8]. According to the guidelines of the International Society for Cellular Therapy, the minimal criteria for human MSCs are the ability to attach and grow on plastic under specific culture conditions, differentiate in vitro into osteoblasts, chondroblasts and adipocytes and express specific surface markers which are >95 % (positive) CD105, CD73 and CD90, and <2 % (negative) CD45, CD34, CD14/CD11b, CD79α/CD19 and HLA-DR [9].

Dental Pulp Stem Cells (DPSCs)

Stem cells from the pulp of permanent teeth are obtained mainly from impacted third molars after enzymatic digestion of the pulp tissue or an explant culture of tissue fragments [7, 10]. As typical MSCs, DPSCs express several surface antigens such as CD73, CD90, CD105, STRO-1, CD29, CD44, CD166, CD271 and CD146 but not CD14, CD34 or CD45 [11]. These cells are highly proliferative, exhibit a faster population doubling time than BMSCs and possess immunosuppressive properties [12, 13].

DPSCs can develop in vitro into odontoblasts, osteoblasts, adipocytes, neural cells, chondrocytes, myocytes, cardiomyocytes, melanocytes, endotheliocytes and hepatocyte-like cells [10, 14–22]. Under specific inductive conditions DPSCs express nestin, a specific marker of functional odontoblasts, and can form mineralized structures that resemble dentin or osteodentin and express alkaline phosphatase (ALP), dentin sialophosphoprotein (DSPP), osteonectin (OCN) and bone sialoprotein (BSP) [7, 23–27].

The osteogenic differentiation capacity of DPSCs seems to be superior compared to BMSCs [12] and has been well demonstrated in vitro in conjunction with several materials [16, 28]. For example, DPSCs that were cultured on different titanium surfaces were differentiated into osteoblasts and produced bone [29]. Immobilization of DPSCs into alginate microspheres resulted in osteo-differentiation and enhanced mineralization as shown by the upregulated osteo-related genomic profile of these cells [30]. DPSCs that were pre-differentiated into osteoblasts and seeded onto collagen-hydroxyapatite-poly (L-lactide-co-ε-caprolactone), provided osteoblast-like cells capable for extracellular matrix mineralization [31].

When DPSCs are transplanted ectopically in experimental animals, they can form dentin-pulp-like complexes [7, 10, 32]. DPSCs in combination with carriers like hydroxyapatite/tricalcium phosphate (HA/TCP) or dentin surfaces, led to the formation of vascularized pulp-like tissue and mineralized nodules resembling dentin [33].

Regarding the osteogenic properties of DPSCs upon ectopic transplantation, numerous studies in animal models have shown bone-like tissue formation after subcutaneous implantation of DPSCs loaded on various carriers [16, 34–40]. More specifically, a subpopulation of DPSCs (c-kit /CD34 positive and CD45 negative,

SBP-DPSCs), able to osteo-differentiate in vitro, generated a fibrous lamellar bone, when transplanted into immunocompromised rats [19]. This bone that contained osteocytes was named LAB (living autologous fibrous bone tissue). Under a similar experimental set-up, the formed bone had its own vasculature, suggesting differentiation of DPSCs into endothelial cells and a tendency to enhance angiogenesis in vivo [16]. In further studies, well-mineralized hard tissue with distinct concentric lamellae and partially developed bone marrow-like haematopoietic tissue was formed after subcutaneous implantation of rat STRO-1-selected DPSCs with a 3D porous HA/TCP carrier [39]. Constructs prepared from DPSCs and an absorbable self-assembling peptide nanofibre hydrogel cultured in an osteogenic medium for 2 weeks and subsequently implanted subcutaneously into nude mice resulted into vascularized mineralized tissue pieces after 4 weeks [34]. The ability to produce ectopic bone-like tissue seems to be enhanced by the addition of the bone morphogenetic protein-2 (BMP-2) [41–43]. Collectively, most studies on bone formation following ectopic transplantation of DPSCs in animal models have provided positive results with very few exceptions [40].

Aspects of DPSCs isolation and handling before implantation can exert a significant influence in bone tissue formation [44]. For example, specific markers have been proposed as indicators of DPSCs' enhanced osteogenic potential such as STRO-1, flk-1, CD34 and very recently a combination of low CD271 with high CD90 and thus have been used for the isolation of specific cell subgroups [19, 45–47]. The number of passages required to generate sufficient cells for clinical application may affect the potency of the cultured cells, while different culture conditions may affect osteogenic lineage differentiation [36]. It is well known that DPSCs are able to differentiate into osteoblasts when cultured for 3–4 weeks in a medium supplemented with dexamethasone, β-glycerophosphate and ascorbic acid [36, 48]. However, the use of standard medium supplemented with 20 % FBS has been proposed as equally effective [19], whereas human serum instead of bovine serum is considered more clinically relevant [49, 50]. Other components or additives of the culture medium can also affect the equilibrium between stemness maintenance and cellular differentiation. Indicatively, human DPSCs exposed in vitro to optimal concentrations of small molecules (pluripotin/SC-1/6-bromoindirubin-3-oxime, rapamysin) showed an increased expression of STRO-1, NANOG, OCT4 and SOX2, but a diminished differentiation into odontogenic/osteogenic, adipogenic and neurogenic lineages [51].

Critical for clinical use is the ability of DPSCs to maintain their characteristics after cryopreservation for years. Several studies reported no changes in cell viability and stem cell marker expression as a result of cryopreservation using controlled cooling rates but no cryopreservant [52–54]. Under such conditions, osteoblasts differentiated from DPSCs, as well as DPSCs themselves were still capable of re-starting proliferation, differentiating and producing mineralized matrix [55, 56]. The choice of the culture medium and the pre-treatment of DPSCs also depend on the type of the defect. Ectopic bone formation in subcutaneous pockets particularly might require osteoinductive supplementation since this model lacks pre-osteogenic signaling cascades that develop due to injury and exerted mechanical forces [44].

Stem Cells from Human Exfoliated Deciduous Teeth (SHEDs)

SHEDs represent a unique and easily accessible source of MSCs. A significant advantage of these cells is that they are taken from deciduous teeth, a temporary tissue that normally is rejected. They are distinct from DPSCs due to their higher proliferation rate, increased population doublings and an ability to form sphere-like clusters in culture. SHEDs express the typical MSCs markers (CD105, CD146, STRO-1, and CD29; are negative for CD31, CD34) but also the embryonic stem cell markers Oct4 and Nanog, the neural stem cells marker nestin and the stage-specific embryonic antigens SSEA-3 and SSEA-4 [57]. SHEDs exhibit higher plasticity than DPSCs and are able to differentiate in vitro into odontoblasts, osteoblasts, adipocytes, myoblasts, chondroblasts, endothelial cells, neural cells and hepatocytes [58–62]. The addition of dexamethasone, L-ascorbate-2-phosphate and inorganic phosphate in the culture medium led to the production of mineralized tissue within 4 weeks, while further addition of human recombinant BMP-4 promoted their osteogenic induction [57].

In numerous studies, SHEDs seeded onto several carriers (tooth slices, HA/TCP or a polyethylene glycol(PEG)ylated fibrin carrier in combination with dentin discs) and subsequently implanted subcutaneously into immunocompromised mice, were able to differentiate in odontoblast-like cells. These cells were associated with a dentin-like structure and produced a tissue very similar to human dental pulp [57, 58, 63–66]. However, they are not able to reconstitute a dentin-pulp-like complex like the one formed by DPSCs. SHEDs in combination with HA/TCP scaffolds are also capable of inducing new bone formation not by differentiating directly into osteoblasts but probably by recruiting host osteogenic cells around the ectopic transplant [57].

Stem Cells from the Apical Papilla (SCAP)

SCAP are a population of MSCs isolated from the apical papilla, a soft tissue located at the root apex of developing teeth [67, 68]. These cells appear to be the source of the primary odontoblasts responsible for the formation of root dentin. They are located mainly around vessels, show higher proliferation rate and mineralization potential than DPSCs, express typical MSC markers (CD73, CD44, CD105, CD146, CD166, CD24, STRO-1, CD34) and exhibit a distinct gene expression profile from that of DPSCs [25, 67]. Like DPSCs and SHEDs, SCAP can differentiate in vitro into odontoblasts and form 3D osteodentin structures. SCAP represent a population of cells from a developing tissue and might thus exhibit greater plasticity than other stem cells of dental origin. In culture, following induction they can differentiate into osteoblasts, adipocytes and hepatocyte-like cells [20, 67]. Additionally, ex vivo expanded SCAP, without any stimulation, express several neural markers that are further increased after neurogenic stimulation [68].

SCAP transplanted subcutaneously in immunocompromised mice lead to the formation of a typical dentin-pulp-like complex [67, 69], while a vascularized pulp-like tissue is formed when these cells are transplanted into root canals [70]. Regarding their in vivo osteogenic potential, SCAP could generate a cement/woven bone-like tissue with embedded cementocyte/osteocyte-like cells when transplanted subcutaneously in combination with porous ceramic discs into immunocompromised mice for 8 weeks, but the authors could not identify if the tissue was bone, dentin or cementum [71].

Dental Follicle Stem Cells (DFSCs)

This population of dental tissue-derived stem cells is isolated from the dental follicle, an ectomesenschymal tissue that surrounds the developing tooth germ prior to eruption. DFSCs are considered the progenitor cells for the formation of all periodontium tissues including cementum, periodontal ligament and alveolar bone [72]. Similar to SCAP, the fact that DFSCs are isolated from developing tissues could provide them with an advantage compared to other stem cells of dental tissue origin [73]. DFSCs exhibit a considerable proliferative ability and express similar surface antigen markers with the other dental stem cells (positive for CD105, CD44, CD29, negative for CD34, CD117), as well as the putative stem cell markers Notch-1 and nestin [74]. Under the appropriate inductive conditions, DFSCs can differentiate in vitro into osteoblasts, cementoblasts, chondrocytes, hepatocytes, adipocytes and neural cells [20, 75–78]. Long-term cultures of DFSCs with dexamethasone produce membrane-like structures containing compact calcified nodules [79, 80]. Ex vivo expanded mouse or bovine DFSCs transplanted into immunosuppressed mice are capable of forming cementum- or PDL-like structures [81, 82].

Hard tissue formation by DFSCs is still a matter of debate. In studies conducted in immunocompromised mice, the subcutaneous transplantation of DFSCs was reported to had no effect in the formation of any type of hard tissue (cementum or bone) [72, 79, 83] or to lead in bone tissue formation when DFCSs were combined with porous ceramic discs [71].

Stem Cells from the Periodontal Ligament (PDLSCs)

Stem cells from the periodontal ligament are isolated from the roots of extracted teeth, mostly impacted third molars, after enzymatic digestion or explant culture of the PDL. They express the typical MSC markers (STRO-1, CD44, CD90, CD105, CD146) and also high levels of the tendon-specific marker scleraxis and several markers related to osteoblasts and cementoblasts [84, 85]. Under the appropriate inductive conditions, PDLSCs have the ability to differentiate into odontoblasts/osteoblasts forming calcified nodules, cementoblast-like cells, adipocytes, chondrocytes, neurons, hepatocytes and also form connective tissue rich in collagen type I [67, 69, 84, 86–90]. They have a similar proliferative potential with DPSCs but an apparent higher ability for differentiation into cementoblasts than in osteoblasts [91]. Formation of a typical cementum/PDL-like complex was reported after subcutaneous transplantation of PDLSCs into the dorsal surfaces of immunocompromised mice [84]. Bone formation has been observed in cultures of human PDLSCs with different 3D-scaffolds like fibrin sponges and bovine-derived substrates [92]. Additionally, PDLSCs isolated from dogs developed well on polymer scaffolds and the constructs generated new bone after subcutaneous transplantation in animals of the same species [93].

Gingival Mesenchymal Stem Cells (GMSCs)

This population of MSCs recently isolated from the human gingiva possesses all the stem cell-like properties such as clonogenicity, self-renewal ability, multipotent differentiation capacity and immunomodulatory properties. These cells display a higher proliferation rate than BMSCs, express typical CD markers of MSCs and also Oct4, Sox2, Nanog, nestin, SSEA-4, STRO-1. They can differentiate in vitro into osteoblasts, chondroblasts, adipocytes and neural cells [94–96]. Under typical osteogenic culture conditions, GMSCs show similar osteogenic ability with the BMSCs [97], while, when encapsulated in a RGD (arginine-glycine-aspartic acid tripeptide)-coupled alginate microencapsulation system, their osteogenic differentiation capability becomes inferior to both BMSCs and PDLSCs [98]. GMSCs have been seeded onto HA/TCP scaffolds and were subsequently incubated in osteogenic medium and transplanted subcutaneously in the dorsal surface of immunocompromised mice. This has led to an increase in the expression levels of osteocalcin, osteopontin and collagen I, but not to the formation of any hard tissue [96, 97].

Alveoral-Bone Derived Mesenchymal Stem Cells (ABMSCs)

Mesenchymal stem cells have been isolated from the alveoral bone and they exhibit a fibroblast-like morphology, plastic adherence and colony formation capacity [99]. They express relevant MSC markers and display osteogenic, chondrogenic and adipogenic differentiation potential [99–101]. Expanded ABMSCs demonstrate in vitro high ALP expression. Several treatments can enhance osteogenesis in these cells [99]. They have also been used to test a new fabric (CS (chitosan)/HA (hydroxyapatite)) as a scaffold for bone tissue engineering and the results were excellent, with the cells showing improved differentiation capacity, increased ALP expression and calcium accumulation in vitro [102]. Following subcutaneous transplantation into immunodeficient mice, human ABMSCs induced new bone formation with a significant activity of cuboidal osteoblasts and osteocytes [99].

Tooth Germ Progenitor Cells (TGPCs)

TGPCs refer to a distinct population of progenitor cells identified in the dental mesenchyme of the third molar tooth germ during the late bell stage [75]. These cells express MSC-associated markers, such as STRO-1 and several CDs and an ability to differentiate in vitro into adipocytes, osteoblasts, odontoblasts, chondrocytes, neurons, hepatocytes, while they also seem to contribute to vascularization [75, 78]. TGPCs/HA subcutaneous implants in immunocompromised rats have shown new bone formation with the presence of osteocytes in the newly formed bone matrix and active osteoblasts lining on the matrix surface [75].

The differentiation properties of dental stem cells mentioned in this chapter are summarized in Table 1.

Table 1 Categories and differentiation potential of dental-tissue derived mesenchymal stem cells

	Source	In vitro multipotency	In vivo ectopic tissue formation
DPSCs [7]	Pulp adult teeth	Osteogenic, dentinogenic, adipogenic, chondrogenic, neurogenic, myogenic, cardiomyogenic, hepatocyte-like cell, melanocyte	Dentin-pulp-like complex
			Bone-like tissues
			Adipose
			Muscle
SHEDs [57]	Pulp deciduous teeth	Osteogenic, dentinogenic, adipogenic, chondrogenic, neurogenic, myogenic, bone induction, endothelial cell	Pulp-like tissues Bone-like tissues
SCAP [67]	apical papilla	Osteogenic, dentinogenic, adipogenic, chondrogenic, neurogenic, hepatocyte-like cell	Dentin-pulp like complex
			Bone-like tissues
DFSCs [72, 80]	Dental follicle	Cementogenic, osteogenic, adipogenic, chondrogenic, neurogenic, hepatocyte-like cell	Periodontal tissues (alveolar bone, PDL, cementum)
PDLSCs [84]	Periodontal ligament	Osteogenic, cementogenic, adipogenic, chondrogenic, neurogenic	Periodontal tissues (cementum, PDL)
GMSCs [96]	Gingiva	Adipogenic, chondrogenic, osteogenic, neurogenic, endoderm cell	Cartilage
			Bone
			Muscle
ABMSCs [99]	Alveolar bone	Adipogenic, chondrogenic, osteogenic	Bone
TPSCs [75]	Dental mesenchyme	Adipogenic, chondrogenic, osteogenic, odontogenic, neurogenic, hepatogenic, endothelial cell	Bone

First reference for each category is provided in brackets in the first column

Dental Stem Cells in Craniofacial Bone Healing

The craniofacial structures constitute the epitome of the vertebrate development as they consist of an incongruous medley of tissues derived from the coordinated integration of each germ layer tissue (i.e., ectoderm, mesoderm, and endoderm) [103]. For example, the anterior section of the calvaria originates from neural crest cells and the parietal bones from mesodermal cells. Specifically, the craniofacial mesenchymal tissues originate from the neural crest, paraxial and lateral mesoderm, with the contribution of the cranial neural crest being the most dominant [104]. Indeed, when the cranial neural crest cells migrate superficially, they demonstrate a stounding multilineage differentiation potential towards cells that generate the majority of the components of the craniofacial complex. Whether the neural crest or mesodermal origin of a bone is related to its function and healing potential remains elusive. However there are convincing insights that the two cell populations exhibit molecular differences that are reflected to their osteogenic potential and regenerative capacity [105, 106]. Further investigation is needed to delineate the possible involvement of these embryological differences to the healing capacity of the discrete craniofacial parts.

Calvarial Bone Defects

Experimental Models of Calvarial Bone Defects

Bone tissue reconstruction in the craniofacial area is typically required in cases of trauma, skeletal diseases, congenital malformations and cancer surgery and affects the life quality of millions of patients worldwide [107]. The 'gold standard' bone grafts, despite their associated limitations (autogenous, allogenous, xenogenous and synthetic) [108], constitute the basic therapeutic modalities. The recently introduced treatment alternatives are based on regenerative medicine and bone tissue engineering [109, 110]

Irrespective of the treatment choice, designing appropriate experimental models to evaluate the efficacy and safety of any regenerative procedure stands out as an absolute necessity. In vivo animal models could provide a link between in vitro studies and clinical trials. Although a single ideal model is currently unavailable, there are specific criteria that a model used for bone tissue engineering should meet [111]. The calvarial bone defect in rodents is frequently used as anorthotopic model to evaluate bone regeneration, as it fulfills the basic requirements of an adequate experimental design [112]. Specifically, the rodent calvarial defect model constitutes a uniform, reproducible and standardized defect that allows radiographic and histological analysis; calvaria is an easily accessible site providing sufficient area for surgical handling; the dura mater and the overlying skin provide adequate support for the grafts without the need of any kind of fixation; includes

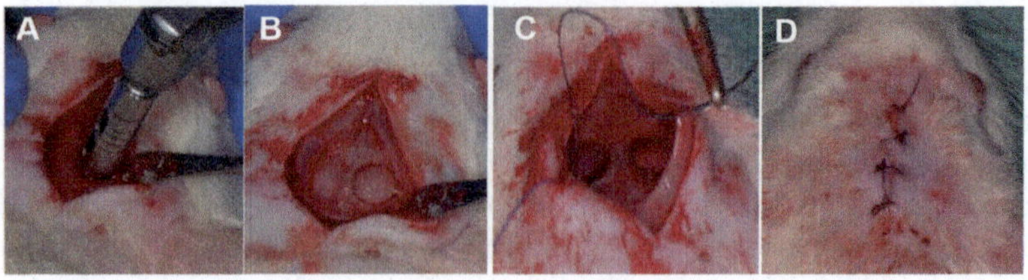

Fig. 1 Surgical procedure in calvaria defects. (**a**) Preparation of the surgical area. (**b**) Creation of the 5-mm critical size defect in the rat calvarium. (**c**) Closure of the periosteum. (**d**) Closure of the overlaying skin [121]. Reproduced upon permission from Elsevier

both cortical and cancellous bone; it is inexpensive and widely used, thus allowing for direct comparisons between the different graft materials; it has a fast healing time [113–116].

An important parameter in the design of the calvarial bone defect is its size. The most widely accepted defect size in assessing the osteogenic capacity of different grafts is the critical-size defect (CSD) [117]. CSDs were originally defined as "the smallest size intraosseous wound in a particular bone and species of animal that will not heal spontaneously during the lifetime of the animal" [118] and were later established at the calvaria of rats as the 8 mm [119] or the paired 5 mm cranial defect [120] (see also Fig. 1). Although the value of the CSDs has been keenly criticized [122], it still remains the most encountered osseous defect in the literature and constitutes the 'gold standard' experimental model in appraising craniofacial regeneration.

Dental Stem Cells and Bone Repair in the Calvarial Experimental Model

The rodent calvarial bone defect model has been employed extensively as an orthotopic experimental design to assess the bone regeneration capacity of the DPSCs in the craniofacial complex. Following the first isolation of MSCs derived from the dental tissues [7], a huge impetus towards the extensive study of their stem cell properties has been noted [69]. Within the context of craniofacial bone engineering, DPSCs constitute a favorable population of adult MSCs for several reasons: they have common developmental origin from the cranial neural crest with the craniofacial bones [123]; they are easily harvested; they exhibit enhanced osteogenic potential in vitro and in vivo compared to other MSC populations [10, 12, 44, 46] and they retain their 'stemness' after cryopreservation [54, 56].

As mentioned previously, a series of in vitro studies have clearly demonstrated the capacity of the c-kit$^+$/CD34$^+$/CD45$^-$ sub-population from the mixed DPSCs (SBP-DPSCs) to produce two- and three-dimensional structures of mature bone [19, 55, 124]. Further research on the regenerative efficacy of a selected subpopulation of DPSCs (cells positive for c-kit$^+$/CD34$^+$/STRO-1$^+$) seeded on collagen or fibroin scaffolds, has demonstrated exceptional bone healing of critical-size calvaria defects in immunosuppressed rats. The human cells were detected in the areas of newly-formed bone and the enhanced bone healing was attributed also to the increased angiogenic potential of the selected stem cell populations [50, 125, 126]. Similar results were obtained recently with a selected $CD271^{low+}/CD90^{high+}$ DPSC population. These cells mixed with Matrigel Matrix and implanted into critical-size calvaria defects in immunodeficient mice survived throughout a 4-week period and promoted significantly bone regeneration [47]. These studies have shown that selective DPSC subpopulations, with enriched osteogenic and angiogenic potential, are of utmost importance for an optimal result. The healing properties of the entire population of human dental pulp cells have also been demonstrated in the rat calvaria model [121]. In our study, dental pulp cells isolated from human third molars were transplanted in the 5 mm calvaria defects of immunocompetent rats following a short osteoinduction period in culture. An ECM-mimicking hyaluronic-based scaffold was used as the cell carrier. Enhanced bone regeneration was detected by histomorphometry in the cell-treated defects, compared to the scaffold-treated and untreated ones (Fig. 2). However, in contrast to the complete bone healing observed in the afore-mentioned studies [50, 125, 126], none of the defects exhibited complete bone regeneration in our study. The mechanisms underlying this incomplete bone regeneration may relate to several factors, including the lower osteo/angiogenic potential of the mixed pulp population used and the type and properties of the scaffold employed that was not degraded properly, thus entrapping many of the pulp cells in it. The latter resulted in inflammatory reactions around the non-degraded scaffold islets that contained dead pulp cells (Fig. 2). Though inflammatory signals have a stimulating role in the beginning of the healing process, the host's response that was detected at a later stage due to pulp cells' entrapment might have hampered the repair procedure. Despite the incomplete bone restoration, the significantly higher percent of new bone formation in the cell-treated group of this study, compared to the control groups, support the osteogenic efficacy of the entire dental pulp cell population. At the same time, these results underline the need for further improvement of our knowledge on the scaffolds' in vivo properties and bio-behaviors.

Stem cell-based bone regeneration is critically relayed on the properties of the stem cell populations used. These properties mostly refer to their osteogenic potential. However, additional characteristics may contribute to their efficiency in bone healing, such as their interplay with the host tissue immune response [127] and their paracrine 'trophic activity' to adjacent cells [128]. Although the bone regenerative properties of DPSCs in vivo have been shown in the calvarial defects of immunosuppressed rats [50, 125, 126], still little is known about the sequence of events following implantation of the DPSCs. More specifically, studies investigating the biological cues in the defect area after the implantation of the biocomplexes are lacking. The DPSCs-induced immunological response and the paracrine action of the DPSCs warrants for further investigation.

Another important parameter that needs to be taken into consideration in bone regeneration is the type of scaffold employed. These three-dimensional matrices accommodate the dental stem cell populations by forming biocomplexes. The

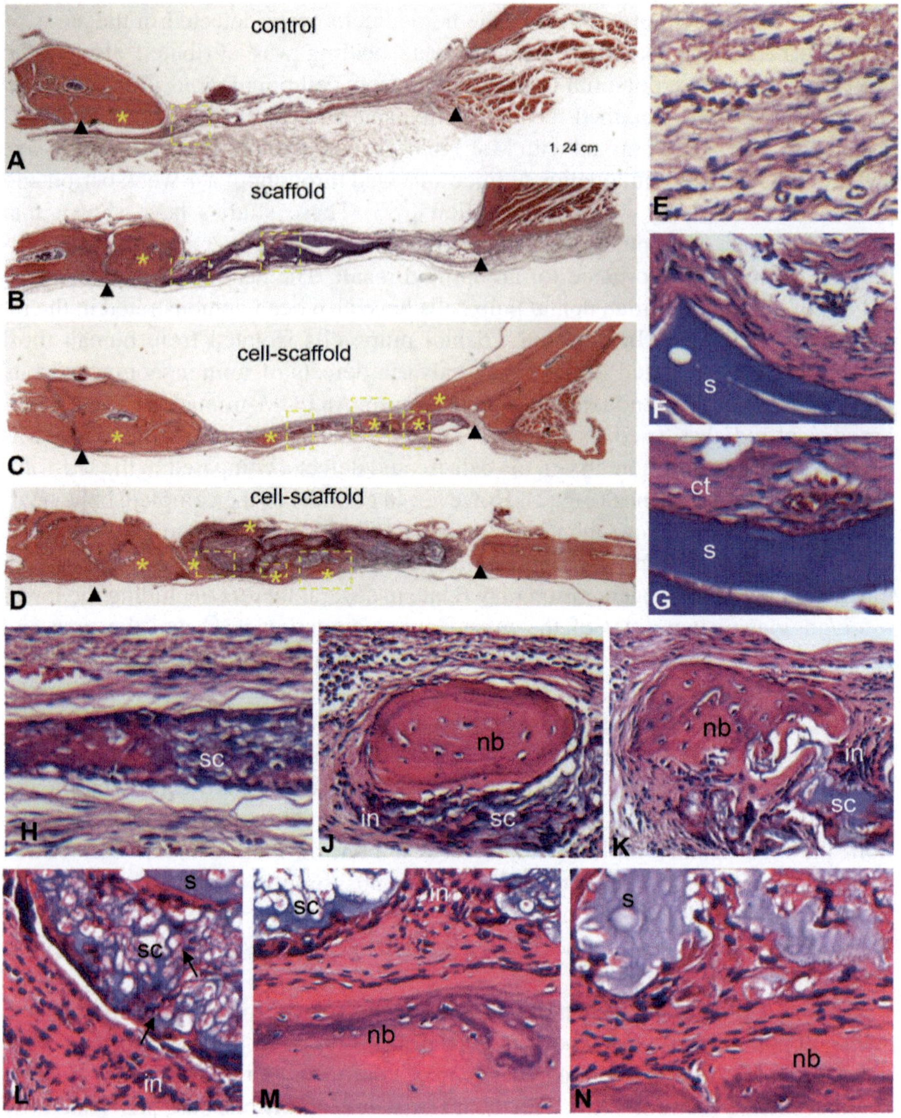

Fig. 2 Histological findings in haematoxylin and eosin stained calvaria sections. Representative low magnification (×1.25) photomicrographs of the defect site in the control (**a**), scaffold (**b**) and cell-scaffold (**c**, **d**) groups. *Arrowheads* indicate defect margins. Areas of new bone formation are marked by *asterisks*. *Rectangles* indicate sites within defect for which higher magnification (×20) images (**e**–**n**) are provided. (**e**): connective tissue (ct) in the framed area of the control group; (**f**, **g**): large areas of scaffold (s) remnants surrounded by dense connective tissue in the framed area of the scaffold group; (**h**–**n**): non-degraded scaffold with entrapped cells (sc), and sites of new bone (nb) formation within the defect in the framed areas of the cell-scaffold group in C (**h**–**k**) and D (**l**–**n**). Note the inflammation (in) around areas of cell-scaffold remnants and dead cells into the scaffold (*arrows* in L) [121]. Reproduced upon permission from Elsevier

constructs are either cultured in vitro leading to the ex vivo production of bone that is subsequently implanted in the osseous defect or they are implanted directly to the site inducing in vivo bone regeneration [129]. In general, the types of scaffolds used to regenerate calvaria bones can be divided into ceramics and polymers (synthetic or natural) and it is noteworthy that the cell-scaffold interactions as well as the interplay between the bioengineered construct and the host tissue can profoundly affect bone regeneration [130]. In a study assessing calvaria bone regeneration after the transplantation of BMSCs-loaded collagen-glycozaminoglycan constructs in rodents, the initial inflammation induced by the biocomplex led to incomplete bone healing [131, 132]. According to the authors, this host response accelerated the degradation of the scaffold, which in turn impaired the bone matrix deposition [132]. In our study, the utterly different host response to the dental pulp cell-biocomplexes, compared to the inflammatory response towards the cell-free scaffolds, was the major cause of the non-optimal repair of the defects [121].

Besides from DPSCs, stem cells from exfoliated deciduous teeth (SHEDs) have also been successfully used to regenerate critical size bone defects in the calvaria rat model [133–135]. Similarly to the DPSCs, the neural crest origin of the SHEDs and their enhanced stemness make them excellent candidates for craniofacial bone tissue engineering.

Bioengineered constructs comprised of stem cells from the periodontal ligament (PDLCs) have been used to regenerate bone in calvarial defects with very encouraging results, thus rendering this treatment modality a viable option in the field of craniofacial bone regeneration [98, 136, 137].

Finally, stem cells from the dental follicle (DFSCs) have shown enhanced bone formation when transferred as scaffoldless transplants in 8 mm critical-size calvarial defects of immunodeficient rats [138].

Overall, the use of dental stem cells in craniofacial bone engineering seems justified. In particular, the results from the transplantation of DPSCs in the calvarial experimental model render this stem cell-based treatment a quite promising future treatment modality for repairing defects in the craniofacial area. Alternatively, stem cell populations from other dental tissues displaying enhanced osteogenic capacity and sharing common developmental origin with the craniofacial bone structures may also prove beneficial in the cranial bone tissue engineering.

Mandibular and Alveolar Bone Defects

Mandibular bone defects due to traumas, neoplastic tumors or other pathological conditions are challenging problems for oral and maxillofacial surgeons. In most cases, repair of such defects requires bone grafts for restoration of continuity, function and aesthetics. In dentistry, bone grafts are mainly applied in order to repair alveolar bone defects caused by periodontal or peri-implant disease and to preserve or reconstruct the alveolar ridge for future implant placement. Dental stem cells and tissue engineering technology in oral and maxillofacial region stand out as new alternatives for repairing mandibular and alveolar defects.

Experimental Models of Mandibular and Alveolar Bone Defects

The majority of in vivo studies evaluating the application of stem cells of dental origin on mandible defects are performed in experimental animals. Various animal models of mandibular bone defects have been used. Dogs, pigs, rabbits and rats are the most frequently used animals in this field, with dogs and pigs possessing anatomical and physiological features more close to the human ones [139]. The rat model, on the other hand, is dissimilar in terms of size and bone structure compared to humans [36], but is easily handled and cost effective [112].

The type of the mandible defect varies from segmental [140] or cylindrical [141] alveolar defects to orofacial defects in the parasymphyseal region of the mandible [142] or in front of the mandibular corner [143]. All of the above defects are considered critical size defects, in which by definition complete bone healing does not spontaneously occur and therapeutical intervention is needed [118]. The efficacy of dental stem cells on bone regeneration has also been tested in a mandibular distraction osteogenesis model in rabbits [144]. In the clinical setting, alveolar defects of the mandible, produced after extraction of impacted third molars, have been treated with autologous DPSCs seeded onto collagen sponge scaffolds [145].

Alveolar defects of the mandible often linked with periodontal degeneration may result in loss of cementum, periodontal ligament and alveolar bone. Animal models simulating these conditions have been also used for evaluating the regenerative efficacy of dental stem cells. Experimental models of periodontitis are surgically created and can be classified to chronic, acute and acute/chronic ones [146]. Models of chronic defects arise from the reproduction of conditions that induce periodontal tissue loss, such as the placement of orthodontic elastics around the teeth at or slightly apical to the gingival margin [147]. The models of acute defects are created by surgical removal of bone, cementum and periodontal ligament, while the insertion of a foreign body in the surgically created acute defect leads to an acute/chronic defect. The type of acute defect displays heterogeneity; furcation defects [148], intrabony defects in the mesial or distal aspect of the tooth [149] or fenestration defects [150] have been used. Acute defect models have been used for testing bone regenerative properties of dental stem cells, due to their cost-effectiveness and the reduction of experimental time. Regarding the acute/chronic defect, Liu et al. [151] created such defects in the mesial region of mandibular first molars of pigs by surgical removal of bone and subsequent silk ligament suture. Similarly, Park et al. [152] created surgically circumferential defects in the mesial root of mandibular premolars of dogs, which were then filled with a rubber base impression material. The acute/chronic model resembles ‘true’ periodontitis and is the one recommended in periodontal regeneration research. In cases of bone regeneration around dental implants, experimental models in dogs have been used to examine the effect of dental stem cells on peri-implant bone regeneration [153–155].

Unfortunately, there is no ideal experimental model and the selection of appropriate animal and defect model for each clinical application is complicated [146]. The establishment of reliable and validated animal models is necessary for evaluating and comparing the effect of stem cell-based therapies on mandibular bone regeneration.

Dental Stem Cells in the Repair of Mandibular and Alveolar Defects

In addition to the proven in vitro and ectopic in vivo osteogenic properties of DPSCs and SHEDs, these cells exhibit great regenerative capacity when transplanted orthotopically in critical-sized mandibular defects. Dental pulp stem cells from deciduous or permanent teeth, enriched with platelet rich plasma (PRP), enhanced bone formation in cylindrical alveolar defects of the mandible in dogs [140–143]. Importantly, the stem cells of dental origin showed bone regeneration capacity similar to that of BMSCs, which have been considered the gold standard in bone regenerative medicine [141].

Growth factors, scaffolds and stem cells, constitute the three key elements for bone tissue engineering. Growth factors, such as the recombinant human bone morphogenetic protein-2 (rhBMP-2,) promote the osteogenic capacity of the DPSCs. Autologous DPSCs combined with rhBMP-2 induced more new bone formation than even the autologous bone, when engrafted in segmental alveolar defects in rabbits [140]. The osteogenic properties of DPSCs appear to be scaffold-dependent. The resorption rate of the scaffold material is of paramount importance and, thus, mixing DPSCs into biomaterials with suitable resorption rates creates a synergetic effect that enhances bone regeneration [143].

In the only human study so far, autologous DPSCs seeded onto collagen sponge scaffolds successfully repaired an alveolar mandibular defect produced after extraction of an impacted third molar [145]. A three-year follow-up study revealed that the regenerated bone was uniformly vascularized and qualitatively compact, though differed from the physiological cancellous (spongy) bone [45]. The formation of compact bone by the DPSC-grafts is alleged to have positive clinical impact, favoring implant stability and rehabilitation of large bone defects in oral cancer patients [36, 45]. However, this alteration in the regenerated bone type raises some concern on the possible diversion of the transplanted DPSCs' fate towards undesirable cells and tissues [156]. Further studies are required to assure the absolute control on the behavior of the transplanted multipotent cells in the host environment.

For periodontal tissue regeneration, PDLSCs are considered to be the best dental stem cell type. Several animal studies have evaluated the potential of PDLSCs for regeneration of alveolar periodontal defects and have shown that these cells induce new bone, cementum and periodontal ligament formation in the treated defects [148–152]. In a clinical case series, the application of autologous PDLSCs for the treatment of periodontal defects led to improved periodontal parameters in all cases [157]. DFSCs may be an alternative to PDLCs for periodontal regeneration, although not an equally effective one [152]. Although DPSCs exhibit great bone regeneration capacity, their potential for periodontal regeneration is doubtful due to their limited capacity to form cementum [152, 158]. Instead, SCAP have shown to possess the ability to form PDL-like, cementum-like and bone-like tissues in an in vivo ectopic transplantation rat model [158], but there is lack of evidence regarding their potential to repair periodontal defects. Similarly, there is no sufficient data supporting the use of SHEDs for periodontal regeneration.

As mentioned above, there are few studies evaluating the effect of stem cells of dental origin on peri-implant bone regeneration. PDLSCs have been shown to enhance alveolar bone regeneration in surgically created peri-implant saddle-like defects, although to a lesser extent than BMSCs [154]. Notably, the osseointegration of dental implants, assessed by measuring bone/implant contact, was higher in DPSCs/PRP-treated alveolar defects, compared to analogous BMSCs/PRP-, periosteal cells/PRP-treated and untreated defects [153]. Furthermore, Omori et al. [155] suggested that treatment of Ti implants with SHED-conditioned medium promotes bone morphogenesis not only around the implant interface, but also at distant locations from the implant surface, thus improving early osseointegration.

Dental Stem Cells in Long Bone Healing

During osteogenesis, the long bones of the body are formed by endochondral ossification through distinct steps that include: differentiation of mesenchymal stem cells into chondrocytes; proliferation, hypertrophy and cartilage production by the chondrocytes; differentiation of osteoblasts and bone matrix deposition. In adult bone repair, both endochondral and intramembranous ossification (i.e. without prior cartilage formation) participate in the healing process depending on the degree of fracture stabilization.

The need to enhance the healing of long bone fractures is well appreciated, especially in cases of unsuccessful unions or non-unions. The use of autologous or allogenic bone grafts is the current gold standard. However, this application is accompanied with severe drawbacks including invasiveness, limited graft volume and bone loss at the donor site. Bone tissue engineering, by using isolated bone marrow stem cells in combination with osteoinductive or osteoconductive scaffolds, emerges as a promising alternative to bone transplants. Animal studies in rodents, rabbits, or bigger mammals have used BMSCs to successfully treat long bone fractures [159]. In these models of bone regeneration, BMSCs have been mostly isolated from same-species donors and the stem cells have been used in combination with scaffolds of various types, including calcifying agents, synthetic polymers and ECM-mimicking biomaterials. In some occasions, scaffold-free approaches have also been used, where the cells are applied as multilayered cell sheets produced in culture prior to transplantation [160, 161]. In the majority of the published animal data, the mesenchymal stem cells enhanced the healing process [6]. The ongoing clinical trials in the field, using therapeutic approaches based on the use of bone marrow stem cells/osteoinductive scaffolds have not yet provided published results [162].

The osteogenic properties of DPSCs are quite comparable with those of stromal BMSCs, both in vitro and upon ectopic transplantation [7, 10]. During the last decade DPSCs have been used in animal models of craniofacial bone healing with promising results. However, neither DPSCs nor other stem cells of dental origin have been tested in long bone repair. Given the mesenchymal origin and properties of DPSCs, we were interested to examine the efficacy of dental pulp cells, isolated from human impacted third molars, in the restoration of osteotomized rat tibia. In humans, the dental pulp from impacted immature third molars that are extracted from healthy young subjects is the preferred source of DPSCs. Additionally, dental pulp contains a substantial population of progenitor cells conferring a niche for the 'true' stem cells [163]. We therefore used the entire population of dental pulp cells instead of isolating the DPSCs. The presence of typical mesenchymal stem cell markers and the osteo-differentiation potential of the entire dental pulp population were verified in vitro by flow cytometry. To avoid reduction of the stemness of the pulp cell population following osteo-differentiation and/or repeated subcultures [164, 165], early passage, undifferentiated, pulp cells were transferred into the fracture site. It was also decided to follow a scaffold-less transfer of the cells in order to clearly detect any stem cell-induced effect. Osteotomies were performed in adult male rats and stabilized with intramedullary nails (Fig. 3). The healing process was monitored radiologically for 8 weeks. Quantification of the X-ray films showed a significantly smaller and more homogenous callus in the cell-treated rats compared to the controls, denoting advanced bone healing in the former group. The radiological results were further supported by the histological observations at 8 weeks post surgery [166]. Cell-treated animals exhibited a remodeled bone marrow cavity and higher percentage of lamellar new bone, compared to the vehicle-treated ones that had a higher score of non-bridging and scar tissue (Fig. 4).

To our knowledge, this is a first approach to examine in vivo the effectiveness of dental pulp cells to heal long bone fractures. The results provide evidence that the whole population of dental pulp cells from adult human dentition can promote bone healing in a rat osteotomy model. The developed model was purposely kept simple (i.e. scaffold-less, without specific cell sorting, without immunosuppression of the

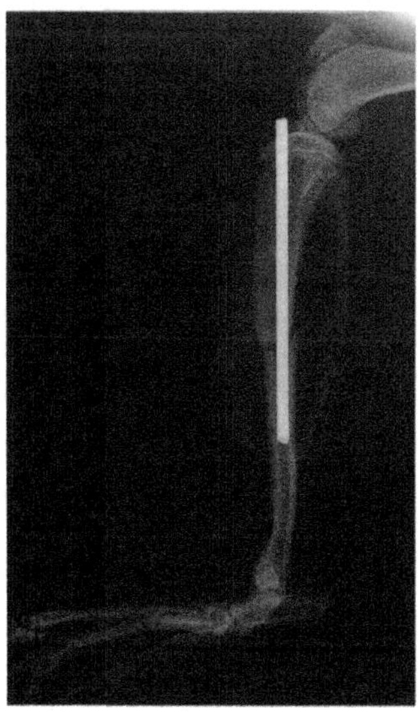

Fig. 3 Representative X-ray of the rat tibia fracture model used in the study of [166]

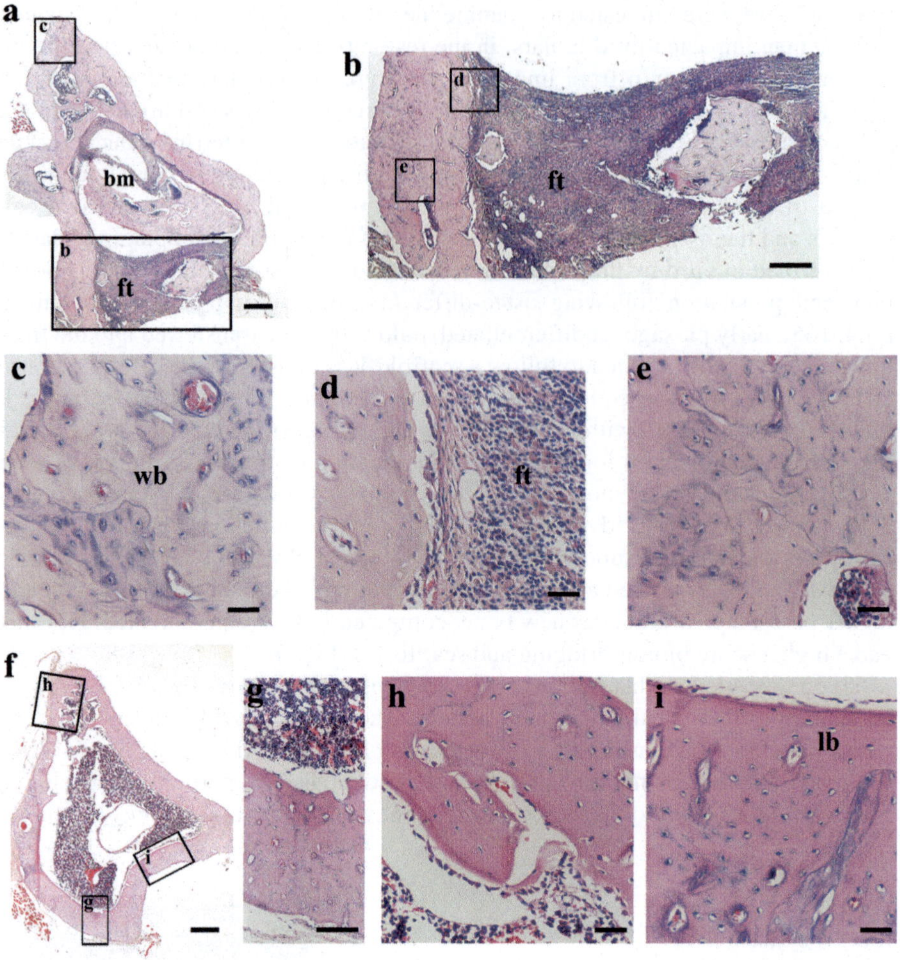

Fig. 4 Representative histological sections stained with haematoxylin and eosin (H&E) from rat tibias at 8 weeks post-fracture of control (**a**–**e**) and dental pulp cell-treated (**f**–**i**) animals [166]. H&E stained sections show smaller bone marrow cavity (**a**), non-bridging (**a**, **b**), extended fibrous tissue (**a**, **b**, **d**) and mostly woven bone (**c**, **e**) in the controls, compared to cell-treated (**f**–**i**). *Framed areas* in **a**, **b** and **f** refer to higher magnification images, according to the letter inside the *upper left side* of the frame. *bm* bone marrow, *ft* fibrous tissue, *lb* lamellar bone, *wb* woven bone. Reproduced upon permission from J Sci Med Central

host that could potentially hinder healing initiation; [2]), in order to investigate a potential cellular effect. Further refinement of the protocol is necessary to delineate the optimal conditions for the best biological outcome. Additionally, there is need for detailed characterization of the cellular interactions (including dental pulp cells' fate in the host tissue) and the molecular responses during the process of bore repair.

Conclusion and Future Perspectives

For successful cell-based bone repair, the candidate stem/progenitor cell population must fulfill the following key criteria: (a) to differentiate into bone forming cells or to stimulate endogenous tissue repair, (b) to impede uncontrolled growth and differentiation, (c) to be tolerated by the host immune system, (d) to support neoangiogenesis and blood supply, (e) to form biocomplexes with scaffolding materials, (f) to be easily harvested in a harmless manner, (g) to retain stemness upon cryopreservation and (h) to be cost effective.

Experimental evidence from mesenchymal stem cells of dental origin (DSCs) suggests that several types of these stem cells can satisfy the above criteria. Indeed, they are responsive to osteogenic signals and exhibit a bone forming potential comparable to that of BMSCs. Their origin (adult mesenchymal stem cells) assures a minimal oncogenic potential compared to pluripotent cells. DSCs are well tolerated in the fracture site and some categories have immunosuppressive properties. Discrete subpopulations of DSCs have been shown to promote new vessel formation at the fracture site. Most types of DSCs can form composite grafts with several types of scaffold materials. Importantly, they can be obtained in a non-invasive and low cost procedure and retain their osteogenic properties upon storage.

The results from the use of DSCs in animal models and the one human study of bone repair are encouraging, especially for the restoration of craniofacial bone defects. However, the existence of different models of bone repair in combination with the different subpopulations of DSCs and scaffolds used, has not allowed all proper comparisons. More studies are thus required to define the most valuable healing approach in terms of the bone defect, and the cell population/scaffold used. Furthermore, there is need to expand our knowledge on the biobehavior of the transplanted cells in the in vivo situation, regarding their interaction with the carrier material, their cell fate and possible induction properties.

References

1. Deschaseaux F, Senséb́é L, Heymann D (2009) Mechanisms of bone repair and regeneration. Trends Mol Med 15(9):417–429
2. Schmidt-Bleek K, Schell H, Schulz N, Hoff P, Perka C, Buttgereit F, Volk HD, Lienau J, Duda GN (2012) Inflammatory phase of bone healing initiates the regenerative healing cascade. Cell Tissue Res 347:567–573
3. Friedenstein AJ, Chailakhjan RK, Lalykina KS (1970) The development of fibroblast colonies in monolayer cultures of guinea-pig bone marrow and spleen cells. Cell Tissue Kinet 3(4):393–403
4. Friedenstein AJ, Piatetzky S II, Petrokova KV (1966) Osteogenesis in transplants of bone marrow cells. J Embryol Exp Morphol 16(3):381–390
5. Rosset P, Deschaseaux F, Layrolle P (2014) Cell therapy for bone repair. Orthop Traumatol Surg Res 100:S107–S112
6. Watson L, Elliman SJ, Coleman CM (2014) From isolation to implantation: a concise review of mesenchymal stem cell therapy in bone fracture repair. Stem Cell Res Ther 5(2):51

7. Gronthos S, Mankani M, Brahim J, Gehron Rodoy P, Shi S (2000) Postnatal human dental pulp stem cells (DPSCs) in vitro and in vivo. Proc Natl Acad Sci U S A 97:13625–13630
8. Liu J, Yu F, Sun Y, Jiang B, Zhang W, Yang J, Xu GT, Liang A, Liu S (2015) Concise reviews: characteristics and potential applications of human dental tissue-derived mesenchymal stem cells. Stem Cells 33:627–638
9. Dominici M, Le Blanc K, Mueller I, Slaper-Cortenbach I, Marini F, Krause D, Deans R, Keating A, Prockop DJ, Horwitz E (2006) Minimal criteria for defining multipotent mesenchymal stromal cells. The International Society for Cellular Therapy position statement. Cytotherapy 8:315–317
10. Gronthos S, Brahim J, Li W, Fisher LW, Cherman N, Boyde A, DenBesten P, Gehron Rodoy P, Shi S (2002) Stem cell properties of human dental pulp cells. J Dent Res 81:531–535
11. Kawashima N (2012) Characterisation of dental pulp stem cells: A new horizon for tissue regeneration? Arch Oral Biol 57:1439–1458
12. Alge DL, Zhou D, Adams LL, Wyss BK, Shadday MD, Woods EJ, Gabriel Chu TM, Goebel WS (2010) Donor-matched comparison of dental pulp stem cells and bone marrow-derived mesenchymal stem cells in a rat model. J Tissue Eng Regen Med 4(1):73–81
13. Peng L, Ye L, Zhou XD (2009) Mesenchymal stem cells and tooth engineering. Int J Oral Sci 1:6–12
14. Armiñán A, Gandía C, Bartual M, García-Verdugo JM, Lledó E, Mirabet V, Llop M, Barea J, Montero JA, Sepúlveda P (2009) Cardiac differentiation is driven by NKX2.5 and GATA4 nuclear translocation in tissue-specific mesenchymal stem cells. Stem Cells Dev 18:907–918
15. Arthur A, Rychkov G, Shi S, Koblar SA, Gronthos S (2008) Adult human dental pulp stem cells differentiate toward functionally active neurons under appropriate environmental cues. Stem Cells 26:1787–1795
16. d'Aquino R, Graziano A, Sampaolesi M, Laino G, Pirozzi G, de Rosa A, Papaccio G (2007) Human postnatal dental pulp cells co-differentiate into osteoblasts and endotheliocytes: a pivotal synergy leading to adult bone tissue formation. Cell Death Differ 14:1162–1171
17. Iohara K, Zheng L, Ito M, Tomokiyo A, Matsushita K, Nakashima M (2006) Side populations isolated from porcine dental pulp tissue with self-renewal and multipotency for dentinogenesis, chondrogenesis, adipogenesis and neurogenesis. Stem Cells 24:2493–2503
18. Király M, Kádár K, Horváthy DB, Nardai P, Rácz GZ, Lacza Z, Varga G, Gerber G (2011) Integration of neuronally predifferentiated human dental pulp stem cells into rat brain in vivo. Neurochem Int 59(3):371–381
19. Laino G, d'Aquino R, Graziano A, Lanza V, Carinci F, Naro F, Pirozzi G, Papaccio G (2005) A new population of human adult dental pulp stem cells: a useful source of living autologous fibrous bone tissue (LAB). J Bone Miner Res 20:1394–1402
20. Patil R, Kumar BM, Lee WJ, Jeon RH, Jang SJ, Lee YM, Park BW, Byun JH, Ahn CS, Kim JW, Rho GJ (2014) Multilineage potential and proteomic profiling of human dental stem cells derived from a single donor. Exp Cell Res 320:92–107
21. Stevens A, Zuliani T, Olejnik C, LeRoy H, Obriot H, Kerr-Conte J, Formstecher P, Bailliez Y, Polakowska RR (2008) Human dental pulp cells differentiate into neural crest-derived melanocytes and have label-retaining and sphere-forming abilities. Stem Cells Dev 17:1175–1184
22. Zhang W, Walboomers XF, van Kuppevelt TH, Daamen WF, Bian Z, Jansen JA (2006) The performance of human dental pulp stem cells on different three-dimensional scaffold materials. Biomaterials 27:5658–5668
23. About I, Bottero MJ, de Denato P, Camps J, Franquin JC, Mitsiadis TA (2000) Human dentin production in vitro. Exp Cell Res 258(1):33–41
24. About I, Laurent-Maquin D, Lendahl U, Mitsiadis TA (2000) Nestin expression in embryonic and adult human teeth under normal and pathological conditions. Am J Pathol 157(1): 287–295
25. Bakopoulou A, Leyhausen G, Volk J, Tsiftsoglou A, Garefis P, Koidis P, Geurtsen W (2011) Comparative analysis of in vitro osteo/odontogenic differentiation potential of human dental

pulp stem cells (DPSCs) and stem cells from the apical papilla (SCAP). Arch Oral Biol 56:709–721
26. Couble ML, Farges JC, Bleicher F, Perrat-Mabillon B, Boudeulee M, Magloire H (2000) Odontoblast differentiation of human dental pulp cells in explant cultures. Calcif Tissue Inter 66:129–138
27. Huang GT, Shagramanova K, Chan SW (2006) Formation of odontoblast-like cells from cultured human dental pulp cells on dentin in vitro. J Endod 32:1066–1073
28. Lindroos B, Mäenpää K, Ylikomi T, Oia H, Suuronen R, Miettinen S (2008) Characterisation of human dental stem cells and buccal mucosa fibroblasts. Biochem Biophys Res Commun 368:329–335
29. Mangano C, De Rosa A, Desiderio V, d'Aquino R, Piattelli A, De Francesco F, Tirino V, Mangano F, Papaccio G (2010) The osteoblastic differentiation of dental pulp stem cells and bone formation on different titanium surface textures. Biomaterials 31:3543–3551
30. Kanafi MM, Ramesh A, Gupta PK, Bhonde RR (2014) Dental pulp stem cells immobilized in alginate microspheres for applications in bone tissue engineering. Int Endod J 47(7):687–697
31. Akkouch A, Zhang Z, Rouabhia M (2014) Engineering bone tissue using human dental pulp stem cells and an osteogenic collagen-hydroxyapatite-poly(-lactide-co-ε-caprolactone) scaffold. J Biomater Appl 28(6):922–936
32. El-Backly RM, Massoud AG, El-Badry AM, Sherif RA, Marei MK (2008) Regeneration of dentine/pulp-like tissue using a dental pulp stem cell/poly (lactic-co-glycolic) acid scaffold construct in New Zealand white rabbits. Aust Endod J 34:52–67
33. Batouli S, Miura M, Brahim J, Tsutsui TW, Fisher LW, Gronthos S, Gehron Robey P, Shi S (2003) Comparison of stem-cell-mediated osteogenesis and dentinogenesis. J Dent Res 82:976–981
34. Chan B, Wong RW, Rabie B (2011) In vivo production of mineralized tissue pieces for clinical use: a qualitative pilot study using human dental pulp cells. Int J Oral Maxillofac Surg 40:612–620
35. Graziano A, d'Aquino R, Laino G, Papaccio G (2008) Dental pulp stem cells: a promising tool for bone regeneration. Stem Cell Rev 4:21–26
36. La Noce M, Paino F, Spina A, Naddeo P, Montella R, Desiderio V, De Rosa A, Papaccio G, Tirino V, Laino L (2014) Dental pulp stem cells: state of the art and suggestions for a true translation of research into therapy. J Dent 42:761–768
37. Li JH, Liu DY, Zhang FM, Wang F, Zhang WK, Zhang ZT (2011) Human dental pulp stem cell is a promising autologous seed cell for bone tissue engineering. Chin Med J 124:4022–4028
38. Morito A, Kida Y, Suzuki K, Inoue K, Kuroda N, Gomi K, Arai T, Sato T (2009) Effects of basic fibroblast growth factor on the development of the stem cell properties of human dental pulp cells. Arch Histol Cytol 72:51–64
39. Yang X, Walboomers XF, Van den Beucken JJ, Bian Z, Fan M, Jansen JA (2009) Hard tissue formation of STRO-1-selected rat dental pulp stem cells in vivo. Tissue Eng Part A 15(2):367–375
40. Zhang W, Walboomers XF, van Osch GJ, van der Dolder J, Jansen JA (2008) Hard tissue formation in a porous HA/TCP ceramic scaffold loaded with stromal cells derived from dental pulp and bone marrow. Tissue Eng Part A 14:285–294
41. Ikeda H, Sumita Y, Ikeda M, Ikeda H, Okumura T, Sakai E, Nishimura M, Asahina I (2011) Engineering bone formation from human dental pulp-and periodontal ligament-derived cells. Ann Biomed Eng 39:26–34
42. Tatullo M, Marrelli M, Shakesheff KM, White LJ (2014) Dental pulp stem cells: function, isolation and applications in regenerative medicine. J Tissue Eng Regen Med. doi:10.1002/term.1899
43. Yang X, van der Kraan PM, Bian Z, Fan M, Walboomers XF, Jansen JA (2009) Mineralized tissue formation by BMP-2-transfected pulp stem cells. J Dent Res 88(11):1020–1025

44. Morad G, Kheiri L, Khojasteh A (2013) Dental pulp stem cells for in vivo bone regeneration: a systematic review of literature. Arch Oral Biol 58:1818–1827
45. Giuliani A, Manescu A, Langer M, Rustichelli F, Desiderio V, Paino F, De Rosa A, Laino L, d'Aquino R, Tirino V, Papaccio G (2013) Three years after transplants in human mandibles, histological and in-line holotomography revealed that stem cells regenerated a compact rather than a spongy bone: biological and clinical implications. Stem Cells Transl Med 2:316–324
46. Graziano A, d'Aquino R, Laino G, Proto A, Giuliano MT, Pirozzi G, De Rosa A, Di Napoli D, Papaccio G (2008) Human CD34+ stem cells produce bone nodules in vivo. Cell Prolif 41:1–11
47. Yasui T, Mabuchi Y, Toriumi H, Ebine T, Niibe K, Hoolihan DD, Morikawa S, Onizawa K, Kawana H, Akazawa C, Suzuki N, Nakagawa T, Okano H, Matsuzaki Y (2015) Purified human dental pulp stem cells promote osteogenic regeneration. J Dent Res. doi:10.1177/0022034515610748
48. Liu L, Ling J, Wei X, Wu L, Xiao Y (2009) Stem cell regulatory gene expression in human adult dental pulp and periodontal ligament cells undergoing odontogenic/osteogenic differentiation. J Endod 35:1368–1376
49. Ferro F, Spelat R, Beltrami AP, Cesselli D, Curcio F (2012) Isolation and characterization of human dental pulp derived stem cells by using media containing low human serum percentage as clinical grade substitutes for bovine serum. PLoS One 7, e48945
50. Pisciotta A, Riccio M, Carnevale G, Beretti F, Gibellini L, Maraldi T, Cavallini GM, Ferrari A, Bruzzesi G, De Pol A (2012) Human serum promotes osteogenic differentiation of human dental pulp cells in vitro and in vivo. PLoS One 7, e50542
51. Al-Habib M, Yu Z, Huang GT (2013) Small molecules affect human dental pulp stem cell properties via multiple signaling pathways. Stem Cells Dev 22(17):2402–2413
52. Lee SY, Chiang PC, Tsai YH, Tsai SY, Jeng JH, Kawata T, Huang HM (2010) Effects of cryopreservation of intact teeth on the isolated dental pulp stem cells. J Endod 36: 1336–1340
53. Woods EJ, Perry BC, Hockema JJ, Larson L, Zhou D, Goebel WS (2009) Optimized cryopreservation method for human dental-pulp derived stem cells and their tissues of origin for banking and clinical use. Cryobiology 59:150–157
54. Zhang W, Walboomers XF, Shi S, Fan M, Jansen JA (2006) Multilineage differentiation potential of stem cells derived from human dental pulp after cryopreservation. Tissue Eng 12:2813–2823
55. Laino G, Carinci F, Graziano A, d'Aquino R, Lanza V, De Rosa A, Gombos F, Caruso F, Guida L, Rullo R, Menditti D, Papaccio G (2006) In vitro bone production using stem cells derived from human dental pulp. J Craniofac Surg 17(3):511–515
56. Papaccio G, Graziano A, d'Aquino R, Graziano MF, Pirozzi G, Menditti D, de Rosa A, Carinci F, Laino G (2006) Long-term cryopreservation of dental pulp stem cells (SBP-DPSCs) and their differentiated osteoblasts: a cell source for tissue repair. J Cell Physiol 208:319–325
57. Miura M, Gronthos S, Zhao M, Lu B, Fisher LW, Robey PG, Shi S (2003) SHED: stem cells from human exfoliated deciduous teeth. Proc Natl Acad Sci U S A 100:5807–5812
58. Cordeiro MM, Dong Z, Kaneko T, Zhang Z, Miyazawa M, Shi S, Smith AJ, Nör JE (2008) Dental pulp tissue engineering with stem cells from exfoliated deciduous teeth. J Endod 34:962–969
59. Daltoé FP, Mendonça PP, Mantesso A, Deboni MCZ (2014) Can SHED or DPSCs be used to repair/regenerate non-dental tissues? A systematic review of in vivo studies. Braz Oral Res 28:1–7
60. Ishkitiev N, Yeagaki K, Calenic B, Nakahara T, Ishikawa H, Mitiev V, Haapasalo M (2010) Deciduous and permanent dental pulp mesenchymal cells acquire hepatic morphologic and functional features in vitro. J Endod 36:469–474
61. Kerkis I, Kerkis A, Dozortsev D, Stukart-Parsons GC, Gomes Massironi SM, Pereira LV, Caplan AI, Cerruti HF (2006) Isolation and characterization of a population of immature

dental pulp stem cells expressing OCT-4 and other embryonic stem cell markers. Cells Tissues Organs 184:105–116
62. Zhao H, Chai Y (2015) Stem cells in teeth and craniofacial bones. J Dent Res 94(11): 1495–1501
63. Casagrande L, Demarco FF, Zhang Z, Araujo FB, Shi S, Nör JE (2010) Dentin-derived BMP-2 and odontoblast differentiation. J Dent Res 89:603–608
64. Demarco FF, Casagrande L, Zhang Z, Dong Z, Tarquinio SB, Zeitlin BD, Shi S, Smith AJ, Nör JE (2010) Effects of morphogen and scaffold porogen on the differentiation of dental pulp stem cells. J Endod 36:1805–1811
65. Galler KM, Cavender AC, Koeklue U, Suggs LJ, Schmaltz G, D'Souza RN (2011) Bioengineering of dental stem cells in a PEGylated fibrin gel. Regen Med 6:191–200
66. Sakai VT, Zhang Z, Dong Z, Neiva K, Machado M, Shi S, Santos C, Nör JE (2010) SHED differentiate into functional odontoblasts and endothelium. J Dent Res 89:791–796
67. Sonoyama W, Liu Y, Fang D, Yamaza T, Seo BM, Zhang C, Liu H, Gronthos S, Wang CY, Shi S (2006) Mesenchymal stem cell-mediated functional tooth regeneration in swine. PLoS One 1, e79
68. Sonoyama W, Liu Y, Yamaza T, Tuan RS, Wang S, Shi S, Huang GT (2008) Characterization of the apical papilla and its residing stem cells from human immature permanent teeth: a pilot study. J Endod 34:166–171
69. Huang GT, Gronthos S, Shi S (2009) Mesenchymal stem cells derived from dental tissues vs. those from other sources: their biology and role in regenerative medicine. J Dent Res 88:792–806
70. Huang GT, Yamaza T, Shea LD, Djouad F, Kuhn NZ, Tuan RS, Shi S (2010) Stem/progenitor cell-mediated *de novo* regeneration of dental pulp with newly deposited continuous layer of dentin in an in vivo model. Tissue Eng Part A 16:605–615
71. Yagyuu T, Ikeda E, Ohgushi H, Tadokoro M, Hirose M, Maeda M, Inagake K, Kirita T (2010) Hard-tissue forming potential of stem/progenitor cells in human dental follicle and dental papilla. Arch Oral Biol 55:68–76
72. Morsczeck C, Moehl C, Götz W, Heredia A, Schäffer TE, Eckstein N, Sippel C, Hoffmann KH (2005) In vitro differentiation of human dental follicle cells with dexamethasone and insulin. Cell Biol Int 29:567–575
73. Estrela C, Alencar AH, Kitten GT, Vencio EF, Gava E (2011) Mesenchymal stem cells in the dental tissues: perspectives for tissue regeneration. Braz Dent J 22:91–98
74. Vollkommer T, Gosau M, Felthaus O, Reichert TE, Morsczeck C, Gotz W (2015) Genome-wide gene expression profiles of dental follicle stem cells. Acta Odontol Scand 73:93–100
75. Ikeda E, Yagi K, Kojima M, Yaguu T, Ohshima A, Sobajima S, Tadokoro M, Katsube Y, Isoda K, Kondoh M, Kawase M, Go MJ, Adachi H, Yokota Y, Kirita T, Ohgushi H (2008) Multipotent cells from the human third molar: feasibility of cell-based therapy for liver disease. Differentiation 76:495–505
76. Kémoun P, Laurencin-Dalicieux S, Rue J, Farges JC, Gennero I, Conte-Auriol F, Briand-Mesange F, Gadelorge M, Arzate H, Narayanan AS, Brunel G, Salles JP (2007) Human dental follicle cells acquire cementoblast features under stimulation by BMP-2/-7 and enamel matrix derivatives (EMD) in vitro. Cell Tissue Res 329:283–294
77. Völlner F, Ernst W, Driemel O, Morsczeck C (2009) A two-step strategy for neuronal differentiation in vitro of human dental follicle cells. Differentiation 77:433–441
78. Yalvac ME, Ramazanoglu M, Rizvanov AA, Sahin F, Bayrak OF, Salli U, Palotás A, Kose GT (2010) Isolation and characterization of stem cells derived from human third molar tooth germs of young adults: implications in neo-vascularization, osteo-, adipo- and neurogenesis. Pharmacogenomics J 10:105–113
79. Lin NH, Gronthos S, Bartold PM (2009) Stem cells and future periodontal regeneration. Periodontol 2000 51:239–251
80. Morsczeck C, Gotz W, Schierholz J, Zeilhofer F, Kuhn U, Mohl C, Sippel C, Hoffmann KH (2005) Isolation of precursor cells (PCs) from human dental follicle of wisdom teeth. Matrix Biol 24:155–165

81. Handa K, Saito M, Tsunoda A, Yamauchi M, Hattori S, Sato S, Toyoda M, Teranaka T, Narayanan AS (2002) Progenitor cells from dental follicle are able to form cementum matrix in vivo. Connect Tissue Res 43:406–408
82. Yokoi T, Saito M, Kiyono T, Iseki S, Kosaka K, Nishida E, Tsubakimoto T, Harada H, Eto K, Noguchi T, Teranaka T (2007) Establishment of immortalized dental follicle cells for regenerating periodontal ligament in vivo. Cell Tissue Res 327:301–311
83. Morsczeck C, Schmaltz G, Reichert T, Völlner F, Galler K, Driemel O (2008) Somatic stem cells for regenerative dentistry. Clin Oral Investig 12:113–118
84. Seo BM, Miura M, Gronthos S, Bartold PM, Batouli S, Brahim J, Young M, Gehron Robey P, Wang CY, Shi S (2004) Investigation of multipotent postnatal stem cells from human periodontal ligament. Lancet 364:149–155
85. Wada N, Menicanin D, Shi S, Bartold PM, Gronthos S (2009) Immunomodulatory properties of human periodontal ligament stem cells. J Cell Physiol 219:667–676
86. Gault P, Black A, Romette JL, Fuente F, Schroeder K, Thillou F, Brune T, Berdal A, Wurtz T (2010) Tissue-engineered ligament: implant constructs for tooth replacement. J Clin Periodontol 37:750–758
87. Gay IC, Chen S, MacDougall M (2007) Isolation and characterization of multipotent human periodontal ligament stem cells. Orthod Craniofac Res 10:149–160
88. Gronthos S, Mrozik K, Shi S, Bartold PM (2006) Ovine periodontal ligament stem cells: isolation, characterization and differentiation potential. Calcif Tissue Int 79:310–317
89. Menicanin D, Mrozik KM, Wada N, Marino V, Shi S, Bartold PM, Gronthos S (2014) Periodontal-ligament-derived stem cells exhibit the capacity for long-term survival, self-renewal and regeneration of multiple tissue types in vivo. Stem Cells Dev 23:1001–1011
90. Xu J, Wang W, Kapila Y, Lotz J, Kapila S (2009a) Multiple differentiation capacity of STRO-1+/CD146+ PDL mesenchymal progenitor cells. Stem Cells Dev 18(3):487–496
91. Shi S, Bartold PM, Miura M, Seo BM, Robey PG, Gronthos S (2005) The efficacy of mesenchymal stem cells to regenerate and repair dental structures. Orthod Craniofac Res 8:191–199
92. Trubiani O, Orsini G, Zini N, Di Iorio D, Piccirilli M, Piattelli A, Caputi S (2008) Regenerative potential of human periodontal ligament derived stem cells on three-dimensional biomaterials: a morphological report. J Biomed Mater Res A 00087:986–993
93. He H, Yu J, Cao J, E LL, Wang D, Zhang H, Liu H (2011) Biocompatibility and osteogenic capacity of periodontal ligament stem cells on nHAC/PLA and HA/TCP scaffolds. J Biomater Sci Polym Ed 22:179–194
94. Marynka-Kalmani K, Treves S, Yafee M, Rachima H, Gafni Y, Cohen MA, Pitaru S (2010) The lamina propria of adult human oral mucosa harbors a novel stem cell population. Stem Cells 28:984–995
95. Tang L, Li N, Xie H, Jin Y (2011) Characterization of mesenchymal stem cells from human normal and hyperplastic gingiva. J Cell Physiol 226:832–842
96. Zhang Q, Shi S, Liu Y, Uyanne J, Shi S, Le AD (2009) Mesenchymal stem cells derived from human gingiva are capable of immunomodulatory functions and ameliorate inflammation-related tissue destruction in experimental colitis. J Immunol 183:7787–7798
97. Tomar GB, Srivastava RK, Gupta N, Barhanpurkar AP, Pote ST, Jhaveri HM, Mishra GC, Wani MR (2010) Human gingiva-derived mesenchymal stem cells are superior to bone marrow-derived mesenchymal stem cells for cell therapy in regenerative medicine. Biochem Biophys Res Commun 393:377–383
98. Moshaverinia A, Chen C, Xu X, Akiyama K, Ansari S, Zadeh HH, Shi S (2014) Bone regeneration potential of stem cells derived from periodontal ligament or gingival tissue sources encapsulated in RGD-modified alginate scaffold. Tissue Eng Part A 20:611–621
99. Matsubara T, Suardita K, Ishii M, Sugiyama M, Igarashi A, Oda R, Nishimura M, Saito M, Nkagawa K et al (2005) Alveolar bone marrow as a cell source for regenerative medicine: differences between alveolar and iliac bone marrow stromal cells. J Bone Miner Res 20(3):399–409

100. Mason S, Tarle S, Osibin W, Kinfu Y, Kaigler D (2014) Standardization and safety of alveolar bone-derived stem cell isolation. J Dent Res 33:55–61
101. Park J, Kim JC, Kim YT, Choi SH, Cho KS, Im GI, Kim BS, Kim CS (2012) Acquisition of human alveolar bone-derived stromal cells using minimally irrigated implant osteotomy: in vitro and in vivo evaluations. J Clin Periodontol 39:495–505
102. Kim BS, Kim JS, Chung YS, Sin YW, Ryu KH, Lee J, You HK (2013) Growth and osteogenic differentiation of alveoral human bone marrow-derived mesenchymal stem cells on chitosan/hydroxyapatite composite fabric. J Biomed Mater Res A 101:1550–1558
103. Arkell RM, Tam PPL (2012) Signalling and transcriptional activity: initiating head development in mouse embryos: integrating. Open Biol 2(3):120030
104. Noden DM, Trainor PA (2005) Relations and interactions between cranial mesoderm and neural crest populations. J Anat 207(5):575–601
105. Behr B, Panetta NJ, Longaker MT, Quarto N (2010) Different endogenous threshold levels of fibroblast growth factor-ligands determine the healing potential of frontal and parietal bones. Bone 47(2):281–294
106. Senarath-Yapa K, Li S, Meyer NP, Longaker MT, Quarto N (2013) Integration of multiple signaling pathways determines differences in the osteogenic potential and tissue regeneration of neural crest-derived and mesoderm-derived calvarial bones. Int J Mol Sci 14(3): 5978–5997
107. Greenwald AS, Boden SD, Goldberg VM, Khan Y, Laurencin CT, Rosier RN (2001) Bone-graft substitutes: facts, fictions and applications. J Bone Joint Surg Am 83A(Suppl 2 Pt 2): 98–103
108. Dinopoulos H, Dimitriou R, Giannoudis PV (2012) Bone graft substitutes: what are the options? Surgeon 10(4):230–239
109. Mao JJ, Giannobile WV, Helms JA, Hollister SJ, Krebsbach PH, Longaker MT, Shi S (2006) Craniofacial tissue engineering by stem cells. J Dent Res 85(11):966–979
110. Zaky SH, Cancedda R (2009) Engineering craniofacial structures: facing the challenge. J Dent Res 88(12):1077–1091
111. Muschler GF, Raut VP, Patterson TE, Wenke JC, Hollinger JO (2010) The design and use of animal models for translational research in bone tissue engineering and regenerative medicine. Tissue Eng Part B Rev 16(1):123–145
112. Gomes PS, Fernandes MH (2011) Rodent models in bone-related research: the relevance of calvarial defects in the assessment of bone regeneration strategies. Lab Anim 45(1):14–24
113. An Y, Friedman R (1999) Animal models of bone defect repair. In: An Y, Friedman R (eds) Animal models in orthopaedic research. CRC Press, Boca Raton, pp 241–260
114. Frame JW (1980) A convenient animal model for testing bone substitute materials. J Oral Surg 38(3):176–180
115. Honma T, Itagaki T, Nakamura M, Kamakura S, Takahashi I, Echigo S, Sasano Y (2008) Bone formation in rat calvaria ceases within a limited period regardless of completion of defect repair. Oral Dis 14(5):457–464
116. Liebschner M (2004) Biomechanical considerations of animal models used in tissue engineering of bone. Biomaterials 25(9):1697–1714
117. Mooney MP, Siegel MI (2005) Animal models for bone tissue engineering. In: Wnek G, Bowlin G (eds) Encyclopedia of biomaterials and biomedical engineering. Marcel Dekker, New York, pp 1–19
118. Schmitz JP, Hollinger JO (1986) The critical size defect as an experimental model for cranio-mandibulofacialnonunions. Clin Orthop Relat Res 205:299–308
119. Hollinger JO, Kleinschmidt JC (1990) The critical size defect as an experimental model to test bone repair materials. J Craniofac Surg 1(1):60–68
120. Bosch C, Melsen B, Vargervik K (1998) Importance of the critical-size bone defect in testing bone-regenerating materials. J Craniofac Surg 9(4):310–316
121. Petridis X, Diamanti E, Trigas GC, Kalyvas D, Kitraki E (2015) Bone regeneration in critical-size calvarial defects using human dental pulp cells in an extracellular matrix-based scaffold. J Craniomaxillofac Surg 43(4):483–490

122. Cooper GM, Mooney MP, Gosain AK, Campbell PG, Losee JE, Huard J (2010) Testing the critical size in calvarial bone defects: revisiting the concept of a critical-size defect. Plast Reconstr Surg 125(6):1685–1692
123. Chai Y, Jiang X, Ito Y, Bringas P Jr, Han J, Rowitch DH, Soriano P, McMahon AP, Sukov HM (2000) Fate of the mammalian cranial neural crest during tooth and mandibular morphogenesis. Development 127(8):1671–1679
124. Riccio M, Resca E, Maraldi T et al (2010) Human dental pulp stem cells produce mineralized matrix in 2D and 3D cultures. Eur J Histochem 54(4), e46
125. Maraldi T, Riccio M, Pisciotta A, Zavatti M, Carnevale G, Beretti F, La Sala GB, Motta A, De Pol A (2013) Human amniotic fluid-derived and dental pulp derived stem cells seeded into collagen scaffold repair critical-size bone defects promoting vascularization. Stem Cell Res Ther 4:53
126. Riccio M, Maraldi T, Pisciotta A, La Sala GB, Ferrari A, Bruzzesi G, Motta A, Migliaresi C, De Pol A (2012) Fibroin scaffold repairs critical-size bone defects in vivo supported by human amniotic fluid and dental pulp stem cells. Tissue Eng Part A 18(9–10):1006–1013
127. Su Y, Shi S, Liu Y (2014) Immunomodulation regulates mesenchymal stem cell-based bone regeneration. Oral Dis 20(7):633–636
128. Caplan AI, Dennis JE (2006) Mesenchymal stem cells as trophic mediators. J Cell Biochem 98(5):1076–1084
129. O'Brien FJ (2011) Biomaterials and scaffolds for tissue engineering. Mater Today 14(3): 88–95
130. Murphy CM, O'Brien FJ, Little DG, Schindeler A (2013) Cell-scaffold interactions in the bone tissue engineering triad. Eur Cell Mater 26:120–132
131. Alhag M, Farrell E, Toner M, Claffey N, Lee TC, O'Brien FJ (2011) Evaluation of early healing events around mesenchymal stem cell-seeded collagen–glycosaminoglycan scaffold. An experimental study in Wistar rats. Oral Maxillofac Surg 15:31–39
132. Alhag M, Farrell E, Toner M, Clive Lee T, O'Brien FJ, Claffey N (2012) Evaluation of the ability of collagen–glycosaminoglycan scaffolds with or without mesenchymal stem cells to heal bone defects in Wistar rats. Oral Maxillofac Surg 16:47–55
133. de Mendonça Costa A, Bueno DF, Martins MT, Kerkis I, Kerkis A, Fanganiello RD, Cerutti A, Alonso N, Passos-Bueno MR (2008) Reconstruction of large cranial defects in nonimmunosuppressed experimental design with human dental pulp stem cells. J Craniofac Surg 19(1):204–210
134. Ma L, Makino Y, Yamaza H, Akiyama K, Hoshino Y, Song G, Kukita T, Nonaka K, Shi S, Yamaza T (2012) Cryopreserved dental pulp tissues of exfoliated deciduous teeth is a feasible stem cell resource for regenerative medicine. PLoS One 7(12), e51777
135. Seo BM, Sonoyama W, Yamaza T, Coppe C, Kikuiri T, Akiyama K, Lee JS, Shi S (2008) SHED repair critical-size calvarial defects in mice. Oral Dis 14(5):428–434
136. Ge S, Zhao N, Wang L, Yu M, Liu H, Song A, Huang J, Wang G, Yang P (2012) Bone repair by periodontal ligament stem cellseeded nanohydroxyapatite-chitosan scaffold. Int J Nanomedicine 7:5405–5414
137. Tour G, Wendel M, Moll G, Tcacencu I (2012) Bone repair using periodontal ligament progenitor cell-seeded constructs. J Dent Res 91(8):789–794
138. Honda MJ, Imaizumi M, Suzuki H, Ohshima S, Tsuchiya S, Satomura K (2011) Stem cells isolated from human dental follicles have osteogenic potential. Oral Surg Oral Med Oral Pathol Oral Radiol Endod 111(6):700–708
139. Pearce AI, Richards RG, Milz S, Schneider E, Pearce SG (2007) Animal models for implant biomaterial research in bone: a review. Eur Cell Mater 13:1–10
140. Liu HC, E LL, Wang DS, Su F, Wu X, Shi ZP, Lv Y, Wang JZ (2011) Reconstruction of alveolar bone defects using bone morphogenetic protein 2 mediated rabbit dental pulp stem cells seeded on nano-hydroxyapatite/collagen/poly(L-lactide). Tissue Eng Part A 17(19–20): 2417–2433
141. Yamada Y, Ito K, Nakamura S, Ueda M, Nagasaka T (2011) Promising cell-based therapy for bone regeneration using stem cells from deciduous teeth, dental pulp, and bone marrow. Cell Transplant 20(7):1003–1013

142. Zheng Y, Liu Y, Zhang CM, Zhang HY, Li WH, Shi S, Le AD, Wang SL (2009) Stem cells from deciduous tooth repair mandibular defect in swine. J Dent Res 88(3):249–254
143. Kuo TF, Lee SY, Wu HD, Poma M, Wu YW, Yang JC (2015) An in vivo swine study for xenografts of calcium sulfate-based bone grafts with human dental pulp stem cells (hDPSCs). Mater Sci Eng C Mater Biol Appl 50:19–23
144. Alkaisi A, Ismail AR, Mutum SS, Ahmad ZA, Masudi S, Abd Razak NH (2013) Transplantation of human dental pulp stem cells: enhance bone consolidation in mandibular distraction osteogenesis. J Oral Maxillofac Surg 71(10):1758.e1-13
145. d'Aquino R, De Rosa A, Lanza V, Tirino V, Laino L, Graziano A, Desiderio V, Laino G, Papaccio G (2009) Human mandible bone defect repair by the grafting of dental pulp stem/progenitor cells and collagen sponge biocomplexes. Eur Cell Mater 18:75–83
146. Liu N, Lyu X, Fan H, Shi J, Hu J, Luo E (2014) Animal models for craniofacial reconstruction by stem/stromal cells. Curr Stem Cell Res Ther 9(3):174–186
147. Caton J, Mota L, Gandini L, Laskaris B (1994) Non-human primate models for testing the efficacy and safety of periodontal regeneration procedures. J Periodontol 65(12):1143–1150
148. Suaid FF, Ribeiro FV, Gomes TR, Silvério KG, Carvalho MD, Nociti FH Jr, Casati MZ, Sallum EA (2012) Autologous periodontal ligament cells in the treatment of Class III furcation defects: a study in dogs. J Clin Periodontol 39(4):377–384
149. Tsumanuma Y, Iwata T, Washio K, Yoshida T, Yamada A, Takagi R, Ohno T, Lin K, Yamato M, Ishikawa I, Okano T, Izumi Y (2011) Comparison of different tissue-derived stem cell sheets for periodontal regeneration in a canine 1-wall defect model. Biomaterials 32(25): 5819–5825
150. Han J, Menicanin D, Marino V, Ge S, Mrozik K, Gronthos S, Bartold PM (2014) Assessment of the regenerative potential of allogeneic periodontal ligament stem cells in a rodent periodontal defect model. J Periodontal Res 49(3):333–345
151. Liu Y, Zheng Y, Ding G, Fang D, Zhang C, Bartold PM, Gronthos S, Shi S, Wang S (2008) Periodontal ligament stem cell-mediated treatment for periodontitis in miniature swine. Stem Cells 26(4):1065–1073
152. Park JY, Jeon SH, Choung PH (2011) Efficacy of periodontal stem cell transplantation in the treatment of advanced periodontitis. Cell Transplant 20(2):271–285
153. Ito K, Yamada Y, Nakamura S, Ueda M (2011) Osteogenic potential of effective bone engineering using dental pulp stem cells, bone marrow stem cells, and periosteal cells for osseointegration of dental implants. Int J Oral Maxillofac Implants 26(5):947–954
154. Kim SH, Kim KH, Seo BM, Koo KT, Kim TI, Seol YJ, Ku Y, Rhyu IC, Chung CP, Lee YM (2009) Alveolar bone regeneration by transplantation of periodontal ligament stem cells and bone marrow stem cells in a canine peri-implant defect model: a pilot study. J Periodontol 80(11):1815–1823
155. Omori M, Tsuchiya S, Hara K, Kuroda K, Hibi H, Okido M, Ueda M (2015) A new application of cell-free bone regeneration: immobilizing stem cells from human exfoliated deciduous teeth-conditioned medium onto titanium implants using atmospheric pressure plasma treatment. Stem Cell Res Ther 6:124
156. Asatrian G, Pham D, Hardy WR, James AW, Peault B (2015) Stem cell technology for bone regeneration: current status and potential applications. Stem Cells Cloning 10(8):39–48
157. Feng F, Akiyama K, Liu Y et al (2010) Utility of PDL progenitors for in vivo tissue regeneration: a report of 3 cases. Oral Dis 16:20–28
158. Xu LL, Liu HC, Wang DS, E LL, Xu L, Jin ZL, Duan YZ (2009b) Effects of BMP-2 and dexamethasone on osteogenic differentiation of rat dental follicle progenitor cells seeded on three-dimensional beta-TCP. Biomed Mater 4(6):065010
159. Peric M, Dumic-Cule I, Grcevic D, Matijasic M, Verbanac D, Paul R, Grgurevic L, Trkulja V, Bagi CM, Vukicevic S (2015) The rational use of animal models in the evaluation of novel bone regenerative therapies. Bone 70:73–86
160. Nakamura A, Akahane M, Shigematsu H, Tadokoro M, Morita Y, Ohgushi H, Dohi Y, Imamura T, Tanaka Y (2010) Cell sheet transplantation of cultured mesenchymal stem cells enhances bone formation in a rat nonunion model. Bone 46:418e24

161. Yiying QI, Tengfei Z, Weiqi Y, Kan XU, Zhongli S, Jianwei W (2013) Mesenchymal stem cell sheet transplantation combined with locally released simvastatin enhances bone formation in a rat tibia osteotomy model. Cytotherapy 15:44–56
162. Gómez-Barrena E, Rosset F, Lozano D, Stanovici J, Ermthaller C, Gerbhard F (2015) Bone fracture healing: cell therapy in delayed unions and nonunions. Bone 70:93–101
163. Sloan AJ, Waddington RJ (2009) Dental pulp stem cells: what, where, how? Int J Paediatr Dent 19(1):61–70
164. Castano-Izquierdo H, Alvarez-Barreto J, van der Dolder J, Jansen JA, Mikos AG, Sikavitsas VI (2007) Pre-culture period of mesenchymal stem cells in osteogenic media influences their in vivo bone forming potential. J Biomed Mater Res A 82:129–138
165. Patel M, Smith AJ, Sloan AJ, Smith G, Cooper PR (2009) Phenotype and behaviour of dental pulp cells during expansion culture. Arch Oral Biol 54:898–908
166. Kitraki E, Zakkas S, Synolaki E, Diamanti E, Tiniakos DG, Stamatakis A, Matsioula C, Stylianopoulou F, Papapolychroniou T (2014) Dental pulp cells enhance bone healing in a rat osteotomy model. Ann Orthop Rheumatol 2(1):1009

Neuronal Properties of Dental Stem Cells

Letizia Ferroni, Sara Ricci, and Chiara Gardin

The neurodegenerative disorders are characterized by loss of nerve structure or function. The prevalent disorders are Alzheimer's disease, frontotemporal dementia, amyotrophic lateral sclerosis, Parkinson's disease, Huntington's disease, and multiple sclerosis. These neurodegenerative diseases are increasing worldwide, not only due to increased incidences of direct or indirect injury to the central nervous system (CNS) but also due to the increase in the percentage of the aging population [1]. Despite great progress in understanding the etiology of these disorders, the underlying mechanisms are still indistinct. Furthermore, no means of treating the underlying cause have been devised. For the past decade, researchers have been interested in stem cells and the prospect of using them for understanding the pathogenesis of disease and for facilitating the development of novel therapeutics [2]. In model organisms, both endogenous and exogenous neural stem cells (NSCs) have been investigated for their capacity to regenerate a damaged nervous system [3]. Due to the low incidence of human adult NSCs and problems with accessibility, the use of exogenous sources of stem cells with neural potential has been suggested as a plausible approach to stem cell therapy. Although embryonic stem cells (ESCs) and bone marrow stem cells (BMSCs) have been assessed as potential candidates for neuronal therapy, adult stem cell populations derived from cranial neural crest cells may possess a greater propensity for neuronal differentiation and repair [4]. Interestingly, the dental pulp tissue, termed "ectomesenchyme", derives from ectodermal cells growing on the periphery of the neural tube during embryonic development, migrating into the oral region, and transdifferentiating into a mesenchymal phenotype [5]. Consequently, Dental Stem Cells (DSCs) have been proposed as a

L. Ferroni • C. Gardin (✉)
Department of Biomedical Sciences, University of Padova, via Ugo Bassi 58/B, 35131 Padova, Italy
e-mail: chiaragardin@gmail.com

S. Ricci
Department of Neurosciences, University of Padova, via Giustiniani 5, 35128 Padova, Italy

B. Zavan, E. Bressan (eds.), *Dental Stem Cells: Regenerative Potential*, Stem Cell Biology and Regenerative Medicine, DOI 10.1007/978-3-319-33299-4_12

promising source of stem cells to treat nerve regeneration due to their close embryonic origin and easy accessibility. Indeed, DSCs can be obtained from extracted teeth and their surrounding tissue that are usually discarded during dental procedures without invasive methods and additional injury.

Preliminary investigations into the neural potential of human adult DSCs have shown that, under non-neuronal inductive conditions, these cells expressed the neural progenitor marker nestin, and the glial marker glial fibrillary acidic protein (GFAP), at both the gene and protein levels. The DSCs retain their neural crest properties following ex vivo expansion, and express the postmitotic neuron-specific marker neuronal nuclei (NeuN), when cultured under neural inductive conditions [4, 6–8]. Therefore, neural crest-derived adult DSCs exposed to the appropriate environmental cues could differentiate into functionally active neurons and could provide an alternative stem cell source for the treatment of neuronal disorders and injury.

The neural differentiation potential of human adult dental pulp stem cells (DPSCs) was demonstrated by Arthur and colleagues [4]. DPSCs were cultured in neuronal inductive conditions with mitogen factors Epidermal Growth Factor (EGF) and basic Fibroblast Growth Factor (FGF) for 3 weeks. Following 3 weeks of induction, DPSCs acquired a bipolar and stellate morphology consistent with that of sensory and motor neurons, respectively. Assessment of the cell proliferation status of DPSCs over the same time period showed that there was a significant decrease in the proliferation rate of DPSCs cultured in neuronal inductive media compared with non-neuronal inductive condition. Immunocytochemical analysis showed that ex vivo expanded DPSCs constitutively express the proteins nestin and GFAP. While nestin protein expression continued to be detected by the majority of DPSCs following neuronal induction, nestin gene transcription was found to be downregulated at 3 weeks after neuronal induction, correlating to neuronal maturation. Also the expression of βIII-tubulin, a neuronal specific marker expressed during early brain development and downregulated in adult brain [9], was evaluated. Whereas βIII-tubulin protein increased over the 3 week induction period, the mRNA level was found to be downregulated, which also correlated to the observed reduction in nestin gene expression. Moreover, the number of DPSCs expressing either neurofilament (NF) medium (NFM) or NF heavy (NFH) gene or protein were found to be significantly higher following neuronal induction compared with non-neuronal inductive condition [4].

Also human exfoliated deciduous teeth (SHED) showed neural differentiation potential. SHED differ from DPSCs with respect to their higher proliferation rate, increased cell-population doublings, sphere-like cell-cluster formation, osteoinductive capacity in vivo, and failure to reconstitute a dentin-pulp-like complex [8]. Therefore, SHED apparently represent a population of multipotent stem cells that are perhaps more immature than previously examined postnatal stromal stem-cell populations. SHED under non-neuronal inductive conditions express a variety of neural cell markers including nestin, βIII-tubulin, glutamic acid decarboxylase (GAD), NeuN, GFAP, NFM, and 2′,3′-cyclic nucleotide-3′-phosphodiesterase (CNPase). After 4 weeks of neural inductive culture, SHED lost their fibroblastic morphology developing multicytoplasmic processes and increased the expression levels of neuronal markers including βIII-tubulin, GAD, and NeuN, whereas the levels of nestin, GFAP, NFM, and CNPase remained unchanged [8]. These findings were confirmed by Govindasamy and co-workers that compared the neural differentiation potential of SHED and DPSCs [10]. They observed that, under non-neuronal inductive conditions, the proliferation rate of SHED was higher than that of DPSCs. The fold expression of several pluripotent markers such as OCT4, SOX2, NANOG, and REX1 were higher in SHED as compared with DPSCs. Conversely, DPSCs showed higher expression of neuro-ectodermal markers PAX6, GBX2, and nestin. These data supported the notion that SHED are more primitive or pluripotent cells than DPSCs. Indeed, the overexpression of the transcription factors OCT4, SOX2, and NANOG are responsible for the maintenance of pluripotency in early embryos and ESCs [11]. However, when SHED and DPSCs were cultured in non-coated dishes containing the neuronal induction factors FGF and EGF for 15 days, both cells were capable of forming distinct neurospheres, in which growing cells aggregate in floating spheres. Once attached in coated dishes, the neurospheres derived from SHED and DPSCs spontaneously showed outgrowth and dendrite-like structure and expressed neuronal markers, such as βIII-tubulin, NF and GFAP. Despite SHED and DPSCs came from the same origin, higher neurosphere formation and neuronal marker expression were found in the differentiated DPSCs into neuron-like cells as compared with SHED. Since nestin is essential for the induction of neurospheres [12], the high level of nestin expressed by undifferentiated DPSCs could potentially enable them to differentiate more efficiently than SHED into neuronal cells. Nevertheless, SHED could also be forced to form neuronal cells under the influence of appropriate microenvironment [10].

In a study of Lee and colleagues the neural differentiation potential of DPSCs was compared to that of stem cells from apical papilla (SCAP), and periodontal ligament stem cells (PDLSCs) [13]. After induction in neurogenic medium for 24 h, the morphology of DPSCs, SCAP, and PDLSCs changed into neurite-like cells, including cell processes. Most of the cells were similar in shape to neuronal cells, with very thin and long cytoplasmic processes, resembling axons and dendrites. The neurogenic potential of DPSCs, SCAP, and PDLSCs was analyzed assessing the transcription levels of βIII-tubulin, microtubule-associated protein 2 (MAP2), and GFAP. In each cell line, mRNA level of the mature neuronal marker MAP2 tended to increase after neurogenic induction. The RNA level of βIII-tubulin significantly increased within SCAP and PDLSCs in neurogenic differentiation medium but not in DPSCs. DPSCs showed a significantly increased GFAP mRNA expression under differentiation conditions, whereas SCAP expressed very little GFAP mRNA in both control and differentiation groups. Interestingly, MAP2 and βIII-tubulin mRNAs were also found in undifferentiated conditions. These results suggest that SCAP might have a neural origin and thus have a capacity for neurogenic differentiation. In terms of protein level, MAP2 was strongly expressed in SCAP, DPSCs, and PDLSCs after neurogenic induction. The protein expression of βIII-tubulin increased in DPSCs under differentiation conditions, whereas GFAP was not detected. However, the expressions of neuronal markers in DPSCs were lower than those in SCAP and PDLSCs. These results indicate that PDLSCs and SCAP, as well as DPSCs, may be appropriate cell sources for neuronal regeneration [13].

Altogether these in vitro studies support the finding that DSCs are capable of differentiating into neuronal cells when cultured under appropriate inductive conditions basing on analyses of cellular morphology and expression of early neuronal markers. However, functional neurons express voltage-gated potassium, sodium, and calcium channels, which are required for the generation and propagation of action potentials [14]. The protocol of Arthur and co-workers demonstrated that DPSCs can be differentiated into neuronal-like cells by means of an inductive medium supplemented with FGF and EGF. Their protocol, however, resulted in an incomplete neuronal differentiation, since only voltage gated sodium channels could be detected without the presence of voltage gated potassium channels which are also regarded as a basic criterion for functional neuronal cell identification [4]. In a later work of Király and colleagues, DPSCs were differentiated into neuronal cells that not only express neuronal markers, but also display simultaneous voltage dependent sodium and potassium channels [15]. Their protocol consists of three steps in 10 days: pretreatment, induction, and maturation. First, DPSCs were seeded onto poly-L-lysine coated surfaces and pretreated with a medium consisting of basic FGF and 5-azacytidine. The 5-azacytidine causes hemi-demethylation of DNA leading to dedifferentiation of partly committed cells to a multipotent state. It was also reported to be a potent maturation inducing factor for neurogenesis [16]. Then, neural induction was performed by basic FGF, nerve growth factor (NGF), and neurotrophin-3 (NT-3), and the simultaneous activation of protein kinase C (PKC) and cyclic adenosine monophosphate (cAMP) pathways. In cell lines of diverse origins, it was demonstrated that neuron and glia differentiation can be promoted by activation of PKC and cAMP pathways [17]. Finally, maturation of the induced cells was achieved by continuous treatment with NT-3, and other neuroprotective factors [15]. The DPSC-derived neuronal cells generated following the protocol of Király et al. gave good results even after the engraftment into rat brain. The DPSC-derived neuronal cells were integrated into the host brain, even into the injured cortex, and showed neuronal properties not only by expressing neuron-specific markers but also by exhibiting voltage dependent sodium and potassium channels [18]. However, Király and co-workers did not observe action potential firing by the differentiated cells, addressing the incomplete differentiation of DPSCs to neuronal cells. Recently, Gervois and colleagues have implemented a two-step protocol to improve the differentiation outcome [19]. First, they adopted the neuronal induction step of Arthur's protocol based on FGF and EGF signaling [4]. These signaling molecules are essential to induce the formation of the three-dimensional neurosphere structures, in which neural progenitor cells are close in a favorable microenvironment for the creation of close physical contacts essential for neuronal commitment [20, 21]. The second step concerning the neuronal maturation is based on activation of cAMP and NT-3 signaling pathways, as previously reported by Király [15]. Elevated intracellular cAMP is essential in sustaining neurogenic differentiation of early neuronal committed cells [22], whereas NT-3 signaling is essential for neurogenic maturation [23]. After neuronal differentiation, patch-clamp analysis demonstrated the functional activity of differentiated DPSCs by the presence of voltage-gated sodium and potassium channels selectively blocked by tetrodotoxin and tetraethylammonium, respectively. Moreover, a subset of neuronal-differentiated DPSCs was able to fire a single action potential. However, a train of repeated action potential firing after stimulation was not observed, which would be the ultimate proof of functional neurons. The failure to fire repeated action potentials might be attributed to the gating kinetics of the delayed rectifier potassium channels, that is resulting in an incomplete repolarization. The incomplete repolarization failed to deactivate sodium channels, which would be necessary for repetitive firing [19].

The above studies have demonstrated the capability of human DSCs to differentiate in neuronal cells when they are cultured in vitro in presence of neuronal inductive conditions. However, interesting results were also achieved after in vivo transplantation of not manipulated DSCs in rodent and avian brain. In particular, neural developmental potential of SHED was studied by injecting them into the dentate gyrus of the hippocampus of immunocompromised mice. Histological examination showed that SHED survived for 10 days inside the mouse brain microenvironment and continued to express neural markers such as NFM [8]. Human DPSCs were found to express neuronal-associated markers (NFM and βIII-tubulin) within 48 h after injection into 4-day-old chicken embryos, indicating a rapid response to endogenous environmental cues [4]. They survived and differentiated into neuronal derivatives, and potentially integrated into host neuronal networks within 7 days. The location of DPSCs into the embryo was essential for their specific neural morphology and differentiation capacity. DPSCs located near sensory trigeminal ganglion neurons displayed a bipolar morphology characteristic of sensory neurons, whereas in the CNS, DPSCs exhibited multidendritic processes associated with motor neurons. These observations suggest that not manipulated DPSCs are able to respond directly to surrounding environment and differentiate into specific neurons when they are transplanted in host brain [4].

The interesting studies of Nosrat and colleagues have demonstrated the capability of DPSCs to promote the survival of sensory neurons of the trigeminal ganglion, and motoneurons of the spinal cord by providing neurotrophic support [24, 25]. It was shown that DPSCs have the ability to produce neurotrophic factors, such as NGF, brain-derived neurotrophic factor (BDNF) and glial cell line-derived neurotrophic factor (GDNF) both in vivo and in vitro. Neurotrophic factors are general modulators of neuronal survival, plasticity, and target innervation and play important roles in axon growth and synaptogenesis [26]. In vivo experiments have demonstrated that NGF, BDNF, and GDNF transcripts are present in the dental pulp at the time of the onset of dental pulp innervation. These neurotrophic factors are crucial in supporting trigeminal nerves as they grow into the developing dental pulp and as the innervation of teeth is established [27]. Primary cultures of DPSCs have also shown the expression of NGF, BDNF, and GDNF mRNA transcripts. Interestingly, the expression levels of these transcripts in DPSCs were upregulated after sub-culturing, suggesting their long-term neurotrophic competence in culture [24]. Generally, the maintenance in vitro of central and peripheral nervous tissues requires the presence of exogenous neurotrophic factors to regulate neuronal survival and differentiation as well as axonal growth and path-finding [28]. However, in vitro experiments proved that neurotrophic factors produced by DPSCs promote survival of trigeminal neurons and neurite outgrowth [24]. Moreover, an in vitro study demonstrated that the neurotrophic factors produced by DPSCs promote the survival of dopaminergic neurons and provide neuroprotection for dopaminergic neurons against 6-hydroxy-dopamine [25]. DPSCs not only produce neurotrophic factors for supporting the survival of neuronal cells, but also support the homing of endogenous neural stem cells to the injury site during transplantation. In an avian embryonic model system, Arthur and coworkers demonstrated that DPSCs are able to coordinate axon guidance within a receptive host nervous system. In particular, they observed that DPSCs chemoattracted avian trigeminal ganglion axons toward them by the release of the chemokine CXCL12 [29]. The neurotrophic support of DPSCs was also confirmed by ectopically transplantation into the anterior chamber of eye in rats. Dental pulp grafts induced collateral sprouting from existing iris nerve fibers and promoted innervation of the ectopically transplanted dental pulp tissue. Additionally, dental pulp grafts increased the density of catecholaminergic nerve fibers of the irises [24]. Interestingly, dental pulp grafting also promoted the survival of injured motoneurons in rat model of spinal cord injury (SCI), indicating a functional bioactivity of the dental pulp-derived neurotrophic factors in vivo by rescuing motoneurons [24].

Similarly, SHED exhibited therapeutic benefits for recovery after SCI [30]. It was shown that SHED inhibited SCI-induced apoptosis of neurons, astrocytes, and oligodendrocytes, which promoted the preservation of neural fibers and myelin sheaths. SHED also regenerated transected axon through the direct inhibition of multiple axon growth inhibitors signals by paracrine mechanisms. Finally, SHED replaced lost or damaged oligodendrocytes after SCI through specific differentiation into mature oligodendrocytes under the extreme conditions of SCI [30].

SHED have also been proposed as a source of stem cells for the treatment of Parkinson's disease, as it has been demonstrated their differentiation into dopaminergic neurons [31]. Wang and colleagues have first induced SHED into neurospheres in serum-free culture medium supplemented with EGF and basic FGF, and then have induced SHED-derived spheres into dopaminergic neurons with a cocktail of sonic hedgehog (SHH), FGF8, GDNF, and forskolin. Under this condition, SHED-derived spheres generated a number of βIII-tubulin- and MAP2-positive neurons, and some of them were positive to tyrosine hydroxylase (TH), the enzyme responsible for synthesis of dopamine. In order to increase the cell survival capacity, Wang and coworkers transplanted into the striatum of 6-hydroxy-dopamine-treated rats the SHED-derived spheres instead of the quite matured differentiation neurons, because the latter has been reported poor survival rate [32]. Intrastriatal transplantation of SHED-derived spheres ameliorated behavioral deficits in parkinsonian rats, and enhanced the dopamine content in the striatum. Probably, this in vitro induction system activated in SHED-derived spheres some transcriptional factors for the regulation of TH expression. Once these cells were transplanted to striatum, they responded to the microenvironmental signals to undergo the committed differentiation [31].

The exposure to a cocktail miming embryonic midbrain cues was also proposed by Kanafi and colleagues to differentiate DPSCs into functional dopaminergic neuron [33]. The cocktail consisted of basic FGF, SHH and FGF8. The last two factors are secreted by the ventral neural tube and at the mid/hindbrain boundary, respectively. The induced cells were characterized by an increase in mature neuronal markers (βIII-tubulin, MAP2) and dopamine-specific markers, such as Engrailed 1 (En1), Nuclear Receptor related protein 1 (Nurr1), paired-like homeodomain transcription factor 3 (Pitx3), and TH. The induced cells presented a simultaneous decrease in early neuronal markers nestin, musashi12, neural crest marker HNK1 and mesenchymal markers CD73, CD90, and CD105. Functional studies indicated that the induced DPSCs secreted dopamine constitutively and upon stimulation with potassium chloride (KCl) or adenosine triphosphate (ATP). Additionally, the induced DPSCs showed intracellular calcium influx in the presence of KCl, and after ATP stimulation. These in vitro findings suggest that DPSCs in the presence of embryonic midbrain cues have efficient propensity towards functional dopaminergic cell-type [33].

In conclusion, in the development of a successful autologous transplantation therapy for repairing and regenerating injured nervous system, accessibility of the cells intended for grafting is a very important issue. This type of therapy should preferably use easily accessible tissues from the patient to be treated. Dental tissue constitutes an interesting source of cells that could be explored for possible autologous transplantation therapy. DSCs are easily available from adult teeth such as wisdom teeth that are sometimes extracted due to mal positioning. This procedure of preparation of DSCs is devoid of any relevant ethical concerns. In addition, DSCs can be easily handled in the laboratory. They are expandable, cryopreservable, and robust in culture. The achieved results support the use of DSCs as promising candidates for cell-based therapies for neuron disorders. DSCs could produce neurotrophic and neuroprotective factors for supporting the survival of neuronal cells, but also could support the homing of endogenous neural stem cells to the injury site during transplantation. Moreover, DSCs exposed to the appropriate environmental cues could differentiate into functionally active neurons and may provide an alternative stem cell source for therapy-based treatments of neuronal disorders and injury.

References

1. Mouhieddine TH, Kobeissy FH, Itani M et al (2014) Stem cells in neuroinjury and neurodegenerative disorders: challenges and future neurotherapeutic prospects. Neural Regen Res 9:901–906
2. Wan W, Cao L, Kalionis B et al (2015) Applications of induced pluripotent stem cells in studying the neurodegenerative diseases. Stem Cells Int 2015:382530
3. Vishwakarma SK, Bardia A, Tiwari SK et al (2014) Current concept in neural regeneration research: NSCs isolation, characterization and transplantation in various neurodegenerative diseases and stroke: a review. J Adv Res 5:277–294

4. Arthur A, Rychkov G, Shi S et al (2008) Adult human dental pulp stem cells differentiate toward functionally active neurons under appropriate environmental cues. Stem Cells 26:1787–1795
5. Chai Y, Jiang X, Ito Y et al (2000) Fate of the mammalian cranial neural crest during tooth and mandibular morphogenesis. Development 127:1671–1679
6. Davidson RM (1994) Neural form of voltage-dependent sodium current in human cultured dental pulp cells. Arch Oral Biol 39:613–620
7. Gronthos S, Brahim J, Li W et al (2002) Stem cell properties of human dental pulp stem cells. J Dent Res 81:531–535
8. Miura M, Gronthos S, Zhao M et al (2003) SHED: stem cells from human exfoliated deciduous teeth. Proc Natl Acad Sci U S A 100:5807–5812
9. Jiang YQ, Oblinger MM (1992) Differential regulation of beta III and other tubulin genes during peripheral and central neuron development. J Cell Sci 103:643–651
10. Govindasamy V, Abdullah AN, Ronald VS et al (2010) Inherent differential propensity of dental pulp stem cells derived from human deciduous and permanent teeth. J Endod 36:1504–1515
11. Niwa H (2007) How is pluripotency determined and maintained? Development 134:635–646
12. Tropepe V, Hitoshi S, Sirard C et al (2001) Direct neural fate specification from embryonic stem cells: a primitive mammalian neural stem cell stage acquired through a default mechanism. Neuron 30:65–78
13. Lee JH, Um S, Song IS et al (2014) Neurogenic differentiation of human dental stem cells in vitro. J Korean Assoc Oral Maxillofac Surg 40:173–180
14. Hille B (2001) Ionic channels of excitable membranes, 3rd edn. Sinauer Associates, Sunderland
15. Király M, Porcsalmy B, Pataki A et al (2009) Simultaneous PKC and cAMP activation induces differentiation of human dental pulp stem cells into functionally active neurons. Neurochem Int 55:323–332
16. Kohyama J, Abe H, Shimazaki T et al (2001) Brain from bone: efficient "meta-differentiation" of marrow stroma-derived mature osteoblasts to neurons with Noggin or a demethylating agent. Differentiation 68:235–244
17. Scintu F, Reali C, Pillai R et al (2006) Differentiation of human bone marrow stem cells into cells with a neural phenotype: diverse effects of two specific treatments. BMC Neurosci 7:14
18. Király M, Kádár K, Horváthy DB et al (2011) Integration of neuronally predifferentiated human dental pulp stem cells into rat brain in vivo. Neurochem Int 59:371–381
19. Gervois P, Struys T, Hilkens P et al (2015) Neurogenic maturation of human dental pulp stem cells following neurosphere generation induces morphological and electrophysiological characteristics of functional neurons. Stem Cells Dev 24:296–311
20. Schlett K, Czirók A, Tárnok K et al (2000) Dynamics of cell aggregation during in vitro neurogenesis by immortalized neuroectodermal progenitors. J Neurosci Res 60:184–194
21. Sasaki R, Aoki S, Yamato M et al (2008) Neurosphere generation from dental pulp of adult rat incisor. Eur J Neurosci 27:538–548
22. Deng W, Obrocka M, Fischer I et al (2001) In vitro differentiation of human marrow stromal cells into early progenitors of neural cells by conditions that increase intracellular cyclic AMP. Biochem Biophys Res Commun 282:148–152
23. Tatard VM, D'Ippolito G, Diabira S et al (2007) Neurotrophin-directed differentiation of human adult marrow stromal cells to dopaminergic-like neurons. Bone 40:360–373
24. Nosrat IV, Widenfalk J, Olson L et al (2001) Dental pulp cells produce neurotrophic factors, interact with trigeminal neurons in vitro, and rescue motoneurons after spinal cord injury. Dev Biol 238:120–132
25. Nosrat IV, Smith CA, Mullally P et al (2004) Dental pulp cells provide neurotrophic support for dopaminergic neurons and differentiate into neurons in vitro; implications for tissue engineering and repair in the nervous system. Eur J Neurosci 19:2388–2398
26. Lewin GR, Barde YA (1996) Physiology of the neurotrophins. Annu Rev Neurosci 19: 289–317

27. Nosrat CA, Fried K, Ebendal T et al (1998) NGF, BDNF, NT3, NT4 and GDNF in tooth development. Eur J Oral Sci 106:94–99
28. Davies AM (1994) The role of neurotrophins in the developing nervous system. J Neurobiol 25:1334–1348
29. Arthur A, Shi S, Zannettino AC et al (2009) Implanted adult human dental pulp stem cells induce endogenous axon guidance. Stem Cells 27:2229–2237
30. Sakai K, Yamamoto A, Matsubara K et al (2012) Human dental pulp-derived stem cells promote locomotor recovery after complete transection of the rat spinal cord by multiple neuro-regenerative mechanisms. J Clin Invest 122:80–90
31. Wang J, Wang X, Sun Z et al (2010) Stem cells from human-exfoliated deciduous teeth can differentiate into dopaminergic neuron-like cells. Stem Cells Dev 19:1375–1383
32. Suon S, Yang M, Iacovitti L et al (2006) Adult human bone marrow stromal spheres express neuronal traits in vitro and in a rat model of Parkinson's disease. Brain Res 1106:46–51
33. Kanafi M, Majumdar D, Bhonde R et al (2014) Midbrain cues dictate differentiation of human dental pulp stem cells towards functional dopaminergic neurons. J Cell Physiol 229: 1369–1377

MSCs and Biomaterials

Adriano Piattelli and Giovanna Iezzi

Introduction

Osseointegration was first observed, although not recognized as such, by Bothe and coworkers in 1940. These researchers found that titanium dental implants placed in animal bone were found to be in close and tight contact with bone tissue [1]. They, moreover, reported that titanium, for its strength and hardness, could have a potential as a prosthesis material. Another researcher, Leventhal, in 1951 placed titanium screws in rat femurs and observed that after 16 weeks, the screws were so strongly apposed to bone that the bone fractured when it was tryied to remove the screws [2]. The peri-implant bone did not show untoward processes and no reaction to the presence of the titanium implants. The bone trabeculae appeared to be absolutely normal. In 1952, Per-Ingvar (PI) Brånemark, a young researcher in anatomy in Lund, Sweden, did a study with a titanium implant chamber used to evaluate the blood flow of bone in rabbits. At the end of the experiment, when trying to remove these chambers from the bone, he found that the bone had so completely and fully integrated with the implant that the titanium chamber could not be removed. Brånemark called this process "osseointegration," from the Latin words os—meaning bone, and integrate—meaning to make whole, in the sense of a combined and close presence between the metal biomaterial and the living bone, and, like the beforementioned authors before him, saw the possibilities for human use [3].

Osseointegration was defined by Branemark as "the formation of a direct interface between the implant and bone, without the presence of an intervening soft tissue". In dental implants, this meant that bone tended to grow right up to the implant metal surface without an interposing layer of soft tissue. This direct contact between bone and the implant metal surface had to be verified histologically [4].

A. Piattelli (✉) • G. Iezzi
Department of Medical, Oral, and Biotechnological Sciences,
University of Chieti-Pescara, Chieti, Italy
e-mail: a.piattelli@unich.it

B. Zavan, E. Bressan (eds.), *Dental Stem Cells: Regenerative Potential*, Stem Cell Biology and Regenerative Medicine, DOI 10.1007/978-3-319-33299-4_13

The concept of osseointegration was applied for the first time in dentistry in the mid-1960s. In 1965 Brånemark, who had become Professor of Anatomy at the University of Gothenburg, inserted dental implants into the first human patient, Gösta Larsson. Mr. Larsson had a cleft palate and he needed dental implants to support a palatal obturator. He died in 2005, and the original implants were still in place, working successfully, after 40 years of function. There are many pictures of him with PI Branemark, and Mr. Larsson became a minor celebrity and he was interviewed many times by television.

In the mid-1970s PI Brånemark started a commercial partnership with a Swedish defense manufacturer called Bofors, to produce dental implants and the metal instruments needed for their insertion. Osseointegrated implantology was criticized by the official dental academia at that time, but, many years and many fights later, the careful documentation of the efficacy and safety of the dental implants originated the widespread and enthusiastic acceptance of implantology as a viable, and in some cases, preferred treatment by the worldwide dental community. Brånemark spent almost three decades trying to get the acceptance of osseointegration from the dental community. In Gothenburg the University stopped the funds for his research, and he was forced to transfer to a private clinic to continue the implant treatment of patients. Toronto's George Zarb, a Canadian prosthodontist, who, by the way speaks Italian fluently, played a pivotal role in presenting the concept of osseointegration to the wider world. The 1983 Toronto Conference was the turning point, and, at long last, the dental scientific community accepted Brånemark's concepts. Today osseointegration is considered by all clinicians and practitioners to be a highly predictable and commonplace treatment modality, with a more than 99.00 % success percentages. The osseointegration concepts have been also transferred in other fields, e.g. orthopedics, where an intramedullary prosthesis was inserted into the residual bone of amputees and then connected to a limb prosthesis. This fact allowed amputees to move in a more comfortable way, and with less energy consumption. Osseointegrated prosthesis can also been combined with replaced joints, allowing, in such a way, below knee amputees with arthritis of the knee or a small residual bone to move without the necessity of a socket prosthesis.

Osseointegration is a dynamic process where the implant features (i.e. macrogeometry, surface properties, mechanical properties, structure of the metal, etc.) play an important and pivotal role in the behavior of molecules and cells. Osseointegration has been observed using different materials (e.g. tantalum, niobium), but most studies concerned the bone reactions to titanium. Titanium implants were found to have mineralized bone tissue at the interface, either through direct contact between calcium and titanium atoms, or by a chemico-physical bonding due to a cement line-like layer located at the implant/bone interface. The healing processes during osseointegration seemed to mimic the mechanisms observed during the healing of bone fractures [5].

For an implant to be osseointegrated, the bone to implant contact (BIC) does not need to be 100 %, and the concept of osseointegration is more related to the stability of the fixation than to the degree of bone to implant contact in histologic terms; we must also consider that about 20–25 % of the bone tissue is composed by marrow spaces, needed to bring oxygen and nutrients to the bone cells. In short, osseointegration is deemed to be successful when the fixation of titanium implants in bone is achieved and, especially, maintained during functional loading in the chewing cycles. After implant insertion, the healing period lasts several weeks or months before the implant is fully integrated into the surrounding bone, and could then be loaded [6]. The first appearance of metal-bone integration occurs usually after a few weeks, while a more strong connection is progressively observed over the next months or years [7]. Although the osseointegrated interface can be resistant to external shocks over time, it may be damaged by prolonged adverse stimuli and overloading forces, which may determine a complete failure with loss of the implant and prosthetic superstructure. Furthermore, clinicians must bear in mind that there is a critical threshold of micromotion above which there is the formation of fibrous tissues at the interface, rather than bone, with subsequent mobility and possible loss of the implant [8].

In the last decade, implants with a porous metal structure have been introduced in dentistry and orthopedics. Porous metals may allow the formation of vascular systems within the porous area and they show high tensile strength and corrosion resistance with excellent biocompatibility. Moreover, the porous structure of the metal has been shown to allow an extensive bone infiltration, allowing osteoblasts to move inside the metal and to form bone there [9–11]. These porous materials are nowadays used in hip replacement, knee replacement and dental implantology.

Other applications of the osseointegration concept have been in:

- Construction of different types of epitheses, used in retention of a craniofacial prosthesis such as an artificial ear (ear prosthesis), maxillofacial reconstruction, eye (orbital prosthesis), or nose (nose prosthesis) [12–14];
- An hearing conduction amplification anchored to the bone (bone anchored hearing aid) [15, 16];
- Replacement of knee and joints.

PI Brånemark has been awarded many prizes for his work, including the Swedish Soederberg Prize, and the Swedish Engineering Academy medal for technical innovations.

Once the dental implants become osseointegrated, a prosthetic suprastructure needs connected to them and loading forces, during the chewing cycles, are transmitted via the implants to the surrounding bone tissue [17, 18]. This peri-implant bone tends to remodel to arrive to a steady state around the implants. When osseointegrated, the implants not only become a part of the body, but also of the mind. Professor Brånemark has called this special type of mental acceptance "osseoperception". Osseoperception seems to be of very important, for example, when bone anchored prosthetic replacements communicate with the mind to restore function, i.e. picking an object using an osseointegrated finger prosthesis [4, 19].

Osseointegration, then, was born and came of age in Dentistry, but has been extremely useful in other medical fields such as orthopedics and maxillofacial reconstruction.

Tissue Engineering

Tissue engineering (TE) is a quite new and very promising approach to obtain the repair and regeneration of tissues and organs lost, damaged or compromised due to trauma, injury, disease or aging [20]. A key component of TE approach to bone regeneration is represented by natural or man-made scaffold that are used as template for the interactions between different types of cells, and the formation of bone extracellular matrix providing structural support to the newly formed tissue. An ideal scaffold should have the following features (1) a three-dimensional (3-D) and highly porous structure with interconnected pores to allow cells migration, flow transport of nutrients, and removal of metabolic waste; (2) biocompatibility and resorbability, with a rate of resorption similar to that of the forming new bone tissues; (3) a surface chemistry that favors cells attachment, proliferation and differentiation; (4) mechanical properties comparable to bone and soft tissues at the implantation site; (5) a possibility to be commercially produced and safely sterilized without any alteration of its properties [21–25]. Several approaches have been used in bone regeneration procedures, and calcium phosphate ceramics are, probably, extremely effective as scaffolds. There is a necessity of further studies of the best ways in which materials, cells and biologically active molecules could interact. Different types of cells and growth factors are two pivotal elements in bone biology/healing, and their interaction is extremely important towards an effective regeneration process. The best combination of materials, cells and growth factors seems to be a must for a very effective bone TE strategy [26]. A system to be used for bone repair and regeneration would ideally require osteoconductive and osteoinductive properties, so that new bone formation can be improved through an adequately shaped three-dimensional (3D) scaffold (osteoconduction) and by a biological stimulus (osteoinduction) [27]. Ceramic materials, e.g. hydroxyapatite, tricalcium phosphate and coralline-derived calcium phosphate, due to their inorganic nature and ionic composition, are extremely useful in several applications. These materials are known for their ability to bond to bone and stimulate new bone formation. 3D systems have been produced with the use of particulates or blocks having a porous interconnected structure [28]. The formation of 3D scaffolds in particulate or block shape creates a potential for their use either without cells (placing of the scaffold in the tissues, and its colonization by surrounding cells) or combining them in vitro with cells, creating a hybrid cell–material construct. These 3D scaffolds can be used also as a delivery system, releasing bioactive agents and enhancing the regenerative potentialities of the system [29]. The ability of micro-CT to evaluate 3-D structures in a non-destructive way has made its use and application extremely wide in several fields, such as physics, materials science, medicine, mineral processing and powder technology. Furthermore, the possibility to use synchrotron radiation X-ray sources has further improved the application of micro-CT due to its numerous advantages compared to conventional X-ray sources, including a higher beam intensity, a higher spatial coherence and the monochromaticity [30]. The monochromaticity property of synchrotron radiation reduces significantly the beam hardening effects, thus allowing to simplify the segmentation process of the whole image analysis. Synchrotron radiation X-ray micro-CT has been used to evaluate the 3-D porous architecture and microstructure of several different calcium phosphate scaffolds after a long-term healing period in humans [31]. In the last decade, bone substitute biomaterials have been used in combination with cells for the fabrication of artificial bone grafts. The use of multipotent mesenchymal stem cells (MSCs) has opened up new therapeutic perspectives for the in situ or in vitro TE of bone. The success of tissue regeneration is related to the structure of the scaffold and its ability to allow invasion by cells and tissues. This construct can then be placed in living tissues to act as replacement tissue after the in-vitro colonization of MSC. Blood vessels [32] begin to grow around and into the construct, and as the scaffold undergoes resorption, the newly formed bone tissue starts blending with the surrounding tissues and finally replaces the scaffold. The scaffolds can be reproducibly manufactured with a specific, desired structure obtained according to stochastic, fractal, or periodic principles. In recent years, the efforts in TE have been focused mainly on the characterization of the regenerative properties of different sources of stem cells (dental pulp, periodontal ligament, amniotic fluid) [33–36]. Amniotic derived stem cells (ADSCs) are an intermediate stage between embryonic stem cells and lineage-restricted adult progenitor cells. Their high proliferation rate together with their differentiation potential into cells of all three embryonic germ layers (ectoderm, endoderm and mesoderm) are important advantages over most of the known adult stem cell sources. In vitro studies and tests are needed to evaluate the attitude of the constructs to support cellular events such as adhesion, proliferation and osteogenic cells differentiation [37]. All the results obtained from in vitro and animal experimentations will give essential informations to try to transfer and apply these novel therapeutic strategies to the field of tissue regeneration.

Approaches to tooth regeneration. The approaches to tooth regeneration are still in their infancy and face many obstacles. The different types of approaches that have been tried include: (a) remineralization of carious dentin by inorganic polyphosphates; (b) calcium phosphate coatings; (c) engineering of bone and tooth root using bioactive materials; (d) regeneration of different dental tissues with the use of different substances, i.e. amelogenins for the regeneration of the periodontal tissues, and calcium-phosphate ceramics and collagen for the reconstruction of the bone [38]. The deep and complete understanding of the principles that support the formation of teeth and periodontium represents the basic foundation to design innovative biomaterials to be used in the possible regeneration of these structures. When enamel undergoes demineralization, the residual mineral crystals can serve as templates for the new formation of apatite crystals [39]. The same process can be used in demineralized dentin, as for example can occur in a carious tooth, where dentin apatite crystals tend to remain and can be used as templates. It is also possible to attempt to remineralize dentin with the use of agents likes polyacrylic or polyaspartic acid [40]. These acids bind to collagen and serve to bind calcium and promoting apatite nucleation. The biomineralization processes, such as the formation of tooth enamel, is under the influence of various proteins such as the amelogenins. Mimicking nature, it has been tried to restore enamel by inducing remineralization of hydroxyapatite on the surface of the tooth. The regeneration of different parts of a tooth by implementation of biomimetic mineralization processes will represent a significant step for the future development of scaffolds for dental regeneration [41]. This will lead to the possible formation of teeth, and this fact could have an enormous benefit to human health with a huge socio-economic impact. Different hybrid composites will be evaluated to mimic the different regions of the tooth and of the periodontal tissues. These composites could have a structure as follows:

1. an unmineralized, collagen central portion, the dental pulp;
2. an unmineralized, layer, the pre-dentin;
3. a highly mineralized layer, the dentin.

The different structure characteristics of the different tooth tissues could be obtaining by different degree of cross-linking. To regenerate the periodontal tissues could be used constructs composed of highly mineralized portions, i.e periodontal bone and cementum, connected by fibrous layers, mimicking the unmineralized periodontal ligament. Tooth regeneration could be a difficult but very important part of regenerative medicine in the future and could have a relevant importance in healthcare.

Graphene

Graphene is a relatively new allotrope of carbon composed of a single layer of monocrystalline graphite with hybridized carbon atoms. Due to its structure, a one-atom-thick two-dimensional (2D) carbon material, graphene has attracted increasing attention in the past several years due to its high surface area, remarkable thermal conductivity, excellent charge mobility, and mechanical properties [42]. The unique structure and outstanding properties render graphene highly promising for a wide range of applications in the fields of electronics, sensor, and energy storage/conversion. Some of the other interesting aspects of graphene include high transparency toward visible light, high values of elasticity, unusual magnetic properties, and charge transfer interaction with molecules, that allowed it to gain a lot of interest in the biomedical field as a new component for biosensors, tissue engineering, and drug delivery. Graphene can be obtained following different approaches [43]. Interestingly, the majority of studies on chemistry of graphene do not involve "pristine" graphene, but rather carbon materials produced upon reduction of graphene oxide (GO). Graphene-based materials show unique interactions with DNA and RNA, which make them attractive in DNA or RNA sensing and delivery. GO shows preferential adsorption of single stranded DNA over double stranded DNA and protects the adsorbed nucleotides from attack by nuclease enzymes opening up a wide range of application opportunities [44]. As opposed to interaction with DNA and RNA, there are only a few data on the interaction of graphene with proteins and lipids. It will be, hoewer, very important to understand interaction of graphene with lipid bilayer in the cell membrane. Protein adsorption on nanomaterials surface has received increasing attention for the past several years. This phenomenon affects in a significant way the behaviour of these materials in biological systems (e.g. cellular uptake and toxic responses). Nanomaterial surfaces are immediately covered by proteins, lipids, enzymes when put in a biological medium. These coated surfaces give new features to the nanosystem, i.e. hydrophilicity/hydrophobicity, surface charges and energy, topography [45, 46]. These new characteristics will produce the responses at the cell/tissue level. Due of their high specific surface area, the carbon family of nanomaterials including graphene has a potentially larger protein adsorption capacity than other nanostructures. After interaction with cells, tissues, or organs, graphene sheets surfaces change and have completely different biological properties. Hydrocarbons, organic molecules, and elements was reported to change the surface composition and surface energy, which affected the protein absorption and cell attachment, proliferation, differentiation, and final integration in the tissues [47–51]. The high active surface area of graphene, over other nanomaterials, is one of the main advantages of graphene based materials, which allows a high-density drug loading. Due to the specific geometry of graphene (2D structure), both sides of a single layer graphene sheet can be used as a substrate for the controlled adsorption of molecules and functional groups for surface modification [52]. For instance, it has been shown that covalent attachment of chitosan, folic acid, and polyethylene glycol (PEG) to GO produces a potential platform for the delivery of anti-inflammatory and water insoluble anticancer drugs such as doxorubicin (Dox) and SN38, a camptothecin analogue [53]. The idea is to study drug-graphene interaction as to be able to deliver, in a controlled manner and exploiting graphene-drug interactions, drug in proper quantity directly to the desired target site. It is important to understand such interactions from two points of view, one for biomedical applications and other for their toxicity and biocompatibility. Like other employed materials in nanomedicine, toxicity of the graphene is strongly dependent to its physicochemical properties (e.g., size and its distribution, surface charge, particulate state, number of layers, surface functional groups and particularly shape). One of the most important issues for biomedical applications of graphene is its short- and long-term toxicity [54–56]. Carbon-based materials (carbon nanotubes or nanocrystalline diamonds) have been widely tested for both their potential toxicological risks and their possible use in biomedical applications. Moreover, the antibacterial activity of graphene-based materials can be used in wound healing applications to prevent infections or to potentiate and protect the integration process of diffeent types of biomaterials biomaterials. Graphene, when used as a delivery vehicle, could, probably, potentiate the effects of antibacterial drugs. Gene therapy to treat genetic disorders and cancer is another area where graphene could be usefully employed. Successful gene therapy requires efficient and safe gene vectors that protect DNA from nuclease degradation as well as facilitate DNA uptake with high transfection efficiency. Graphene has been explored for applications in gene delivery, gene–drug co-delivery and protein delivery [43, 55, 57]. Osteogenic differentiation of Mesenchymal Stem Cells was enhanced on titanium surfaces coated with GO carrying BMP-2, compared to titanium surfaces coated only with BMP-2 [58]. In vivo studies in mouse also showed a higher new bone formation when using titanium–GO–BMP2 implants compared to only titanium or titanium–GO or titanium–BMP2 implants; these new composites could be very effective carriers for delivery of drugs. Several studies have emphasized the potential of graphene-based materials as drug and gene delivery vehicles in vitro; however, there is a need to demonstrate their potential in vivo with particular focus on safety, biodistribution and efficacy [59]. Therefore, graphene is an ideal model material for experiments with adherent (anchorage-dependent) cells (e.g. osteoblasts, mesenchymal stromal cells (MSCs), etc.). Adhesion of osteoblasts is a crucial prerequisite to subsequent cell functions, such as proliferation, synthesis of proteins (e.g. proteins of extracellular matrix (ECM), morphogenic factors and osteoinductive molecules) and formation of mineral deposits [43, 60, 61]. Adhesion is generally dependent on time, adhesive forces at the cell/material interface, and surface topography. Cell adhesion is primarily mediated by integrins, a widely expressed family of transmembrane adhesion receptors. Upon ligand binding, integrins rapidly associate with the actin cytoskeleton and cluster together to form focal adhesions (FAs), which are discrete complexes that contain structural (e.g. vinculin) and signalling molecules (e.g., focal adhesion kinase). FAs are central elements in the adhesion process because they function as structural links between the cytoskeleton and ECM to mediate stable adhesion and migration. Furthermore, in combination with growth factor receptors, FAs activate signalling pathways that regulate transcription factor activity and direct cell growth and differentiation [62]. Human MSCs are mononuclear cell population adherent to tissue culture plastic and have been isolated from adult bone marrow. They are capable of further proliferation as well as differentiation into multiple lineages involved with connective tissue (osteoblasts, adipocytes and chondrocytes) when exposed to various growth factor combinations or substrates with different topography and rigidity). Thus, these cells may serve as a good model for testing the possible increased/accelerated differentiation induced by adhesion onto graphene surfaces. Recently, the utilization of graphene foam, a 3D porous structure, showed to be succesfully used as a novel scaffold for Neural Stem Cells (NSCs) in vitro. It was found that three-dimensional graphene foams (3D-GFs) can not only support NSC growth, but also keep cell at an active proliferation state. These findings show that 3D-GFs could offer a powerful platform for NSC research and neural tissue engineering [63].

Digital Dentistry

It has been said that by the end of the decade there will be as many digital bits as there are stars in the universe, some 44 trillion gigabytes. With that degree of digitization it is no wonder that dentistry has been highly influenced by the presence of digital media and digital Technologies [63]. The challenge for the dental community is to try to understand and use at best what the digital world has to offer. In order to do so we need to understand the nuances and complexities of the digital world and with the high speed of its advances, this could be really a major challenge. Probably digital technologies can and will be used in all dental field, e.g. digital data capture systems such as X-ray, CT and MRI [64]. More recently, there have been really interesting and quite exciting new developments in the production and use of intra-oral cameras. The latter will change in a profound way the daily practice of restorative dentistry as it provides us with digital data of the oral cavity that can then be processed by a digital dental laboratory. This will provide an even greater incentive to develop new design software that will allow us to manipulate all the data that we are able to obtain from our patients. Another aspect of the digital world is the transition from traditional manufacturing routes to the use of CAD-CAM and more recently additive manufacturing. In the immediate future there will be a need to explore the extent to which new manufacturing technologies and new materials could be extremely useful to treat our patients.

References

1. Ananth H, Kundapur V, Mohammed HS, Anand M, Amarnath GS, Mankar S (2015) Review on biomaterials in dental implantology. Int J Biomed Sci 11(3):113–120
2. Novaes AB Jr, de Souza SL, de Barros RR, Pereira KK, Iezzi G, Piattelli A (2010) Influence of implant surfaces on osseointegration. Braz Dent J 21(6):471–481
3. Piattelli A, Pontes AE, Degidi M, Iezzi G (2011) Histologic studies on osseointegration: soft tissues response to implant surfaces and components. A review. Dent Mater 27(1):53–60
4. Bhatnagar VM, Karani JT, Khanna A, Badwaik P, Pai A (2015) Osseoperception: an implant mediated sensory motor control—a review. J Clin Diagn Res 9(9):ZE18–ZE20
5. Jemat A, Ghazali MJ, Razali M, Otsuka Y (2015) Surface modifications and their effects on titanium dental implants. Biomed Res Int 2015:791725
6. Santos PL, Gulinelli JL, Telles Cda S, Betoni Júnior W, Okamoto R, Chiacchio Buchignani V, Queiroz TP (2013) Bone substitutes for peri-implant defects of postextraction implants. Int J Biomater 2013:307136
7. Razzouk S, Schoor R (2012) Mesenchymal stem cells and their challenges for bone regeneration and osseointegration. J Periodontol 83(5):547–550
8. Nishimura I (2013) Genetic networks in osseointegration. J Dent Res 92(12 Suppl): 109S–118S
9. Thalji G, Cooper LF (2014) Molecular assessment of osseointegration in vitro: a review of current literature. Int J Oral Maxillofac Implants 29(2):e171–e199
10. Piattelli A, Artese L, Penitente E, Iaculli F, Degidi M, Mangano C, Shibli JA, Coelho PG, Perrotti V, Iezzi G (2014) Osteocyte density in the peri-implant bone of implants retrieved after different time periods (4 weeks to 27 years). J Biomed Mater Res B Appl Biomater 102(2):239–243
11. Ozdemir T, Higgins AM, Brown JL (2013) Osteoinductive biomaterial geometries for bone regenerative engineering. Curr Pharm Des 19(19):3446–3455, Review
12. Mangano FG, Zecca PA, van Noort R, Apresyan S, Iezzi G, Piattelli A, Macchi A, Mangano C (2015) Custom-made computer-aided-design/computer-aided-manufacturing biphasic calcium-phosphate scaffold for augmentation of an atrophic mandibular anterior ridge. Case Rep Dent 2015:941265
13. Narasimhan HV, Dheeraj D, Darshana D, Sreekumar AV (2012) Contemporary concepts in treatment planning, reconstruction and rehabilitation of the maxillofacial region: an overview of methods used in practice. J Contemp Dent Pract 13(6):938–944, Review

14. Verstraaten J, Kuijpers-Jagtman AM, Mommaerts MY, Bergé SJ, Nada RM, Schols JG, Eurocran Distraction Osteogenesis Group (2010) A systematic review of the effects of bone-borne surgical assisted rapid maxillary expansion. J Craniomaxillofac Surg 38(3):166–174
15. Reinfeldt S, Håkansson B, Taghavi H, Eeg-Olofsson M. New developments in bone-conduction hearing implants: a review. Med Devices (Auckl). 2015;8:79–93
16. Reinfeldt S, Håkansson B, Taghavi H, Eeg-Olofsson M (2015) New developments in bone-conduction hearing implants: a review. Med Devices (Auckl) 8:79–93
17. Monument MJ, Lerman DM, Randall RL (2015) Novel applications of osseointegration in orthopedic limb salvage surgery. Orthop Clin North Am 46(1):77–87
18. Van Steenberghe D, Jacobs R (2006) Jaw motor inputs originating from osseointegrated oral implants. J Oral Rehabil 33(4):274–281, Review
19. Wiskott HW, Belser UC (1999) Lack of integration of smooth titanium surfaces: a working hypothesis based on strains generated in the surrounding bone. Clin Oral Implants Res 10(6):429–444, Review
20. Jacobs R, Van Steenberghe D (2006) From osseoperception to implant-mediated sensory-motor interactions and related clinical implications. J Oral Rehabil 33(4):282–292, Review
21. Larsson L, Decker AM, Nibali L, Pilipchuk SP, Berglundh T, Giannobile WV (2016) Regenerative medicine for periodontal and peri-implant diseases. J Dent Res 95(3):255–266
22. Rai R, Raval R, Khandeparker RV, Chidrawar SK, Khan AA, Ganpat MS (2015) Tissue engineering: step ahead in maxillofacial reconstruction. J Int Oral Health 7(9):138–142, Review
23. Gardin C, Ferroni L, Bressan E, Calvo-Guirado JL, Degidi M, Piattelli A, Zavan B (2014) Adult stem cells properties in terms of commitment, aging and biological safety of grit-blasted and acid-etched ti dental implants surfaces. Int J Mol Cell Med 3(4):225–236
24. Gardin C, Bressan E, Ferroni L, Nalesso E, Vindigni V, Stellini E, Pinton P, Sivolella S, Zavan B (2012) In vitro concurrent endothelial and osteogenic commitment of adipose-derived stem cells and their genomical analyses through comparative genomic hybridization array: novel strategies to increase the successful engraftment of tissue-engineered bone grafts. Stem Cells Dev 21(5):767–777
25. Bressan E, Sbricoli L, Guazzo R, Tocco I, Roman M, Vindigni V, Stellini E, Gardin C, Ferroni L, Sivolella S, Zavan B (2013) Nanostructured surfaces of dental implants. Int J Mol Sci 14(1):1918–1931
26. Aurrekoetxea M, Garcia-Gallastegui P, Irastorza I, Luzuriaga J, Uribe-Etxebarria V, Unda F, Ibarretxe G (2015) Dental pulp stem cells as a multifaceted tool for bioengineering and the regeneration of craniomaxillofacial tissues. Front Physiol 6:289. doi:10.3389/fphys.2015.00289
27. Zhao H, Chai Y (2015) Stem cells in teeth and craniofacial bones. J Dent Res 94(11): 1495–1501
28. Gardin C, Ricci S, Ferroni L, Guazzo R, Sbricoli L, De Benedictis G, Finotti L, Isola M, Bressan E, Zavan B (2015) Decellularization and delipidation protocols of bovine bone and pericardium for bone grafting and guided bone regeneration procedures. PLoS One 10(7), e0132344
29. Cecchinato F, Karlsson J, Ferroni L, Gardin C, Galli S, Wennerberg A, Zavan B, Andersson M, Jimbo R (2015) Osteogenic potential of human adipose-derived stromal cells on 3-dimensional mesoporous TiO2 coating with magnesium impregnation. Mater Sci Eng C Mater Biol Appl 52:225–234
30. Tatullo M, Marrelli M, Falisi G, Rastelli C, Palmieri F, Gargari M, Zavan B, Paduano F, Benagiano V (2016) Mechanical influence of tissue culture plates and extracellular matrix on mesenchymal stem cell behavior: a topical review. Int J Immunopathol Pharmacol 29(1):3–8. doi:10.1177/0394632015617951, Epub 2015 Nov
31. Obregon F, Vaquette C, Ivanovski S, Hutmacher DW, Bertassoni LE (2015) Three-dimensional bioprinting for regenerative dentistry and craniofacial tissue engineering. J Dent Res 94(9 Suppl):143S–152S
32. Giuliani A, Manescu A, Mohammadi S, Mazzoni S, Piattelli A, Mangano F, Iezzi G, Mangano C (2016) Quantitative kinetics evaluation of blocks versus granules of biphasic calcium phos-

phate scaffolds (HA/β-TCP 30/70) by synchrotron radiation X-ray microtomography: a human study. Implant Dent 25(1):6–15. doi:10.1097/ID.0000000000000363
33. Manescu A, Giuliani A, Mohammadi S, Tromba G, Mazzoni S, Diomede F, Zini N, Piattelli A, Trubiani O (2015) Osteogenic potential of dualblocks cultured with human periodontal ligament stem cells: in vitro and synchrotron microtomography study. J Periodontal Res. doi:10.1111/jre.12289
34. Sunil PM, Manikandan R, Muthumurugan, Yoithapprabhunath TR, Sivakumar M (2015) Harvesting dental stem cells—overview. J Pharm Bioallied Sci 7(Suppl 2):S384–S38
35. Barboni B, Mangano C, Valbonetti L, Marruchella G, Berardinelli P, Martelli A, Muttini A, Mauro A, Bedini R, Turriani M, Pecci R, Nardinocchi D, Zizzari VL, Tetè S, Piattelli A, Mattioli M (2013) Synthetic bone substitute engineered with amniotic epithelial cells enhances bone regeneration after maxillary sinus augmentation. PLoS One 8(5), e63256
36. Berardinelli P, Valbonetti L, Muttini A, Martelli A, Peli R, Zizzari V, Nardinocchi D, Vulpiani MP, Tetè S, Barboni B, Piattelli A, Mattioli M (2013) Role of amniotic fluid mesenchymal cells engineered on MgHA/collagen-based scaffold allotransplanted on an experimental animal study of sinus augmentation. Clin Oral Investig 17(7):1661–1675
37. Antonucci I, Pantalone A, Tete S, Salini V, Borlongan CV, Hess D, Stuppia L (2012) Amniotic fluid stem cells: a promising therapeutic resource for cell-based regenerative therapy. Curr Pharm Des 18(13):1846–1863, Review
38. Black CR, Goriainov V, Gibbs D, Kanczler J, Tare RS, Oreffo RO (2015) Bone tissue engineering. Curr Mol Biol Rep 1(3):132–140, Epub 2015 Aug 15. Review
39. Bertolai R, Catelani C, Aversa A, Rossi A, Giannini D, Bani D (2015) Bone graft and mesenchimal stem cells: clinical observations and histological analysis. Clin Cases Miner Bone Metab 12(2):183–187
40. Morcos MW, Al-Jallad H, Hamdy R (2015) Comprehensive review of adipose stem cells and their implication in distraction osteogenesis and bone regeneration. Biomed Res Int 2015:842975
41. Rathinam E, Rajasekharan S, Chitturi RT, Martens L, De Coster P (2015) Gene expression profiling and molecular signaling of dental pulp cells in response to tricalcium silicate cements: a systematic review. J Endod 41(11):1805–1817
42. Cao Y, Song M, Kim E, Shon W, Chugal N, Bogen G, Lin L, Kim RH, Park NH, Kang MK (2015) Pulp-dentin regeneration: current state and future prospects. J Dent Res 94(11):1544–1551
43. Dubey N, Bentini R, Islam I, Cao T, Castro Neto AH, Rosa V (2015) Graphene: a versatile carbon-based material for bone tissue engineering. Stem Cells Int 2015:804213
44. Mudedla SK, Balamurugan K, Kamaraj M, Subramanian V (2016) Interaction of nucleobases with silicon doped and defective silicon doped graphene and optical properties. Phys Chem Chem Phys 18(1):295–309. doi:10.1039/c5cp06059a, Epub 2015 Nov 26
45. Qiu H, Sarathy A, Leburton JP, Schulten K (2015) Intrinsic stepwise translocation of stretched ssDNA in graphene nanopores. Nano Lett 15(12):8322–8330. doi:10.1021/acs.nanolett, 5b03963. Epub 2015 Nov 25
46. Kang SH, Hwang WS, Lin Z, Kwon SH, Hong SW (2015) A robust highly aligned DNA nanowire array-enabled lithography for graphene nanoribbon transistors. Nano Lett 15(12):7913–7920. doi:10.1021/acs.nanolett.5b02946, Epub 2015 Nov 19
47. Mao Y, Chen Y, Li S, Lin S, Jiang Y (2015) A graphene-based biosensing platform based on regulated release of an aptameric DNA biosensor. Sensors (Basel) 15(11):28244–28256
48. Choi M, Kim KG, Heo J, Jeong H, Kim SY, Hong J (2015) Multilayered graphene nano-film for controlled protein delivery by desired electro-stimuli. Sci Rep 5:17631
49. Sharma VK, McDonald TJ, Kim H, Garg VK (2015) Magnetic graphene-carbon nanotube iron nanocomposites as adsorbents and antibacterial agents for water purification. Adv Colloid Interface Sci 225:229–240
50. Scarpa F, Chowdhury R, Kam K, Adhikari S, Ruzzene M (2011) Dynamics of mechanical waves in periodic grapheme nanoribbon assemblies. Nanoscale Res Lett 6:430

51. Zhao C, Andersen H, Ozyilmaz B, Ramaprabhu S, Pastorin G, Ho HK (2015) Spontaneous and specific myogenic differentiation of human mesenchymal stem cells on polyethylene glycol-linked multi-walled carbon nanotube films for skeletal muscle engineering. Nanoscale 7(43):18239–18249
52. Tiwari JN, Vij V, Kemp KC, Kim KS (2016) Engineered carbon-nanomaterial-based electrochemical sensors for biomolecules. ACS Nano 10(1):46–80. doi:10.1021/acsnano.5b05690, Epub 2015 Nov 30
53. Zhang C, Wang L, Zhai T, Wang X, Dan Y, Turng LS (2016) The surface grafting of graphene oxide with poly(ethylene glycol) as a reinforcement for poly(lactic acid) nanocomposite scaffolds for potential tissue engineering applications. J Mech Behav Biomed Mater 53:403–413
54. Zhao X, Yang L, Li X, Jia X, Liu L, Zeng J, Guo J, Liu P (2015) Functionalized graphene oxide nanoparticles for cancer cell-specific delivery of antitumor drug. Bioconjug Chem 26(1):128–136. doi:10.1021/bc5005137, Epub 2015 Jan 7
55. Xu C, Wang J, Xu Y, Shang G, Wang R, Lin Y (2013) Review of and perspectives on the toxicology of graphene-based materials. Curr Drug Metab 14(8):863–871, Review
56. Chatterjee N, Eom HJ, Choi J (2014) A systems toxicology approach to the surface functionality control of graphene-cell interactions. Biomaterials 35(4):1109–1127
57. Wu SY, An SS, Hulme J (2015) Current applications of graphene oxide in nanomedicine. Int J Nanomedicine 10(Spec Iss):9–24
58. Jin G, Li K (2014) The electrically conductive scaffold as the skeleton of stem cell niche in regenerative medicine. Mater Sci Eng C Mater Biol Appl 45:671–681
59. La WG, Jin M, Park S, Yoon HH, Jeong GJ, Bhang SH, Park H, Char K, Kim BS (2014) Delivery of bone morphogenetic protein-2 and substance P using graphene oxide for bone regeneration. Int J Nanomedicine 9(Suppl 1):107–116
60. La WG, Park S, Yoon HH, Jeong GJ, Lee TJ, Bhang SH, Han JY, Char K, Kim BS (2013) Delivery of a therapeutic protein for bone regeneration from a substrate coated with graphene oxide. Small 9(23):4051–4060. doi:10.1002/smll.201300571, Epub 2013 Jul 10
61. Bressan E, Ferroni L, Gardin C, Sbricoli L, Gobbato L, Ludovichetti FS, Tocco I, Carraro A, Piattelli A, Zavan B (2014) Graphene based scaffolds effects on stem cells commitment. J Transl Med 12:296
62. Gu M, Liu Y, Chen T, Du F, Zhao X, Xiong C, Zhou Y (2014) Is graphene a promising nanomaterial for promoting surface modification of implants or scaffold materials in bone tissue engineering? Tissue Eng Part B Rev 20(5):477–491
63. Durán N, Martinez DS, Silveira CP, Durán M, de Moraes AC, Simões MB, Alves OL, Fávaro WJ (2015) Graphene oxide: a carrier for pharmaceuticals and a scaffold for cell interactions. Curr Top Med Chem 15(4):309–327, Review
64. Feng ZQ, Wang T, Zhao B, Li J, Jin L (2015) Soft graphene nanofibers designed for the acceleration of nerve growth and development. Adv Mater 27(41):6462–6468

DSC for Ocular Regeneration

Fatima N. Syed-Picard

Introduction

The eye is a complex organ that converts light into visual signal accomplishing one of our five senses, sight. Multiple ocular tissues must function in chorus to properly orchestrate the overarching goal of vision. As light travels through the eye, it first enters through the transparent cornea that focuses the light through the pupil. It is then further focused by the lens onto the retina. The retina converts the light into a decipherable visual signal that is transported to the brain via the optic nerve (Fig. 1). This simplistic description of the pathway of light through the eye that doesn't convey the multiple intricate components involved in ensuring proper function of the eye. For instance, the cornea itself comprises three distinct cellular layers serving multiple functions such as controlling corneal hydration and matrix structure, and preventing vascular infiltration; the disruption of any of these cellular processes results in corneal opacity obstructing the passage of light into the eye. The retina contains multiple types of neurons involved the converting light into visual signal and transporting the signal to the brain, diseases causing neuronal death or axon damage prevent this final necessary step for acquiring visual signal. Cellular therapies could provide mechanisms to repair or replace cells in these types of ocular pathologies.

This chapter reviews ongoing research involved in assessing the potential of dental stem cells for regenerative ocular therapy. Several populations of adult stem cells have be identified in dental tissues, and due to their neural crest developmental origins, maybe be advantageous for use in ocular cellular therapy. These populations of stem cells include those isolated from the dental pulp of adult of exfoliated deciduous teeth, or periodontal ligament. Portions of the cornea are derived from the cranial neural crest; therefore due to their similarity in developmental origins, dental

F.N. Syed-Picard (✉)
Department of Bioengineering, University of Pittsburgh,
302 Benedum Hall, 3700 O'Hara St., Pittsburgh, PA 15260, USA
e-mail: fns4@pitt.edu

B. Zavan, E. Bressan (eds.), *Dental Stem Cells: Regenerative Potential*, Stem Cell Biology and Regenerative Medicine, DOI 10.1007/978-3-319-33299-4_14

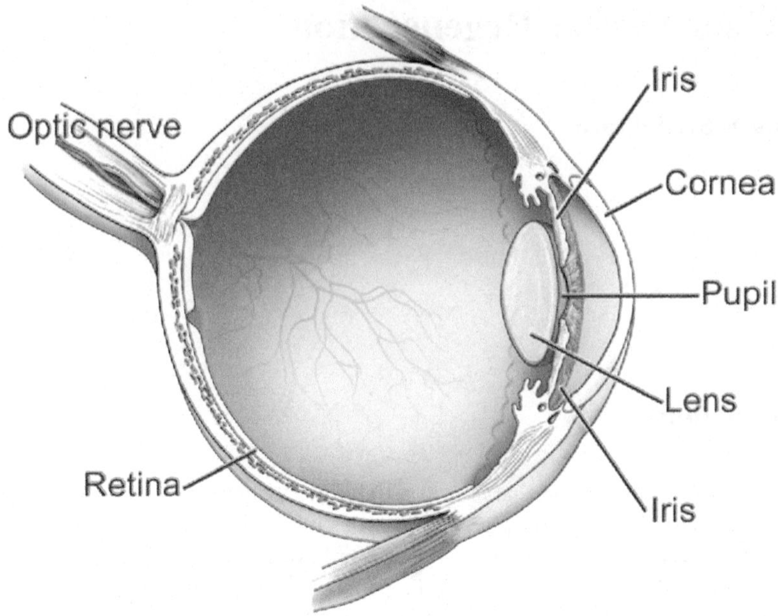

Fig. 1 Drawing of a section of the human eye. Image courtesy of the National Eye Institute, National Institutes of Health

stem cells may potentially possess a stronger affinity to differentiate towards corneal cell lineages. Additionally, the neural crest gives rise to the neurons of the peripheral nervous system, so dental stem cells may have the capacity to differentiate into neurons for replacing damaged cells of the retina. An additional benefit of dental stem cells is their accessibility. These cells can be easily collected from patients for autologous use. These attributes make dental stem cells a potential viable option for developing cellular ocular therapies.

Dental Stem Cells to Treat Corneal Blindness

The cornea is the anterior most tissue of the eye that serves multiple functions such as allowing the passage of light with minimal scatter, acting as a barrier to prevent penetration or infiltration into ocular tissues, and providing two-thirds of the refractive power of the eye to focus light onto the retina [1]. The complex structure of the cornea allows it to facilitate these multiple functions. The cornea comprises three cellular layers: the epithelium is the anterior-most, the stroma is central, and the posterior-most layer is the endothelium. Millions of individuals worldwide suffer from bilateral corneal blindness due to diseases associated with these cellular layers or trauma to the cornea [2]. Stem cell research shows promising results for developing cellular therapies to restore vision to patients suffering from corneal disorders.

Corneal Epithelium

The anterior surface of the cornea comprises a stratified squamous epithelium that acts as a barrier to the eye. The corneal epithelium is a self-renewing structure where the superficial terminal cells are naturally shed and repopulated by a population of stem cells found in the limbus, the transition zone between the cornea and the sclera, termed limbal epithelial stem cells [3]. In addition to functioning in the natural process of epithelial self-renewal, these stem cells are capable of regenerating the entire corneal epithelium following injury.

Disease or injury leading to total limbal stem cell deficiency (LSCD) results in the inability of the corneal epithelium to renew and regenerate itself and therefore leads to corneal opacity due to corneal inflammation and vascular infiltration. Unilateral LSCD is currently treated by transplanting limbal epithelial cells from the contralateral eye [4]. However, treating bilaterial LSCD requires allogeneic tissue transplantation, which is subject to failure due to immunological rejection [5]. Therefore the development of autologous stem cell therapies utilizing cells from an alternative tissue source to repopulate the limbal epithelial stem cells would bypass limitations of current treatments.

Stem cells from exfoliated deciduous teeth (SHED) are a highly investigated population of stem cells since they can be easily isolated from autologous tissues and have the capacity to differentiate into multiple cell types [6]. Recent studies have shown promising results on the ability of SHED to regenerate corneal epithelial cells. Epithelial and stem cells gene expression profiles of SHED were shown to be more similar to that of limbal epithelial stem cells than differentiated corneal epithelial cells for genes such as ABCG2, K12, and K3 [7]. SHED were able to restore corneal epithelial tissue in an LCSD animal model in vivo. LSCD was induced in New Zealand white rabbits through chemical burning. One month following injury, keratectomy was performed to remove pannus formed from the chemical burn, and cell sheets comprising SHED were transplanted onto the exposed rabbit stroma. To prevent damage to the SHED sheet, acellular human amniotic membrane was placed over the cell sheet and sutured to the episclera. Without intervention, the chemical assault to rabbit corneas resulted in corneal vascularization, conjunctivalization, and opacification. Histological analysis of untreated eyes lacked the presence of a stratified epithelium and confirmed the presence of a disorganized, neovascularized conjunctival tissue [7, 8]. Alternatively, the introduction of sheets of SHED to chemically injured rabbit eyes led to clearer corneas with less neovascularization than the control eyes. Furthermore, histological analysis showed that injured eyes receiving SHED had a regenerated corneal epithelium containing cuboidal basal cells, intermediate flattened cells, and polygonal surface cells, similar to nature corneal epithelial tissue [3, 7, 8]. These studies provide promising results for the potential use of SHED in cellular or tissue engineering therapies to regenerate corneal epithelial tissue.

Corneal Stroma

The corneal stroma is the central-most tissue of the cornea and makes up the bulk of the cornea comprising about 90 % of its structure [9]. The stroma has a complex, lamellar structure that facilitates its mechanical integrity and transparency. Each stromal lamella contains long fibrils of collagen that are organized in parallel, have approximately uniform diameter, and are uniformly spaced. The direction of collagen fibril orientation rotates orthogonally between adjacent lamellae [10]. The lattice structure formed by these tightly packed collagen fibrils minimizes light scatter facilitating the transparency of the cornea [11]. Keratocytes are the cells that maintain the stroma. These are quiescent cells serve multiple functions including generating specific molecules that regulate collagen fiber spacing to maintain corneal transparency.

Trauma or disease can cause disruptions to the collagen organization within the corneal stroma diminishing its transparency. Keratocytes respond to stromal trauma by differentiating into fibroblasts and depositing an unorganized scar tissue. This scar tissue does not have the highly organized collagen structure characteristic of healthy stromal tissue and is consequently opaque [12]. Millions of individuals suffer from bilateral corneal blindness resulting from the formation of corneal scars [2]. The current method of treatment is a keratoplasty, which involves using cadaveric grafts to replace the damaged tissues. However, there is a worldwide shortage of donor tissue and this treatment has a failure rate of 38 % after 10 years due in part to immune rejection [13].

The development of a cellular therapy or engineered tissues using autologous stem cells to treat stromal scarring could bypass the limitations associated with current allogeneic grafting treatment. Researchers have started investigating adult stem cell populations such as those derived from adipose tissue or corneal stromal tissue [14–16]. Dental pulp isolated from adult teeth contains a population of multipotent stem cells that are easily accessible and may have a higher affinity for corneal stromal differentiation since, similar to the corneal stroma, they are neural crest-derived.

Human dental pulp stem cells (DPSCs) isolated from adult third molar cells were shown to be able to differentiate into keratocytes in vitro and maintain this phenotype in vivo [17]. The DPSCs were cultured as pellets in a keratocyte differentiation medium, in a similar manner as the culture treatment used to induce keratocyte differentiation from corneal stromal stem cells [18]. After in vitro differentiation, DPSCs had increased expression of molecules characteristic of keratocytes at both the gene and protein levels [17]. Of particular significance was the upregulation in the expression of the proteoglycan keratocan. This molecule is critical in maintaining proper collagen fibril spacing required for corneal transparency. Importantly, keratocan is only found in the corneal stroma and is therefore considered a keratocyte specific marker [19]. The expression of keratocan at the gene and protein levels verifies that DPSCs differentiated into keratocytes in vitro.

DPSCs differentiated into keratocytes in vitro were injected into mouse corneal stroma to assess the behavior of the cells in the in vivo environment. The DPSCs remained in the mouse stroma for up to 5 weeks and produced corneal stromal matrix molecules such as type I collagen and keratocan in vivo. Furthermore, the mouse stroma remained transparent with the injected DPSCs indicating that the DPSCs were not inducing any adverse reaction vivo [17].

The ability of DPSCs to generate tissue engineered corneal stromal constructs with similar structure as native tissues was assessed. DPSCs were cultured on aligned nanofibers to direct cell and matrix organization. In these engineered tissues, DPSCs aligned in parallel and produced a parallel-aligned collagenous matrix, similar to native corneal stromal tissue. Furthermore, the direction of collagen fibrils rotated at different depths within engineered tissue indicating the formation of multiple lamellae. The collagen fibrils were approximately uniform in diameter similar to engineered tissues generated by CSSC [17]. This study showed that DPSCs can differentiation in keratocytes in vitro, maintain this phenotype in vivo, and be used to generate engineered corneal stromal tissue. This shows promise for the development of DPSC-based cellular or tissue engineering therapies to treat corneal stromal scarring.

Corneal Endothelium

The corneal endothelium is the posterior most cellular layer of the cornea. It is a single layer of neural crest derived cells that function to regulate hydration of the stroma through barrier and pump functions [9]. Correct water content is critical in controlling the stromal collagen fibril spacing needed to maintain corneal transparency. Therefore, hydration defects such as edema cause increased light scatter resulting in corneal haze [1]. Corneal endothelial cells have limited proliferative capacity and severe decreases in cellularity due to trauma or diseases such as Fuch's dystrophy results in an edematous cornea with reduced transparency [20]. More than half of corneal transplants occur due to damage or disease to the corneal endothelium [21, 22] indicating a need for the development of cellular therapies. A recent study showed that neural crest derived stem cells isolated from the corneal stroma maintain the capacity to differentiate into corneal endothelial cells [23]. However, few, if any, additional studies have identified alternative adult stem cell populations with this ability. Currently, there are no studies reporting the differentiation competence of dental stem cells into corneal endothelial cells. The capacity of dental stem cells to differentiate into corneal endothelial cells seems plausible since both tissues share similar developmental origins. Potentially, the methods developed to induce corneal endothelial cells from corneal stromal stem cells could be adapted to dental cells. This is still an open scientific question that needs to be answered since, if possible, the use of dental stem cells to treat disorders of the corneal endothelium would greatly impact the field of corneal regeneration.

Dental Stem Cells to Treat Retinal Degeneration

The retina is a complex tissue in the posterior region of the eye involved in converting light into visual signal and transmitting it to the brain. This process is orchestrated by multiple, specialized neuronal cells organized into nuclear layers. The photoreceptor cells, the rods and cones, are located in the outer nuclear layer and are the light-sensitive cells that first detect light and initiate the process of translating it into visual signal. The signal is then processed by neurons in the inner nuclear layer, including horizontal cells, bipolar cells, and amacrine cells. Finally, the signal is received by the retinal ganglion cells (RGCs), which transmit the signal to the brain. The RGC somas are positioned in the ganglion cell layer and their axons stretch towards to posterior of the eye, along the surface of the retina in the retinal nerve fiber layer. The RGC axons collect at the optic nerve head, and assemble to form the optic nerve [24, 25].

Retinal degeneration due to genetic disease or age-related disorders can lead to retinal cell damage or death resulting in vision impairment or loss. Common genetic disorders affecting retinal function result from mutations of the gene for rhodopsin, the light-sensitive protein expressed by the photoreceptor cells [24]. Additional types of retinopathies include glaucoma, a group of diseases resulting from optic nerve degeneration due to RGC loss or damage. Over 60 million individuals worldwide suffer from glaucoma, and this disease has caused bilateral blindness in approximately eight million of these people [26]. Glaucoma is the second leading cause of blindness [27], and currently a cure for this disease is not available. The development of therapies to enhance cell viability and restore cell function in these types of retinopathies could be used to treat these blinding conditions.

Dental Stem Cells for Retinal Cell Regeneration

Multiple groups have shown that dental pulp stem cells have an affinity to differentiate into mature neuronal cells both in vitro and in vivo [28–30]. This characteristic is attributed to the neural crest origins of these cells, since the neural crest gives rise to the nerves of the peripheral nervous system [28]. Under appropriate in vitro conditions, DPSC develop the characteristic stellate neuronal morphology, express neuronal markers at both the gene and protein levels, and produce voltage-dependent current similar to mature neurons [28, 30]. Furthermore, in vivo, neuronally- differentiated DPSC maintained their neuronal phenotype and integrate into brain tissue after transplantation [28, 29]. Because of these capabilities, researchers are investigating the capacity of DPSC to differentiate into retinal neurons.

A recent study has shown that human DPSC cultured with conditioned media from rat retinal explants differentiated into retinal neurons [31]. Conditioned medium was collected from cultures of normal rat retinal explants or explants treated with *N*-methyl-*N*-nitrosourea (MNU), which chemically damages the retina by inducing apoptosis of the photoreceptor cells. Researchers found that DPSC adopt a neuronal cell morphology when cultured with either normal or damaged retinal explant conditioned medium [31]. However, DPSC only expressed rhodopsin after being cultured with conditioned medium from damaged retinal explants [31]. This shows promising data that DPSC have the capacity to differentiate into retinal neuron-like cells, and damaged retinal cells secrete signals to induce this differentiation. Elucidating the specific signals necessary to induce retinal cell differentiation from DPSCs would be additionally powerful and allow for the development of an efficient mechanism to develop a cellular therapy to treat retinal disorders.

Similar to DPSC, periodontal ligament stem cells (PDLSC) are able to differentiate down neuronal lineages [32], and this also is attributed to the neural crest origins of this tissue. Research is now also emerging on the use of PDLSC for retinal cell differentiation [33]. To induce differentiation, PDLSCs were first cultured as neuorospheres and then plated on Matrigel-coated dishes in differentiation medium containing Noggin and Dkk-1, inhibitors of the bone morphogenic protein and Wnt/β-catenin pathways, respectively [33]. Antagonizing these pathways mimicked developmental processes necessary for anterior neural plate formation and retinogenesis. Differentiated PDLSCs expressed genes and proteins characteristic of retinal fate induction such as Pax6 and Rx, and markers of photoreceptors such as rhodopsin and Nrl [33]. This study shows promising data on the potential use of PDLSCs for retinal cell regeneration. Additional animal studies showing the in vivo behavior of these differentiated PDLSCs such as cell integration into retinal structures and cell functionality would provide further support on the potential use of PDLSCs for regeneration therapies to treat retinopathies.

Dental Stem Cells to Facilitate Neuroprotection

Neurotrophic factors (NTFs) are a family of proteins that promote neuronal cell survival and function, and the regeneration of damaged axons. This class of proteins includes nerve growth factor (NGF), brain-derived neurotrophic factor (BDNF), and neurotrophin-3 (NT-3). During development, the expression of NTFs by dental pulp cells is critical for pulp innervation [34–37]. Cultured dental pulp cells maintain the ability to express these factors in vitro and through this production of NTFs promote neuron survival [38]. These studies give promise to the development of cellular therapies utilizing dental pulp cell to treat disorders involving the neuron degeneration.

The introduction of neurotrophic factors is being considered as a strategy to repair damaged RGCs and prevent RGC death in glaucoma [26]. Intraocular injections of recombinant neurotrophic factors results in significant increases in RGC survival following optic nerve injuries in animals [39–43], however, this type of therapy would require repeated injections in patients. Intravitreal injections of viral-mediated NTF genes or cells genetically modified to express NTFs has shown promising results as a method of NTF delivery for optic nerve repair [44, 45]. Alternatively, cellular therapies involving the delivery of adult stem cell populations that naturally express NTFs are being investigated. The intravitreal injection of NTF-secreting bone marrow stromal cells facilitated a neuroprotective response as seen by a significant increase in RGC survival following optic nerve transection [46]. Recently, a study comparing the NTF secretion among adult stem cell populations isolated from different tissues showed that DPSCs produce significantly greater amounts of NTFs when compared to stem cells derived from adipose or bone morrow [47] suggesting that DPSC may be a favorable cell choice to provide NTFs for retinal cell repair.

Research investigating the ability of DPSC to provide neuroprotection and axon regeneration to RGCs after optic nerve injury showed promising results. In vitro, retinal cells cultured with DPSC-conditioned medium had increased neurite number and neurite length when compared to retinal cells cultured with untreated medium or BMSC conditioned medium [48]. Furthermore, this affect was lost by blocking tropomyosin receptor kinase receptors, which bind NTFs, confirming that the increase in neuritogenesis was due to NTF production by the DPSCs [48].

The ability of DPSC to restore retinal cell function in vivo was evaluated by injecting suspensions of DPSC or BMSC intravitreally into rats after optic nerve crush. One observable effect of optic nerve crush is decreased thickness of the RNFL due to RGC atrophy, which can be quantified from optical coherence topography images. Unlike control eyes, DPSC treated eyes did not have reduced RNFL thickness after optic nerve crush [48]. Furthermore, injured eyes with transplanted DPSC had significantly greater RGC survival, determined by positive Brn3a staining, than control or BMSC injected injured eyes, indicating a neuroprotective effect of DPSCs [48]. Additionally, DPSCs promoted a significant increase in RGC axon repair in the optic nerve at distances of up to 1200 μm distal to the crush site as compared to BMSC injected eyes or control eyes [48]. This data shows promising results that through the expression of neurotrophic signals, DPSC could provide a cellular therapy to promote RGC survival and repair to treat glaucoma.

Conclusions and Future Prospects

Populations of multipotent stem cells have been identified in several dental tissues. These stem cells have been shown to have the competence to differentiate down multiple lineages and are therefore being considered for numerous regenerative therapies. This chapter highlighted the exciting current work on investigating the potential of dental stem cells for ocular regeneration. Due to their neural crest origins, researchers postulated that dental stem cells may have a strong affinity to differentiation down ocular lineages, and these recent studies provide strong data to support this. Future studies could also reveal alternative dental stem cell populations for treating the ocular disorders described here or for use in additional ocular diseases not yet investigated. Dental stem cells can be easily isolated from autologous sources and therefore present a viable cell source for the development of new dental stem cell based cellular or tissue engineering therapies to treat blinding ocular disorders.

References

1. Meek KM, Leonard DW (1993) Ultrastructure of the corneal stroma: a comparative study. Biophys J 64:273–280
2. Whitcher JP, Srinivasan M, Upadhyay MP (2001) Corneal blindness: a global perspective. Bull World Health Organ 79:214–221
3. DelMonte DW, Kim T (2011) Anatomy and physiology of the cornea. J Cataract Refract Surg 37:588–598
4. Kenyon KR, Tseng SC (1989) Limbal autograft transplantation for ocular surface disorders. Ophthalmology 96:709–722, discussion 22–3
5. Santos MS, Gomes JA, Hofling-Lima AL, Rizzo LV, Romano AC, Belfort R Jr (2005) Survival analysis of conjunctival limbal grafts and amniotic membrane transplantation in eyes with total limbal stem cell deficiency. Am J Ophthalmol 140:223–230
6. Miura M, Gronthos S, Zhao M, Lu B, Fisher LW, Robey PG et al (2003) SHED: stem cells from human exfoliated deciduous teeth. Proc Natl Acad Sci U S A 100:5807–5812
7. Monteiro BG, Serafim RC, Melo GB, Silva MC, Lizier NF, Maranduba CM et al (2009) Human immature dental pulp stem cells share key characteristic features with limbal stem cells. Cell Prolif 42:587–594
8. Gomes JA, Geraldes Monteiro B, Melo GB, Smith RL, Cavenaghi Pereira da Silva M, Lizier NF et al (2010) Corneal reconstruction with tissue-engineered cell sheets composed of human immature dental pulp stem cells. Invest Ophthalmol Vis Sci 51:1408–1414
9. Levin LA, Kaufman PL (2011) Adler's physiology of the eye: clinical application, 11th edn. Saunders/Elsevier, Edinburgh/New York
10. Smolin G, Foster CS, Azar DT, Dohlman CH (2005) Smolin and Thoft's the cornea: scientific foundations and clinical practice, 4th edn. Lippincott Williams & Wilkins, Philadelphia
11. Hassell JR, Birk DE (2010) The molecular basis of corneal transparency. Exp Eye Res 91:326–335
12. Fini ME (1999) Keratocyte and fibroblast phenotypes in the repairing cornea. Prog Retin Eye Res 18:529–551
13. Williams KA, Esterman AJ, Bartlett C, Holland H, Hornsby NB, Coster DJ (2006) How effective is penetrating corneal transplantation? Factors influencing long-term outcome in multivariate analysis. Transplantation 81:896–901
14. Du Y, Carlson EC, Funderburgh ML, Birk DE, Pearlman E, Guo N et al (2009) Stem cell therapy restores transparency to defective murine corneas. Stem Cells 27:1635–1642
15. Du Y, Funderburgh ML, Mann MM, SundarRaj N, Funderburgh JL (2005) Multipotent stem cells in human corneal stroma. Stem Cells 23:1266–1275
16. Du Y, Roh DS, Funderburgh ML, Mann MM, Marra KG, Rubin JP et al (2010) Adipose-derived stem cells differentiate to keratocytes in vitro. Mol Vis 16:2680–2689
17. Syed-Picard FN, Du Y, Lathrop KL, Mann MM, Funderburgh ML, Funderburgh JL (2015) Dental pulp stem cells: a new cellular resource for corneal stromal regeneration. Stem Cells Transl Med 4:276–285
18. Du Y, Sundarraj N, Funderburgh ML, Harvey SA, Birk DE, Funderburgh JL (2007) Secretion and organization of a cornea-like tissue in vitro by stem cells from human corneal stroma. Invest Ophthalmol Vis Sci 48:5038–5045
19. Carlson EC, Liu CY, Chikama T, Hayashi Y, Kao CW, Birk DE et al (2005) Keratocan, a cornea-specific keratan sulfate proteoglycan, is regulated by lumican. J Biol Chem 280:25541–25547
20. Mimura T, Yamagami S, Amano S (2013) Corneal endothelial regeneration and tissue engineering. Prog Retin Eye Res 35:1–17
21. Arjamaa O (2012) Corneal reconstruction by stem cells and bioengineering. Clin Ophthalmol 6:1407–1409
22. Patel SV (2012) Graft survival and endothelial outcomes in the new era of endothelial keratoplasty. Exp Eye Res 95:40–47

23. Hatou S, Yoshida S, Higa K, Miyashita H, Inagaki E, Okano H et al (2013) Functional corneal endothelium derived from corneal stroma stem cells of neural crest origin by retinoic acid and Wnt/beta-catenin signaling. Stem Cells Dev 22:828–839
24. Karl MO (2013) The potential of stem cell research for the treatment of neuronal damage in glaucoma. Cell Tissue Res 353:311–325
25. Kolb H (1995) Simple anatomy of the retina. In: Kolb H, Fernandez E, Nelson R (eds) Webvision. Moran Eye Center. Web. (May 12, 2016) http://webvision.med.utah.edu/book/part-i-foundations/simple-anatomy-of-the-retina/
26. Almasieh M, Wilson AM, Morquette B, Cueva Vargas JL, Di Polo A (2012) The molecular basis of retinal ganglion cell death in glaucoma. Prog Retin Eye Res 31:152–181
27. Stone EM, Fingert JH, Alward WL, Nguyen TD, Polansky JR, Sunden SL et al (1997) Identification of a gene that causes primary open angle glaucoma. Science 275:668–670
28. Arthur A, Rychkov G, Shi S, Koblar SA, Gronthos S (2008) Adult human dental pulp stem cells differentiate toward functionally active neurons under appropriate environmental cues. Stem Cells 26:1787–1795
29. Kiraly M, Kadar K, Horvathy DB, Nardai P, Racz GZ, Lacza Z et al (2011) Integration of neuronally predifferentiated human dental pulp stem cells into rat brain in vivo. Neurochem Int 59:371–381
30. Kiraly M, Porcsalmy B, Pataki A, Kadar K, Jelitai M, Molnar B et al (2009) Simultaneous PKC and cAMP activation induces differentiation of human dental pulp stem cells into functionally active neurons. Neurochem Int 55:323–332
31. Bray AF, Cevallos RR, Gazarian K, Lamas M (2014) Human dental pulp stem cells respond to cues from the rat retina and differentiate to express the retinal neuronal marker rhodopsin. Neuroscience 280:142–155
32. Bueno C, Ramirez C, Rodriguez-Lozano FJ, Tabares-Seisdedos R, Rodenas M, Moraleda JM et al (2013) Human adult periodontal ligament-derived cells integrate and differentiate after implantation into the adult mammalian brain. Cell Transplant 22:2017–2028
33. Huang L, Liang J, Geng Y, Tsang WM, Yao X, Jhanji V et al (2013) Directing adult human periodontal ligament-derived stem cells to retinal fate. Invest Ophthalmol Vis Sci 54: 3965–3974
34. Luukko K, Arumae U, Karavanov A, Moshnyakov M, Sainio K, Sariola H et al (1997) Neurotrophin mRNA expression in the developing tooth suggests multiple roles in innervation and organogenesis. Dev Dyn 210:117–129
35. Nosrat CA, Fried K, Ebendal T, Olson L (1998) NGF, BDNF, NT3, NT4 and GDNF in tooth development. Eur J Oral Sci 106(Suppl 1):94–99
36. Nosrat CA, Fried K, Lindskog S, Olson L (1997) Cellular expression of neurotrophin mRNAs during tooth development. Cell Tissue Res 290:569–580
37. Nosrat CA, Tomac A, Lindqvist E, Lindskog S, Humpel C, Stromberg I et al (1996) Cellular expression of GDNF mRNA suggests multiple functions inside and outside the nervous system. Cell Tissue Res 286:191–207
38. Nosrat IV, Smith CA, Mullally P, Olson L, Nosrat CA (2004) Dental pulp cells provide neurotrophic support for dopaminergic neurons and differentiate into neurons in vitro; implications for tissue engineering and repair in the nervous system. Eur J Neurosci 19:2388–2398
39. Chen H, Weber AJ (2001) BDNF enhances retinal ganglion cell survival in cats with optic nerve damage. Invest Ophthalmol Vis Sci 42:966–974
40. Mansour-Robaey S, Clarke DB, Wang YC, Bray GM, Aguayo AJ (1994) Effects of ocular injury and administration of brain-derived neurotrophic factor on survival and regrowth of axotomized retinal ganglion cells. Proc Natl Acad Sci U S A 91:1632–1636
41. Mey J, Thanos S (1993) Intravitreal injections of neurotrophic factors support the survival of axotomized retinal ganglion cells in adult rats in vivo. Brain Res 602:304–317
42. Parrilla-Reverter G, Agudo M, Sobrado-Calvo P, Salinas-Navarro M, Villegas-Perez MP, Vidal-Sanz M (2009) Effects of different neurotrophic factors on the survival of retinal ganglion cells after a complete intraorbital nerve crush injury: a quantitative in vivo study. Exp Eye Res 89:32–41

43. Peinado-Ramon P, Salvador M, Villegas-Perez MP, Vidal-Sanz M (1996) Effects of axotomy and intraocular administration of NT-4, NT-3, and brain-derived neurotrophic factor on the survival of adult rat retinal ganglion cells. A quantitative in vivo study. Invest Ophthalmol Vis Sci 37:489–500
44. Di Polo A, Aigner LJ, Dunn RJ, Bray GM, Aguayo AJ (1998) Prolonged delivery of brain-derived neurotrophic factor by adenovirus-infected Muller cells temporarily rescues injured retinal ganglion cells. Proc Natl Acad Sci U S A 95:3978–3983
45. Leaver SG, Cui Q, Plant GW, Arulpragasam A, Hisheh S, Verhaagen J et al (2006) AAV-mediated expression of CNTF promotes long-term survival and regeneration of adult rat retinal ganglion cells. Gene Ther 13:1328–1341
46. Levkovitch-Verbin H, Sadan O, Vander S, Rosner M, Barhum Y, Melamed E et al (2010) Intravitreal injections of neurotrophic factors secreting mesenchymal stem cells are neuroprotective in rat eyes following optic nerve transection. Invest Ophthalmol Vis Sci 51: 6394–6400
47. Mead B, Logan A, Berry M, Leadbeater W, Scheven BA (2014) Paracrine-mediated neuroprotection and neuritogenesis of axotomised retinal ganglion cells by human dental pulp stem cells: comparison with human bone marrow and adipose-derived mesenchymal stem cells. PLoS One 9, e109305
48. Mead B, Logan A, Berry M, Leadbeater W, Scheven BA (2013) Intravitreally transplanted dental pulp stem cells promote neuroprotection and axon regeneration of retinal ganglion cells after optic nerve injury. Invest Ophthalmol Vis Sci 54:7544–7556

DSC-Differentiated Hepatocytes for Treatment of Liver Diseases

Francesco Paduano, Massimo Marrelli, Akhilesh K. Gaharwar, and Marco Tatullo

Abbreviations

ADSCs	Adipose stem cells
AHF	Acute hepatic failure
ALB	Albumin
ALF	Acute liver failure
ALT	Alanine aminotransferase
AST	Aminotransferase
BM-MSCs	Bone marrow stem cells
CHC	Carboxymethyl-hexanol chitosan
CHC/PU-miR122	miR122 complexed with PU-PEI in nanostructured CHC
CK-18	Cytokeratin 18
CPS-1	Carbamoyl phosphate synthetase
Dex	Dexamethasone
DP-iPSC-Heps	Hepatocytes-like cells from dental pulp derived-iPSCs
DP-iPSCs	Dental pulp-derived iPSC
DPSCs	Dental pulp stem cells

F. Paduano • M. Tatullo (✉)
Research and Development in Biomedicine – Stem Cells Unit, TECNOLOGICA Research Institute, st. E. Fermi Loc. Passovecchio I. Z., 88900 Crotone (KR), Italy
e-mail: marco.tatullo@tecnologicasrl.com

M. Marrelli
Department of Maxillofacial Surgery, Calabrodental, Crotone, Italy

A.K. Gaharwar
Department of Biomedical Engineering, Texas A&M University, College Station, TX 77843, USA

Department of Materials Science and Engineering, Texas A&M University, College Station, TX 77843, USA

B. Zavan, E. Bressan (eds.), *Dental Stem Cells: Regenerative Potential*, Stem Cell Biology and Regenerative Medicine, DOI 10.1007/978-3-319-33299-4_15

DSCs	Dental stem cells
DTCPs	Deciduous tooth pulp cells
EGF	Epidermal growth factor
ESC-Heps	Embryonyc stem cell-derived hepatocytes
ESCs	Embryonic stem cells
ETF	Embryotrophic factor
FBS	Fetal bovine serum
FGF-1	Fibroblast growth factor 1
HCC	Hepatocellular carcinoma
HCV	Hepatitis C virus
hdSHED	Human hepatically differentiated SHED
HGF	Hepatocyte growth factor
HTCP	Tooth-pulp stem cells
IL-6	Interleukin 6
iPSC-Heps	iPSCs-derived hepatocyte-like cells
iPSCs	Induced pluripotent stem cells
ITS	Insulin-transferrin selenium
MAPS	Multipotent adult progenitor cells
miR122	MicroRNA122
OSM	Oncostatin M
PU-PEI	Polyurethane-graft-short-branch polyethylenimine copolymer
SHED	Exfoliated deciduous teeth stem cells
STAT3	Signal transducer and activator of transcription 3
TAA	Thioacetamide
TGPCs	Tooth germ progenitor cells
WJ-MSCs	Wharton's jelly-derived MSCs
WTPCs	Wisdom tooth pulp cells
α-FP	Alpha-fetoprotein

Introduction

Chronic liver disease or acute liver failure are caused by several factors including viral infections, genetic disorders, toxic injury and auto-immune defects [1]. Chronic liver injury, such as that triggered by hepatitis C virus (HCV), induces inflammation and fibrosis, often followed by the development of liver cirrhosis, which the most advanced stage of hepatic fibrosis [2]. Current pharmaceutical treatments are able to cure patients with hepatitis virus C thought the administration of direct acting-antiviral agents (DAAs) such as *sofosbuvir* in combination with *simeprevir* [3]. However, the drug therapy alone is not enough to treat the liver fibrosis, especially in patients with decompensated cirrhosis [4]. Consequently, the development of effective treatments for liver fibrosis are necessary for HCV-infected patients. Cirrhosis generally progress to hepatocellular carcinoma (HCC), resulting in a liver failure without the typical self-regenerative capability of liver. Therefore, the inhibition of hepatic fibrogenesis and liver inflammation could prevent the progression of cirrhosis and reduce the incidence of HCC. Similarly, Acute hepatic failure (AHF) is another severe liver injury that sustained liver damage and deterioration of liver functions, leading complications including hepatic encephalopathy and multiorgan failure. The management of patients with AHF continues to be one of the most interesting obstacles in clinical medicine.

Currently clinical treatment for cirrhosis and AHF include liver transplantation. However, transplantation of liver is hampered due to lack of donors and graft rejection. Therefore, there is an unmet need to develop an alternative treatment for acute and chronic liver diseases. Recently, to alleviate the excessive demand for liver transplants, alternative cellular therapy are investigated [5].

Hepatocyte transplantation has been considered a therapeutic alternative for the treatment of patients having hepatic disorders such as cirrhosis and HCC. However, the source of hepatocytes is limited due to scarcity of suitable liver donors. In addition, it has been shown that the hepatic transplantation of hepatocytes could provoke injury to both recipient and donor. Therefore, the researchers have put great effort into obtaining hepatocyte-like cells from sources other than the human adult liver. Consequently, stem cells have been investigated as an alternative source of hepatocytes to overcome these limitations.

Stem cell therapy may represent a new treatment method for tissue and organ regeneration. In fact, one of the main focuses of regenerative medicine in recent years has been to obtain large population of hepatocyte-like cells derived from stem cells for transplantation. In the beginning, the strategies were focused principally on embryonic stem cells (ESCs) as they can successfully differentiate into three germ layers and have limitless capacity of self-renewal [6]. However, safety conditions and ethical issues are some of the obstacle to clinical translations. The difficulties regarding the use of ESCs can be avoided by using the adult stem cells found in different tissue.

Cellular Therapies for Liver Disease

Recently, cell-based therapies and hepatic tissue engineering have been explored as an alternative for the clinical treatment of liver-related diseases [7]. Mesenchymal stem cells (MSCs) have recently attracted considerable attention as promising a promising cell source for the treatment of liver disease because they can be easily isolated, expanded and cryopreserved. MSCs obtained from various tissue sources include human bone marrow mesenchymal stem cells (BM-MSCs) [8], human adipose stem cells (ADSCs) [9], Wharton's Jelly-derived MSCs (WJ-MSCs) [10] and multipotent stem cells from amniotic fluid (AF-MSCs) [11]. These MSCs have been shown to differentiate into endodermal lineages such as hepatocytes. In addition, several clinical phase I, I/II, and II trials have demonstrated that human MSCs transplantation recovers hepatic function in patients with liver cirrhosis [12–14]. This demonstrates that human MSCs might be strong candidate for the treatment of liver dysfunction.

In comparison with other MSCs, dental stem cells (DSCs) offer important advantages of lower ethical controversies and easily accessibilities. DSCs maintain stemness characteristics such as cell proliferation, multi-lineage differentiation potential, and immunomodulatory functions [15], and can be cryopreservation without losing these functions [16]. Thus, DSCs are considered an attractive alternative cell source for cell-based therapies for the treatment of liver-related diseases. In addition, DSCs are good cell-source to generate hepatocyte-like cells as they possess multipotent characteristics. DSC can be expanded allowing autologous transplantation and, most importantly without the need of genetic manipulation.

Effect of Soluble Signals on Hepatic Differentiation of Stem Cells

Several studies demonstrated that DSCs such as tooth germ progenitor cells (TGPCs) [4], dental pulp stem cells (DPSCs) [17–19], tooth-pulp stem cells (HTCP) CD117+ [19] and exfoliated deciduous teeth (SHED) [20] or SHED-CD117+ [21] can differentiate into functional hepatocytes in vitro (Table 1). DSCs under definite stimulation, are able to generate hepatocyte-like cells that resemble hepatocyte morphology, express liver-specific markers at mRNA and protein levels and perform functions typical of hepatocytes. For the hepatic maturation of DSCs, use of fibroblast growth factor (FGF), hepatocyte growth factor (HGF), Dexamethasone (Dex) and Oncostatin M (OSM) as hepatic inductors (Table 1) are reported. However, the mechanism behind the differentiation of DSCs into hepatocytes has not yet been completely known.

Several studies have showed that biochemical factors that have an essential role in liver development are also shown to induce hepatic differentiation of DSCs. Moreover, those biochemical factors have also been useful for maintaining hepatocytes cultures in vitro [26]. For example, the use of HGF and FGF-4 as inductors for hepatic differentiation of multipotent adult progenitor cells (MAPS) [27] and human umbilical cord blood-derived mesenchymal stem cells [28] are reported. HGF is considered a key growth factor for liver growth and function that controls different processes, including morphogenesis, motogenesis and mitogenesis.

HGF initiates multifunctional regulatory roles by interaction with HGF receptor (c-Met) [29], a transmembrane protein-tyrosine kinase expressed in normal hepatocytes, intestinal, ovarian, epithelium and endometrial endothelia [7]. HGF possesses strong antifibrotic activity and contributes to the onset or progress of liver fibrosis/cirrhosis and directly promotes the differentiation of MSCs into hepatic lineages, and transplantation of these cells improves liver injury in rats [30]. Therefore, HGF can potentially be used for the treatment of acute hepatic failure (AHF) [31] and CCl_4 induced liver necrosis [32]. FGF is another growth factor that works as mitotic agent for a variety of cells and tissues [33]. It stimulates differentiation and

Table 1 Studies that assessed the use of human dental stem cells (DSCs) for liver regeneration

Dental stem cells (DSCs)	In vitro hepatic differentiation	Animal model	Remarks	Reference
Human tooth germ progenitor cells (TGPCs)	DMEM low glucose, 2 % FBS, 100 ng/mL a-FGF (5 days); DMEM low glucose, 2 % FBS, 20 ng/mL HGF (5 days); 20 ng/mL DMEM low glucose, 2 % FBS, 20 ng/mL HGF, 10 nM Dexamehtasone, ITS-X, 10 ng/mL oncostatin M (11 days)	Liver fibrosis in CCl_4 treated rats	TGPCs subjected to in vitro hepatic differentiation had therapeutic effect on CCl_4-induced liver injury (portal veins)	Ikeda E. et al. [4]
Human dental pulp stem cells (DPSCs)	20 ng/mL HGF (5 days), 2 % FCS; 10 ng/mL oncostatin M, 10 nM Dex, 1 % ITS-X (15 days)	No in vivo model	Differentiation of DPSCs into hepatocyte like-cells	Ishkitiev et al. [22]
Human dental pulp stem cells (DPSCs)	DMEM low glucose, 1 % FCS, 20 ng/mL HGF, 10 ng/mL oncostatin, 10 mM nicotinammide, 1.25 μg/mL LDL,, 10 ng/mL FGF-4, 4 μg/mL insulin, 1.25 g/L glucose, 180 mg/L linoleic acid (40 day)	No in vivo model	Hepatic function of differentiated DPSCs	Ferro et al. [18]
Human tooth-pulp stem cells (HTCP) $CD117^+$	Serum free-DMEM, 1 % ITS-X, 100 μg/mL embryotrophic factor (ETF) (until 70 % confluence); 20 ng/mL HGF (5 days); 10 ng/mL oncostatin M, 10 nM Dex (15 days)	No in vivo model	H_2S increased the ability of human tooth-pulp cells to undergo hepatogenic differentiation	Ishkitiev et al. [19]
Human dental pulp stem cells (DPSCs) and human exfoliated deciduous teeth (SHED) $CD117^+$	Serum free DMEM, 1 % ITSX, 100 μg/mL embryotrophic factor (ETF) (until 70 % confluence); 20 ng/mL HGF (5 days); 10 ng/mL oncostatin M, 10 nM Dex (15 days)	No in vivo model	Hepatic differentiation of DPSCs and SHED in serum free medium	Ishkitiev et al. 2012 [17]

(continued)

Table 1 (continued)

Dental stem cells (DSCs)	In vitro hepatic differentiation	Animal model	Remarks	Reference
Cryopreserved human dental pulp stem cells (hDPSCs) isolated from vital extracted teeth with disease.	Serum deprived IMDM medium, 20 ng/mL HGF, 10 ng/ml β-FGF, nicodidamine 0.61 g/L (7 days). IMDM, 20 ng/mL oncostatin M, 1 μM Dex, 50 mg/mL ITS (34 days)	No in vivo model	Hepatic differentiation of hDPSCs from cryopreserved dental pulp tissues of vital extracted teeth with disease	Chen et al. [23]
Human exfoliated deciduous teeth (SHED) CD117+	DMEM serum free, 1 % ITS-x 100 μg/mL embryotrophic factor (ETF) (until 70 %); 20 ng/mL HGF (5 days); 10 ng/mL Oncostatin M, 10 nM Dex (15 days)	No in vivo model	The level of hepatic differentiation in SHED compared with BM-MSCs was the same or higher. H_2S increased the level of hepatic differentiation.	Okada et al. [24]
Human exfoliated deciduous teeth (SHED) CD117+	Serum free medium, 20 ng/mL HGF (5 days); 10 ng/mL oncostatin M, 10 nM Dex, HGF 10 ng/mL (15 days)	Acute liver injury (ALI) and secondary biliary cirrhosis in rats	hdSHED engraft morphologically and functionally into the livers of rats (transplantation to the spleen)	Ishkitiev et al. [21]
Human dental pulp stem cells (DPSCs)	DMEM 2 % FCS, 20 ng/mL HGF (5 days); DMEM 2 % FCS, 20 ng/mL HGF, 10 ng/mL oncostatin M, 10 nM Dex (15 days)	CCl_4-induced liver fibrosis in mice	Combined treatment of DPSCs and melatonin for the treatment of liver fibrosis (teil veins)	Cho et al. [25]
Human exfoliated deciduous teeth (SHED)	No in vitro hepatic differentiation	CCl_4-induced liver fibrosis model mice	SHED transplantation recovered liver dysfunction and led anti-fibrotic and anti-inflammatory effects (intrasplenically transplanted into mice)	Yamanaza et al. [20]

proliferation by binding to its receptor epidermal growth factor receptor (EGFR) [34]. Both, HGF and FGF have important roles in liver development, but it is unclear if they can induce definitive endoderm without the contribution of additional factors.

Recently, Lin et al. showed that hepatocytes differentiation of BM-MSCs is induced by cooperation of HGF and FGF-4 through the HGF/c-Met, a transmembrane receptor endowed with tyrosine kinase activity [35]. In some studies, in order to induce the late hepatic differentiation, the medium was supplemented with OSM [36], insulin-transferrin selenium (ITS) [37], Dex [38] and embryotrophic factor (ETF) [39]. OSM is a cytokine belonging to the interleukin 6 (IL-6) family and is recognized as having an essential role in liver development, it is essential for maturation of hepatocytes in combination with Dex [40]. Moreover, OSM plays a crucial role in progression of hepatocyte development through the signaling pathway of activator factor and signal transducer of transcription 3 (STAT 3) [41]. Some studies have shown that Dex is able to preserve the expression of transcription factors essential for the regulation of liver specific genes [38, 42]. Whereas, insulin-transferrin selenium (ITS) have been shown to induce survival and proliferation of primary hepatocytes [43].

In summary, HGF, EGF and OSM have important effects on the maintenance of primary human hepatocytes in vitro [44] and several studies have proposed that there is a need to expose cells to HGF and FGF4 in a sequential approach in order to increase the hepatic differentiation efficacy of BM-MSCs [45].

BM-MSCs and DPSCs have similar function and morphology, and thus several studies speculated that DSCs have the ability to differentiate not only in tissue of ectodermal and mesodermal origin, but also in those of endodermal origin. As a consequence, the combination of HGF, EGF, OSM, Dex and ITX previously established for ESCs and BMSCs was widely used in protocols to differentiate DSCs to hepatocytes [4, 17–19, 21, 22, 25].

Various Tissue Sources of DSCs

DSCs obtained from various sources can be directed towards hepatic differentiation using soluble factors as listed in Table 1. For example, Ikeda et al. demonstrated that stem cells derived from the neural crest-derived dental tissues can differentiate into endoderm cell lineages [4]. In particular, they showed that the tooth germ progenitor cells (TGPCs) were able to differentiate into functional hepatocytes. The differentiated hepatocytes were able to suppress inflammation and liver fibrosis in carbon tetrachloride (CCl_4) treated rats [4]. They further showed that the hepatic differentiated TGPCs can improve to the restoration of liver function as evaluated by the expression of hepatic serum marker such as albumin, bilirubin, aspartate aminotransferase (AST), and alanine aminotransferase (ALT). In their protocol, TGPCs were first treated with a-FGF and HGF followed by the addition of the maturation factors OSM and Dex when the TGPCs already were committed to the hepatic fate.

The changes in morphology were evident and the production of albumin was observed starting from day 10.

In a similar study, Ishkitiev et al. showed that the differentiation of DSCs into hepatic-like cells was possible from DSCs obtained from full-grown wisdom or primary tooth pulp [22]. In fact, they showed the differentiation of wisdom tooth pulp cells (WTPCs) and deciduous tooth pulp cells (DTCPs) into hepatocytes-like cells using HGF, Dex, OSM and ITS-x. The differentiated DSCs produced proteins specific to hepatocytes and acquired hepatic specific functions including the storing of glycogen and the production of urea [22]. Subsequently, the same research group showed that hydrogen sulfide (H_2S) at physiological concentrations improved the capacity of human tooth-pulp stem cells (HTCP) to undergo hepatogenic differentiation [19]. The CD117-positive cells ($CD117^+$) were isolated from deciduous HTCP by using a magnetic cell sorting system and then cells were grown in medium supplemented with HGF, ETF, OSM and insulin-transferrin-selenium-x (ITS-x) with or without H_2S. The results demonstrated that the hepatic markers albumin (ALB), α-fetoprotein (αFP), and carbamoyl phosphate synthetase (CPS-1) were expressed more in H_2S treated HTCP with respect to control. This study demonstrate that H_2S at physiological concentrations is able to improve hepatogenic differentiation of HTCP in vitro [19].

More recently, Okada et al. have compared the hepatic differentiation of SHEDs with those of BM-MSCs in vitro with or without H_2S [24]. In their experiments they observed that either SHEDs or BM-MSCs $CD117^+$ without H_2S expressed similar levels of stem cell transcription factors. However, after differentiation, the expression of hepatic markers such as ALB, αFP, CPS-1 was significantly higher in SHEDs compared to BM-MSCs. Unlike the previous results, no significance differences were observed between H_2S-treated SHEDs and BM-MSCs after hepatic maturation in terms of ALB, αFP, CPS-1 expression, glycogen production and urea concentration. These results strongly suggest the possibility to use SHEDs in combination with H_2S for clinical treatment of liver disease.

In another study, Ferro et al. further demonstrated the hepatic-differentiation potential of DSCs [18]. They reported that under hepatic stimuli, DSCs increased their mRNA expression of albumin and hepatocyte growth factor receptor (C-met/HGF-r). Moreover, they showed that DSCs acquired the pattern of cytokeratin expression (CK8, CK18 and CK19) of normal hepatocytes and exhibited some hepatic functions such as the production and secretion of albumin [18].

The previous studies used fetal bovine serum (FBS) complemented medium for hepatic differentiation. The use of animal products in the differentiation medium represents a enormous obstacle for the application of in vitro cell-therapy methods in clinical applications due to the fact that FBS can cause hypersensitivity reaction and can be a potential vector for prion transmission [46]. Therefore, in order to overcome these limitations, Ishkitiev et al. described the differentiation of DPSCs $CD117^+$ and SHEDs $CD117^+$ into a high-purity hepatic lineage using serum-free medium [17].

Recently, Chen et al. have studied the hepatic potential after cryopreservation of DSCs isolated from human dental pulp tissues of teeth affected by disease such as periodontitis and pericoronitis [23]. In their experiments they observed that hepatically differentiated DSC had a polygonal shape and expressed the specific liver genes including αFP, ALB, HNF1α and CK18.

Moreover, urea production and glycogen storage results denoted that the differentiated human DPSCs were functionally similar to normal hepatocytes-like cells. Although research studies regarding hepatic differentiation of DSCs are at beginning, the use of cryopreserved tissue to generate hepatocytes-like cells represents an encouraging alternative for the treatment of liver diseases.

To highlight the clinical potential of DSCs for the treatment of hepatic diseases, Ishkitiev et al. showed that human hepatically differentiated $CD117^+$ SHED (hdSHED) were able to engraft functionally in the liver of experimental rat models of hepatic diseases [21]. In fact, human specific markers such as albumin and αFP were found in the serum of the rats, confirming that transplanted hdSHED can be incorporated into structural components of the liver of rats having acute injury. This study showed that in vitro differentiation of DSCs in hepatocyte-like cells and the following transplantation could be a useful method to treat human liver diseases.

As described above, the transplantation of human TGPS in the liver of CCl_4 injured mice were able to inhibit fibrosis and inflammation [4]. Using the same animal model, Cho et al. recently reported that the use of melatonin in combination with the transplantation of human DPSCs drastically decrease liver fibrosis demonstrating that the treatment of DPSCs transplantation combined with melatonin administration could be used to improve liver functions [25]. Melatonin promoted hepatic differentiation by regulating the NF-κB pathway, p38, BMP, and ERK. Moreover, the grade of liver fibrosis measured as restoration of ALT, ASP and levels of ammonia was lower in CCl_4-injured mice treated with a combination of DPSCs transplantation and melatonin administration compared to either melatonin administration or DPSCs transplantation alone [25].

In the earlier studies, DSCs were pre-differentiated in vitro before transplantation in animal model [4, 21, 25]. Recently, Yamanaza et al. demonstrated that naïve SHEDs, not previously in vitro differentiated, were able to improve hepatic dysfunction and directly transform into hepatocytes in CCl_4 treated mice [20]. Moreover, these in vivo SHEDs converted hepatocytes participated in the hepatic recovery via tissue replacement and possessed anti-fibrotic and anti-inflammatory effects.

DSCs-Derived Induced Pluripotent Stem Cells (iPSCs) as Potential Cell Source

Recently, human somatic cells have been reprogramming into induced pluripotent stem cells (iPSCs) by addition of four transcription factors [47, 48]. These results offer the possibility of using adult somatic cells to obtain cells that shared many features with ESCs but with the advantage of being obtained from an adult source. As a consequence, in the last years, it became possible to obtain iPSCs from several adult tissues, including the hepatic tissue. iPSCs are typically generated by ectopic expression of four transcription factors, *OCT3/4, Sox2, Klf4* and *c-Myc* in fibroblasts. Several researchers have demonstrated that iPSCs can be derived from different cell sources including skin fibroblasts [47], keratinocytes [49], blood progenitor cells [50], pancreatic cells [51] and primary human hepatocytes [52].

It has recently been showed that iPSCs can be considered a new cell source for stem cell therapy of liver diseases because they were able to differentiate into functional hepatocytes [53, 54]. Because of its accessibility and differentiation potential, DSCs has gradually drawn attention in the applications of iPSCs in regenerative medicine, in fact, DSC has been widely demonstrated as a source of iPSCs [55, 56]. The knowledge previously acquired in hepatic-differentiation of DSCs was successfully applied in iPSCs and two reports of efficient methods to obtain hepatocytes-like cells from dental pulp derived-iPSCs (DP-iPSC-Heps) have been published [57, 58]. A summary of these reports is shown in Table 2.

Interestingly, Chiang et al. reprogrammed human dental pulp fibroblasts into iPSCs and induced the differentiation of these dental pulp-derived iPSC (DP-iPSCs)

Table 2 Studies that assessed the use of human dental pulp derived iPSCs (DP-iPSCs) for liver regeneration

Dental stem cells (DSCs)	In vitro hepatic differentiation	Animal model	Remarks	Reference
Induced pluripotent stem cells (iPSCs) from dental pulp stromal cells Transduction of pMX vectors encoding the transcription factors: OCT4, Sox-2, Klf4 and c-Myc	RPMI 0.5 mg/ml Albumin fraction V, 100 ng/mL activin A (1 day); 0.15 and 1 % ITS (2 days); 30 ng/mL FGF-4, 20 ng/mL BMP-2 (4 days); 20 ng/mL HGF, 20 ng/mL KGF, oncostatin M, 0.1 μM Dex (5 days); DMEM N2 B27, glutamax, 0.1 mM β-mercaptoethanol (3 days)	Acute hepatic failure (AHF) mouse model induced by thioacetamide (TAA)	Engraftment of iPSC-Heps with injectable hydrogel (CHC) containing HGF reduced the TAA-induce hepatic necrosis and rescued liver function	Chiang et al. [58]
Induced pluripotent stem cells derived from human dental pulp (DP-iPSCs) Transduction of pMX vectors encoding the transcription factors: OCT4, Sox-2, Klf4 and c-Myc	Nanostructured amphiphatic carboxymethyl–hexanoyl chitosan (CHC)/ Polyurethane-graft-short-branch polyethylenimine copolymer/ miR122 mixture to accelerate the hepatic differentiation of iPSCs	Acute hepatic failure (AHF) mouse model induced by thioacetamide (TAA)	miR122-iPSC-Heps improved liver functions and rescued recipient survival	Chien et al. [57]

into functional iPSC-Heps [57]. Microarray analysis indicated that DP-iPSC-Heps, normal liver and ESCs derived hepatocytes (ESC-Heps) possessed a similar gene expression profile. Moreover, similar to mature ESC-Heps, DP-iPSCs-Heps acquired regular functions for glycogen storage and LDL uptake during hepatic differentiation. Moreover, DP-iPSCs-Heps expressed several hepatic markers including αFP, hepatocyte nuclear factor 3β (HBF-3β) and cytochrome c P450 enzyme CYP2E1. In their in vivo studies, Chiang et al. used an injectable carboxymethyl-hexanol chitosan hydrogel (CHC) having a continuous release of HGF (HGF-CHC), as delivery vehicle for transplantation of DP-iPSCs-Hep in immunocompromised acute hepatic failure (AHF) mouse model induced by thioacetamide (TAA) [58]. They observed that intrahepatic delivery of CHC-HGF hydrogel coupled with iPSC-Heps decreased the TAA-induced hepatic necrotic area and recovered liver functions, suggesting that HGF-CHC is an excellent vehicle for iPSCs-Heps engraftment in iPSCs based therapy against AHF [58].

In another study from the same group, Chien et al. investigated an alternative strategy to obtain mature hepatocytes-like cells from DP-iPSCs using a nanotechnology based delivery microRNA (miRNA) system. At this purpose, they used CHC to encapsulated the complex of polyurethane-graft-short-branch polyethylenimine copolymer (PU-PEI) and microRNA122 (miR122), a hepatocyte-specific miRNA [57]. Subsequently, PU-PEI complex was used as vehicle to deliver miR122 into DP-iPSCs. miR122 has been chosen because it is associated with some biological functions of the liver [59]. They observed that combination of miR122 plasmids with PU-PEI in nanostructured CHC (CHC/PU-miR122) were able to enhance miR122 delivery into human DP-iPSCs. As consequence, after the entry of miR122 into DP-iPSCs, these cells were able to successfully differentiate into iPSC-Heps having mature functions of hepatocytes [57]. In this work, using microarray analyses it has been observed that CHC/PU-PEI-miR122 shifted gene expression profile of DP-iPSCs toward that of the liver. Moreover, they observed that the delivery of a construct containing CHC/PU-miR122 and iPSCs-Heps into thioacetamide (TAA)-injured rat liver was able to significantly rescued liver functions. These results demonstrated for the first time that miR122-iPSCs-Heps transplantation could be used as new therapeutic approach to obtain liver regeneration of recipients with acute hepatic failure (AHF).

Conclusions and Future Directions

Several groups reported successful differentiation of DSCs into hepatocyte-like cells expressing defined hepatocellular antigens and functional properties that include the following: (a) phenotypic changes leading the acquisition of polygonal and polarized morphology; (b) expression of the specific proteins like α-FP, ALB, CPS-1, CK-18 (c) the capacity to synthetize urea and glycogen and (d) expression of protein or enzymes, such as AST, ALT, HNF-4. In all studies described above, in vitro hepatic differentiation was established on a defined scheme of experimental protocol, which usually required first a differentiation step followed then by maturation step. The differentiation step was sustained by the treatment of DSCs with HGF and EGF, whereas in the maturation step DSCs were exposed to medium containing OSM, Dex and ITS-x. Once hepatocytes-like cells were obtained, several laboratories transplanted in vivo these DSCs to analyse their efficiency in animal models of CCl_4 induced liver injury or acute liver injury (ALI). Most of these experimental studies have reported that transplanted DSCs can effectively engraft in injured liver. These studies have provided evidence indicating that DSCs engrafting the liver during the course of experimental model of liver injury have potential to differentiate in hepatic like cells [4, 20, 21, 25]. All studies described in this chapter lead us to believe that DSCs is one of the promising cell-source to generate functional hepatocyte-like cells in appropriate quantities and purity for cell replacement therapy. However, for clinical applications there are yet several studies that need to be done to better understand the capacity of these cells. In vivo methodologies to check the comportment of the hepatocyte-like cells in humans will be essential to verify their potential application in cell therapy.

A new approach for generating hepatocyte-like cells from DSCs has taken advantage of induce pluripotent stem cells (iPSCs), one of the greatest inventions in the field of regenerative medicine. Human dental pulp stromal cells can be reprogrammed into iPSCs (DP-iPSCs) and can be differentiated into functional iPSCs-derived hepatocyte-like cells (iPSC-Heps) by using constructs of HGF-CHC or CHC/PU-miR122 [54, 58]. HGF-CHC and CHC/PU-miR122 are excellent vehicles for iPSCs-Heps engraftment in iPSCs based therapy against AHF.

In conclusion, this chapter shows the differentiation of DSCs into hepatocyte-like cells, indicating that MSCs of dental tissues have a hepatic potential and can be considered a new cell source for liver cell therapy. The described models offer suitable and convenient cell source of non-embryonic origin to obtain functional hepatocyte-like cells. Wisdom teeth are regularly extracted and deciduous teeth fall out naturally. This potentially makes them exceptionally useful as a simply accessible cell source. Furthermore, dental tissues can be easily manipulated and stored in stem-cell banks. The possibility of preserving DSCs, so that they can be later differentiated into hepatocytes and used for transplantation without risk of immunologic reaction could have an extraordinary effect on the cell therapy of liver deceases in the future. In the past, liver diseases were often considered irreversible, but today progress of research in cellular and molecular biology has demonstrated that hepatic cellular recovery is possible. At this purpose, the prospect of using DSCs as cell therapy for treating liver diseases is very encouraging. However, further studies are necessary to demonstrate long-tem safety of DSCs-based transplantation and moreover, clinical trials are necessary to verify their efficacy to recover liver functions in human patients affected by liver diseases.

References

1. Liu W, Song F, Ren L et al (2015) The multiple functional roles of mesenchymal stem cells in participating in treating liver diseases. J Cell Mol Med 19:511–520. doi:10.1111/jcmm.12482
2. Karidis NP, Delladetsima I, Theocharis S (2015) Hepatocyte turnover in chronic HCV-induced liver injury and cirrhosis. Gastroenterol Res Pract 2015:654105. doi:10.1155/2015/654105
3. Lam BP, Jeffers T, Younoszai Z et al (2015) The changing landscape of hepatitis C virus therapy: focus on interferon-free treatment. Ther Adv Gastroenterol 8:298–312. doi:10.1177/1756283X15587481
4. Ikeda E, Yagi K, Kojima M et al (2008) Multipotent cells from the human third molar: feasibility of cell-based therapy for liver disease. Differentiation 76:495–505. doi:10.1111/j.1432-0436.2007.00245.x
5. Nussler A, Konig S, Ott M et al (2006) Present status and perspectives of cell-based therapies for liver diseases. J Hepatol 45:144–159. doi:10.1016/j.jhep.2006.04.002
6. Baharvand H, Hashemi SM, Kazemi Ashtiani S, Farrokhi A (2006) Differentiation of human embryonic stem cells into hepatocytes in 2D and 3D culture systems in vitro. Int J Dev Biol 50:645–652. doi:10.1387/ijdb.052072hb
7. Ye J-S, Su X-S, Stoltz J-F et al (2015) Signalling pathways involved in the process of mesenchymal stem cells differentiating into hepatocytes. Cell Prolif 48:157–165. doi:10.1111/cpr.12165
8. Pournasr B, Mohamadnejad M, Bagheri M et al (2011) In vitro differentiation of human bone marrow mesenchymal stem cells into hepatocyte-like cells. Arch Iran Med 14:244–249. doi:0011144/AIM.004
9. Lee JH, Lee KH, Kim MH et al (2012) Possibility of undifferentiated human thigh adipose stem cells differentiating into functional hepatocytes. Arch Plast Surg 39:593–599. doi:10.5999/aps.2012.39.6.593
10. Prasajak P, Leeanansaksiri W (2013) Developing a new two-step protocol to generate functional hepatocytes from Wharton's jelly-derived mesenchymal stem cells under hypoxic condition. Stem Cells Int 2013:762196. doi:10.1155/2013/762196
11. Miki T, Lehmann T, Cai H et al (2005) Stem cell characteristics of amniotic epithelial cells. Stem Cells 23:1549–1559. doi:10.1634/stemcells.2004-0357
12. Mohamadnejad M, Alimoghaddam K, Mohyeddin-Bonab M et al (2007) Phase 1 trial of autologous bone marrow mesenchymal stem cell transplantation in patients with decompensated liver cirrhosis. Arch Iran Med 10:459–466. doi:07104/AIM.008
13. Kharaziha P, Hellström PM, Noorinayer B et al (2009) Improvement of liver function in liver cirrhosis patients after autologous mesenchymal stem cell injection: a phase I-II clinical trial. Eur J Gastroenterol Hepatol 21:1199–1205. doi:10.1097/MEG.0b013e32832a1f6c
14. El-Ansary M, Abdel-Aziz I, Mogawer S et al (2012) Phase II trial: undifferentiated versus differentiated autologous mesenchymal stem cells transplantation in Egyptian patients with HCV induced liver cirrhosis. Stem Cell Rev 8:972–981. doi:10.1007/s12015-011-9322-y
15. Yamaza T, Kentaro A, Chen C et al (2010) Immunomodulatory properties of stem cells from human exfoliated deciduous teeth. Stem Cell Res Ther 1:5. doi:10.1186/scrt5
16. Ma L, Makino Y, Yamaza H et al (2012) Cryopreserved dental pulp tissues of exfoliated deciduous teeth is a feasible stem cell resource for regenerative medicine. PLoS One 7, e51777. doi:10.1371/journal.pone.0051777
17. Ishkitiev N, Yaegaki K, Imai T et al (2012) High-purity hepatic lineage differentiated from dental pulp stem cells in serum-free medium. J Endod 38:475–480. doi:10.1016/j.joen.2011.12.011
18. Ferro F, Spelat R, D'Aurizio F et al (2012) Dental pulp stem cells differentiation reveals new insights in Oct4A dynamics. PLoS One 7, e41774. doi:10.1371/journal.pone.0041774
19. Ishkitiev N, Calenic B, Aoyama I et al (2012) Hydrogen sulfide increases hepatic differentiation in tooth-pulp stem cells. J Breath Res 6:017103. doi:10.1088/1752-7155/6/1/017103
20. Yamaza T, Alatas FS, Yuniartha R et al (2015) In vivo hepatogenic capacity and therapeutic potential of stem cells from human exfoliated deciduous teeth in liver fibrosis in mice. Stem Cell Res Ther 6:171. doi:10.1186/s13287-015-0154-6

21. Ishkitiev N, Yaegaki K, Imai T et al (2015) Novel management of acute or secondary biliary liver conditions using hepatically differentiated human dental pulp cells. Tissue Eng Part A 21:586–593. doi:10.1089/ten.TEA.2014.0162
22. Ishkitiev N, Yaegaki K, Calenic B et al (2010) Deciduous and permanent dental pulp mesenchymal cells acquire hepatic morphologic and functional features in vitro. J Endod 36: 469–474. doi:10.1016/j.joen.2009.12.022
23. Chen YK, Huang AHC, Chan AWS, Lin LM (2013) Human dental pulp stem cells derived from cryopreserved dental pulp tissues of vital extracted teeth with disease demonstrate hepatic-like differentiation. J Tissue Eng Regen Med. doi:10.1002/term.1763
24. Okada M, Ishkitiev N, Yaegaki K et al (2014) Hydrogen sulphide increases hepatic differentiation of human tooth pulp stem cells compared with human bone marrow stem cells. Int Endod J 47:1142–1150. doi:10.1111/iej.12262
25. Cho Y-A, Noh K, Jue S-S et al (2015) Melatonin promotes hepatic differentiation of human dental pulp stem cells: clinical implications for the prevention of liver fibrosis. J Pineal Res 58:127–135. doi:10.1111/jpi.12198
26. Sun G-Y, Dong L-Y, An W (2014) Involvement of hepatic stimulator substance in the regulation of hepatoblast maturation into hepatocytes in vitro. Stem Cells Dev 23:1675–1687. doi:10.1089/scd.2013.0468
27. Schwartz RE, Reyes M, Koodie L et al (2002) Multipotent adult progenitor cells from bone marrow differentiate into functional hepatocyte-like cells. J Clin Invest 109:1291–1302. doi:10.1172/JCI15182
28. Kang X-Q, Zang W-J, Bao L-J et al (2005) Fibroblast growth factor-4 and hepatocyte growth factor induce differentiation of human umbilical cord blood-derived mesenchymal stem cells into hepatocytes. World J Gastroenterol 11:7461–7465
29. Neuss S, Becher E, Wöltje M et al (2004) Functional expression of HGF and HGF receptor/c-met in adult human mesenchymal stem cells suggests a role in cell mobilization, tissue repair, and wound healing. Stem Cells 22:405–414. doi:10.1634/stemcells.22-3-405
30. Oyagi S, Hirose M, Kojima M et al (2006) Therapeutic effect of transplanting HGF-treated bone marrow mesenchymal cells into CCl4-injured rats. J Hepatol 44:742–748. doi:10.1016/j.jhep.2005.10.026
31. Kosai K, Matsumoto K, Nagata S et al (1998) Abrogation of Fas-induced fulminant hepatic failure in mice by hepatocyte growth factor. Biochem Biophys Res Commun 244:683–690
32. Burr AW, Toole K, Chapman C et al (1998) Anti-hepatocyte growth factor antibody inhibits hepatocyte proliferation during liver regeneration. J Pathol 185:298–302. doi:10.1002/(SICI)1096-9896(199807)185:3<298::AID-PATH88>3.0.CO;2-B
33. Salehinejad P, Alitheen NB, Mandegary A et al (2013) Effect of EGF and FGF on the expansion properties of human umbilical cord mesenchymal cells. In Vitro Cell Dev Biol Anim 49:515–523. doi:10.1007/s11626-013-9631-3
34. Doetzlhofer A, White PM, Johnson JE et al (2004) In vitro growth and differentiation of mammalian sensory hair cell progenitors: a requirement for EGF and periotic mesenchyme. Dev Biol 272:432–447. doi:10.1016/j.ydbio.2004.05.013
35. Lin N, Lin J, Bo L et al (2010) Differentiation of bone marrow-derived mesenchymal stem cells into hepatocyte-like cells in an alginate scaffold. Cell Prolif 43:427–434. doi:10.1111/j.1365-2184.2010.00692.x
36. Teramoto T, Matsunaga T, Toba M et al (2009) Role of dexamethasone and oncostatin M on the formation of vacuoles in human fetal liver cells. Biol Pharm Bull 32:209–212
37. Das ZC, Gupta MK, Uhm SJ, Lee HT (2014) Supplementation of insulin-transferrin-selenium to embryo culture medium improves the in vitro development of pig embryos. Zygote 22:411–418. doi:10.1017/S0967199412000731
38. Al-Adsani A, Burke ZD, Eberhard D et al (2010) Dexamethasone treatment induces the reprogramming of pancreatic acinar cells to hepatocytes and ductal cells. PLoS One 5, e13650. doi:10.1371/journal.pone.0013650
39. Ishiwata I, Tokieda Y, Kiguchi K et al (2000) Effects of embryotrophic factors on the embryogenesis and organogenesis of mouse embryos in vitro. Hum Cell 13:185–195

40. Liang H, Block TM, Wang M et al (2012) Interleukin-6 and oncostatin M are elevated in liver disease in conjunction with candidate hepatocellular carcinoma biomarker GP73. Cancer Biomark 11:161–171. doi:10.3233/CBM-2012-00276
41. Kamiya A, Kinoshita T, Miyajima A (2001) Oncostatin M and hepatocyte growth factor induce hepatic maturation via distinct signaling pathways. FEBS Lett 492:90–94
42. Sidhu JS, Liu F, Omiecinski CJ (2004) Phenobarbital responsiveness as a uniquely sensitive indicator of hepatocyte differentiation status: requirement of dexamethasone and extracellular matrix in establishing the functional integrity of cultured primary rat hepatocytes. Exp Cell Res 292:252–264
43. Hong SH, Gang EJ, Jeong JA et al (2005) In vitro differentiation of human umbilical cord blood-derived mesenchymal stem cells into hepatocyte-like cells. Biochem Biophys Res Commun 330:1153–1161. doi:10.1016/j.bbrc.2005.03.086
44. Dong J, Mandenius C-F, Lübberstedt M et al (2008) Evaluation and optimization of hepatocyte culture media factors by design of experiments (DoE) methodology. Cytotechnology 57:251–261. doi:10.1007/s10616-008-9168-6
45. Snykers S, Vanhaecke T, Papeleu P et al (2006) Sequential exposure to cytokines reflecting embryogenesis: the key for in vitro differentiation of adult bone marrow stem cells into functional hepatocyte-like cells. Toxicol Sci 94:330–341. doi:10.1093/toxsci/kfl058, discussion 235–9
46. Kuznetsov SA, Mankani MH, Robey PG (2000) Effect of serum on human bone marrow stromal cells: ex vivo expansion and in vivo bone formation. Transplantation 70:1780–1787
47. Takahashi K, Tanabe K, Ohnuki M et al (2007) Induction of pluripotent stem cells from adult human fibroblasts by defined factors. Cell 131:861–872. doi:10.1016/j.cell.2007.11.019
48. Park I-H, Zhao R, West JA et al (2008) Reprogramming of human somatic cells to pluripotency with defined factors. Nature 451:141–146. doi:10.1038/nature06534
49. Chien Y, Liao Y-W, Liu D-M et al (2012) Corneal repair by human corneal keratocyte-reprogrammed iPSCs and amphiphatic carboxymethyl-hexanoyl chitosan hydrogel. Biomaterials 33:8003–8016. doi:10.1016/j.biomaterials.2012.07.029
50. Yamanaka S (2009) A fresh look at iPS cells. Cell 137:13–17. doi:10.1016/j.cell.2009.03.034
51. Nukaya D, Minami K, Hoshikawa R et al (2015) Preferential gene expression and epigenetic memory of induced pluripotent stem cells derived from mouse pancreas. Genes Cells 20:367–381. doi:10.1111/gtc.12227
52. Liu H, Ye Z, Kim Y et al (2010) Generation of endoderm-derived human induced pluripotent stem cells from primary hepatocytes. Hepatology 51:1810–1819. doi:10.1002/hep.23626
53. Rashid ST, Corbineau S, Hannan N et al (2010) Modeling inherited metabolic disorders of the liver using human induced pluripotent stem cells. J Clin Invest 120:3127–3136. doi:10.1172/JCI43122
54. Chen Y-F, Tseng C-Y, Wang H-W et al (2012) Rapid generation of mature hepatocyte-like cells from human induced pluripotent stem cells by an efficient three-step protocol. Hepatology 55:1193–1203. doi:10.1002/hep.24790
55. Tamaoki N, Takahashi K, Tanaka T et al (2010) Dental pulp cells for induced pluripotent stem cell banking. J Dent Res 89:773–778. doi:10.1177/0022034510366846
56. Oda Y, Yoshimura Y, Ohnishi H et al (2010) Induction of pluripotent stem cells from human third molar mesenchymal stromal cells. J Biol Chem 285:29270–29278. doi:10.1074/jbc.M109.055889
57. Chien Y, Chang Y-L, Li H-Y et al (2015) Synergistic effects of carboxymethyl-hexanoyl chitosan, cationic polyurethane-short branch PEI in miR122 gene delivery: accelerated differentiation of iPSCs into mature hepatocyte-like cells and improved stem cell therapy in a hepatic failure model. Acta Biomater 13:228–244. doi:10.1016/j.actbio.2014.11.018
58. Chiang C-H, Wu W-W, Li H-Y et al (2015) Enhanced antioxidant capacity of dental pulp-derived iPSC-differentiated hepatocytes and liver regeneration by injectable HGF-releasing hydrogel in fulminant hepatic failure. Cell Transplant 24:541–559. doi:10.3727/096368915X686986
59. Doddapaneni R, Chawla YK, Das A et al (2013) Overexpression of microRNA-122 enhances in vitro hepatic differentiation of fetal liver-derived stem/progenitor cells. J Cell Biochem 114:1575–1583. doi:10.1002/jcb.24499

Index

B. Zavan, E. Bressan (eds.), *Dental Stem Cells: Regenerative Potential*, Stem Cell Biology and Regenerative Medicine, DOI 10.1007/978-3-319-33299-4

E

F

G

R

S

Printed by Printforce, the Netherlands